T0329913

Hydrometeorology

Advancing Weather and Climate Science Series

Series editor:
John A. Knox
University of Georgia, USA

Other titles in the series:

Meteorological Measurements and Instrumentation
Giles Harrison
Published: December 2014

Fluid Dynamics of the Mid-Latitude Atmosphere
Brian J. Hoskins, Ian N. James
Published: October 2014

OperationalWeather Forecasting
Peter Inness, University of Reading, UK and
Steve Dorling, University of East Anglia, UK
Published: December 2012
ISBN: 978-0-470-71159-0

Time-Series Analysis in Meteorology and Climatology: An Introduction
Claude Duchon, University of Oklahoma, USA and
Robert Hale, Colorado State University, USA
Published: January 2012
ISBN: 978-0-470-97199-4

The Atmosphere and Ocean: A Physical Introduction, 3rd Edition
Neil C. Wells, Southampton University, UK
Published: November 2011
ISBN: 978-0-470-69469-5

Thermal Physics of the Atmosphere
Maarten H.P. Ambaum, University of Reading, UK
Published: April 2010
ISBN: 978-0-470-74515-1

Mesoscale Meteorology in Midlatitudes
Paul Markowski and Yvette Richardson, Pennsylvania State University, USA
Published: February 2010
ISBN: 978-0-470-74213-6

Hydrometeorology

Christopher G. Collier
University of Leeds, UK

WILEY Blackwell

This edition first published 2016
© 2016 by John Wiley & Sons, Ltd.

Registered Office
John Wiley & Sons, Ltd, The Atrium, Southern Gate, Chichester, West Sussex, PO19 8SQ, UK

Editorial Offices
9600 Garsington Road, Oxford, OX4 2DQ, UK
The Atrium, Southern Gate, Chichester, West Sussex, PO19 8SQ, UK
111 River Street, Hoboken, NJ 07030-5774, USA

For details of our global editorial offices, for customer services and for information about how
to apply for permission to reuse the copyright material in this book please see our website at
www.wiley.com/wiley-blackwell.

The right of the author to be identified as the author of this work has been asserted in accordance with
the UK Copyright, Designs and Patents Act 1988.

All rights reserved. No part of this publication may be reproduced, stored in a retrieval system, or
transmitted, in any form or by any means, electronic, mechanical, photocopying, recording or otherwise,
except as permitted by the UK Copyright, Designs and Patents Act 1988, without the prior permission of
the publisher.

Designations used by companies to distinguish their products are often claimed as trademarks. All brand
names and product names used in this book are trade names, service marks, trademarks or registered
trademarks of their respective owners. The publisher is not associated with any product or vendor
mentioned in this book.

Limit of Liability/Disclaimer of Warranty: While the publisher and author(s) have used their best efforts
in preparing this book, they make no representations or warranties with respect to the accuracy or
completeness of the contents of this book and specifically disclaim any implied warranties of
merchantability or fitness for a particular purpose. It is sold on the understanding that the publisher is
not engaged in rendering professional services and neither the publisher nor the author shall be liable for
damages arising herefrom. If professional advice or other expert assistance is required, the services of a
competent professional should be sought.

Library of Congress Cataloging-in-Publication Data

Names: Collier, C. G., author.
Title: Hydrometeorology / Christopher G. Collier.
Description: Chichester, UK ; Hoboken, NJ : John Wiley & Sons, 2016. | Includes index.
Identifiers: LCCN 2016007238| ISBN 9781118414989 (cloth) | ISBN 9781118414972 (pbk.)
Subjects: LCSH: Hydrometeorology. | Hydrodynamic weather forecasting.
Classification: LCC GB2801.7 .C65 2016 | DDC 551.57–dc23
LC record available at http://lccn.loc.gov/2016007238

A catalogue record for this book is available from the British Library.

Wiley also publishes its books in a variety of electronic formats. Some content that appears in print may
not be available in electronic books.

Set in 10/12pt Palatino by SPi Global, Pondicherry, India

1   2016

Dedication

This book is dedicated with love and gratitude to Cynthia for all the support she has given to me over many years.

Contents

Series Foreword

Advancing Weather and Climate Science

Meteorology is a rapidly moving science. New developments in weather forecasting, climate science and observing techniques are happening all the time, as shown by the wealth of papers published in the various meteorological journals. Often these developments take many years to make it into academic textbooks, by which time the science itself has moved on. At the same time, the underpinning principles of atmospheric science are well understood but could be brought up to date in the light of the ever increasing volume of new and exciting observations and the underlying patterns of climate change that may affect so many aspects of weather and the climate system.

In this series, the Royal Meteorological Society, in conjunction with Wiley Blackwell, is aiming to bring together both the underpinning principles and new developments in the science into a unified set of books suitable for undergraduate and postgraduate study as well as being a useful resource for the professional meteorologist or Earth system scientist. New developments in weather and climate sciences will be described together with a comprehensive survey of the underpinning principles, thoroughly updated for the 21st century. The series will build into a comprehensive teaching resource for the growing number of courses in weather and climate science at undergraduate and postgraduate level.

Series Editors

Peter Inness
University of Reading, UK

John A. Knox
University of Georgia, USA

Preface

The Earth, referred to as the blue planet, has three-quarters of its surface covered by water, which is essential to life. However, excessive variations bring disasters in the form of floods and droughts. Water is unevenly distributed in both time and space, and its circulation within the global atmosphere and oceans, the hydrological cycle, is a vital component of the earth's energy system. Water is the medium through which the atmosphere has most influence on human wellbeing, and terrestrial surfaces have significant influence on the atmosphere. Early knowledge of water developed through local attempts to manage and control it.

Although atmospheric and hydrologic science and practice have largely developed separately, meteorological forecasts beyond a few days, and climate predictions, require numerical models that include realistic representations of surface hydrology and associated energy exchanges. Hence it is essential that hydrologists and meteorologists work together. Therefore the discipline of Hydrometeorology is important, and is addressed in this book. However, it is so wide ranging that this book cannot hope to cover everything, and at best I hope it stimulates the reader to investigate areas further.

Rainfall-runoff modelling at scales of interest (small to large catchments) is not able to reproduce all the details of flow processes that give rise to stream hydrology. Indeed it is essential to understand and articulate the uncertainties when addressing modelling problems. A wide range of numerical models have been developed to address river, surface and sewer flows. Also forecasts of rainfall and climate change are made using comprehensive models of the atmosphere at a range of grid scales depending upon the application from those appropriate to urban drainage systems to those appropriate to large continental river catchments. This work has been stimulated by the rapid advances in computer power over the last 30 years or so.

Remote sensing, both surface and space-based has been used for almost 80 years as a practical tool to aid mapping of river flood plain inundation areas and the earth's surface. For many years most of the work has been qualitative. However the growth of both meteorological and hydrological sciences has demanded more comprehensive quantitative measurements. A range of instrumentation from simple raingauges and sophisticated weather radar to satellite passive radiometers and active radars underpin operational systems I examine how these trends have led to advances in hydro-meteorological studies.

I have included a wide range of both recent and historical references. Many of the earlier references remain very relevant to modern applications. The access to a wide range of literature via the internet and electronic databases enables the reader of this book to develop the knowledge contained therein. The book contains 14 chapters with each chapter ending with a summary of the main points in the chapter, a list of problems which readers may wish to use to test their appreciation of the contents and references. Each chapter also includes one or more appendices containing some additional information.

I wish to thank friends and colleagues who have encouraged me to work in the hydro-meteorological field. It is always difficult to engage with more than one discipline. However, I would highlight the scientific and practical benefits of the cross fertilization of ideas, and encourage young scientists in particular to accept the challenges that are offered by Hydrometeorology.

Chris G. Collier
Leeds

Acknowledgements

I am very grateful to Audrie Tan, Delia Sandford, Brian Goodale and the editorial staff of Wiley for all their help in the preparation of this book. Appreciation of all the organisations and individuals who gave permission to use their material is acknowledged.

About the Companion Website

This book is accompanied by a companion website

www.wiley.com/go/collierhydrometerology

The website includes:
- Powerpoints of all figures from the book for downloading
- PDFs of tables from the book

1

The Hydrological Cycle

1.1 Overview

The hydrological cycle describes the continuous movement of water above, on and below the surface of the Earth. It is a conceptual model that describes the storage and movement of water between the *biosphere* (the global sum of all ecosystems, sometimes called the zone of life on Earth), the *atmosphere* (the air surrounding the Earth, which is a mixture of gases, mainly nitrogen (about 80%) and oxygen (about 20%) with other minor gases), the *cryosphere* (the areas of snow and ice), the *lithosphere* (the rigid outermost shell of the Earth, comprising the crust and a portion of the upper mantle), the *anthroposphere* (the effect of human beings on the Earth system) and the *hydrosphere* (see Table 1.1).

Models of the biosphere are often referred to as *land surface parameterization schemes* (LSPs) or *soil–vegetation–atmosphere transfer schemes* (SVATs). An example of an SVAT is described by Sellers et al. (1986). The water on the Earth's surface occurs as streams, lakes and wetlands in addition to the sea. Surface water also includes the solid forms of precipitation, namely snow and ice. The water below the surface of the Earth is ground water.

Most of the energy leaves the ocean surface in the form of latent heat in water vapour, but this is not necessarily the case for land surfaces. Hence maritime air masses are different to continental air masses. The atmosphere and oceans are strongly coupled by the exchange of energy, water vapour, momentum at their interface, and precipitation. The oceans represent an enormous reservoir for stored energy and are denser than the atmosphere, having a larger mechanical inertia. Therefore ocean currents are much slower than atmospheric flows. The atmosphere is heated from below by the Sun's energy intercepted by the underlying surface, whereas the oceans are heated from above. Lakes, rivers and underground water can have significant hydrometeorological and hydroclimatological significance in continental regions. The hydrological cycle is represented by the simplified diagram in Figure 1.1.

Hydrometeorology, First Edition. Christopher G. Collier.
© 2016 John Wiley & Sons, Ltd. Published 2016 by John Wiley & Sons, Ltd.
Companion website: www.wiley.com/go/collierhydrometerology

Table 1.1   Water in the hydrosphere and the distribution of fresh water on the Earth (from Martinec, 1985)

(a) Distribution of water in the hydrosphere

Forms of water present	Water volume (10^6 km³)	As %
Oceans, seas	1348	97.4
Polar ice, sea ice, glaciers	28	2.0
Surface water, ground water, atmospheric water	8	0.6
Total	1384	100.0
Total fresh water	36	2.6

(b) Distribution of fresh water on Earth

Forms of water present	Water volume (10^6km³) *	Water volume (10^6km³) †	As % *	As % †
Polar ice, glaciers	24.8	27.9	76.93	77.24
Soil moisture	0.09	0.06	0.28	0.17
Ground water within reach	3.6	3.56	11.17	9.85
Deep ground water	3.6	4.46	11.17	12.35
Lakes and rivers	0.132	0.127	0.41	0.35
Atmosphere	0.014	0.014	0.04	0.04
Total	32.236	36.121	100.0	100.0

* Based on Volker (1970).
† Based on Dracos (1980), referred to in Baumgartner and Reichel (1975).

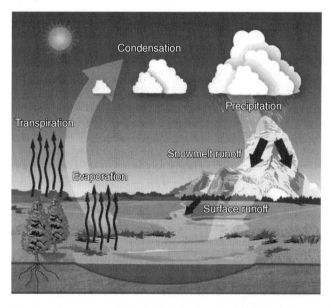

Figure 1.1   Simplified representation of the hydrological cycle (NWS Jetstream NOAA, USA, www. srh.noaa.gov/jetstream/atmos/hydro.htm) (*see plate section for colour representation of this figure*)

1.2   Processes comprising the hydrological cycle

There are many processes involved in the hydrological cycle, the most important of which are as follows:

- *Evaporation* is the change of state from liquid water to vapour. The energy to achieve this may come from the Sun, the atmosphere itself, the Earth or human activity.
- *Transpiration* is the evaporation of water from plants through the small openings found on the underside of leaves (known as stomata). In most plants, transpiration is a passive process largely controlled by the humidity of the atmosphere and the moisture content of the soil. Only 1% of the transpired water passing through a plant is used by the plant to grow, with the rest of the water being passed into the atmosphere. Evaporation and transpiration return water to the atmosphere at rates which vary according to the climatic conditions.
- *Condensation* is the process whereby water vapour in the atmosphere is changed into liquid water as clouds and dew. This depends upon the air temperature and the dew point temperature. The dew point temperature is the temperature at which the air, as it is cooled, becomes saturated and dew can form. Any additional cooling causes water vapour to condense. When the air temperature and the dew point temperature are equal, mist and fog occur. Since water vapour has a higher energy level than liquid water, when condensation occurs the excess energy is released in the form of heat. When tiny condensation particles, through collision or coalescence with each other, grow too large for the ascending air to support them, they fall to the surface of the Earth as precipitation (Chapter 2). Precipitation is the primary way fresh water reaches the Earth's surface, and on average the Earth receives about 980 mm each year over both the oceans and the land.
- *Infiltration* of water into the land surface occurs if the ground is not saturated, or contains cracks or fissures. The flow of water into the ground may lead to the recharge of aquifers, or may move through unsaturated zones to discharge into rivers, lakes or the seas. Between storm or snowmelt periods, stream flow is sustained by discharge from the ground water systems. If storms are intense, most water reaches streams rapidly. Indeed, if the water table – the boundary between the saturated and unsaturated zones – rises to the land surface, overland flow may occur.
- The *residence time* of water in parts of the hydrological cycle is the average time a water molecule will spend in a particular area. These times are given in Table 1.2. Note that ground water can spend over 10,000 years beneath the surface of the Earth before leaving, whereas water stored in the soil remains there very briefly. After water evaporates, its residence time in the atmosphere is about nine days before it condenses and falls to the surface of the Earth as precipitation. Residence times can be estimated in two ways. The first and more common method is to use the principle of conservation of mass, assuming the amount of water in a given store is roughly constant. The residence time is derived by dividing the volume of water in the store by the rate by which water either enters or leaves the store. The second method, for ground water, is via isotropic techniques. These techniques use either in-stream tracer injection combined with modelling, or measurements of naturally occurring tracers such as radon-222 (see for example Lamontagne and Cook, 2007).

Table 1.2 Average residence times for specific stores (see for example www.physicalgeography.net/fundamentals/8b.html)

Reservoir	Average residence time
Antarctica	20,000 years
Oceans	3,200 years
Glaciers	20 to 100 years
Seasonal snow cover	2 to 6 months
Soil moisture	1 to 2 months
Ground water shallow	100 to 200 years
Ground water deep	10,000 years
Lakes	50 to 100 years
Rivers	2 to 6 months
Atmosphere	9 days

- *Human activities* release tiny particles (aerosols) into the atmosphere, which may enhance scattering and absorption of solar radiation. They also produce brighter clouds that are less efficient at releasing precipitation. These aerosol effects can lead to a weaker hydrological cycle, which connects directly to the availability and quality of fresh water (see Ramanathan et al., 2001).

1.3 Global influences on the hydrological cycle

Differential heating by the Sun is the primary cause of the general circulation of the atmosphere. There are a number of regional differences which influence the hydro-logical cycle in addition to the relative position of the Sun and the Earth. There are latitudinal differences in solar input brought about by the rotation of the Earth. These cause seasonal changes in the circulation patterns, and produce a regular diurnal cycle in the longitude.

The fluxes of surface latent and sensible heat are different between continental and oceanic surfaces. Also continental topography varies in ways which significantly impact the atmosphere. Changes in land cover, atmospheric composition and sea surface temperature alter the hydrological cycle both persistently and on a temporary basis.

Clouds transport substantial amounts of water around the atmosphere. Consequently they have a major impact on the absorption of solar radiation and modify surface energy balance. Some tall clouds tend to shade and inhibit solar radiation reaching the ground, whereas high clouds inhibit the loss of long wave radiation.

The peak zonal mean precipitable water is situated at 10°N for the overall mean (Figure 1.2a) and for the mean over sea (Figure 1.2c). The zonal mean precipitable water over land (Figure 1.2b) is nearly symmetrical about the equator for the annual mean, and the winter (DJF) mean is close to the mirror image of the summer (JJA) mean between 40°S and 40°N. The water vapour content of the atmosphere is highly sensitive to the temperature through the saturation vapour pressure, and the temperature decreases towards the upper atmosphere. The SI unit of pressure is the

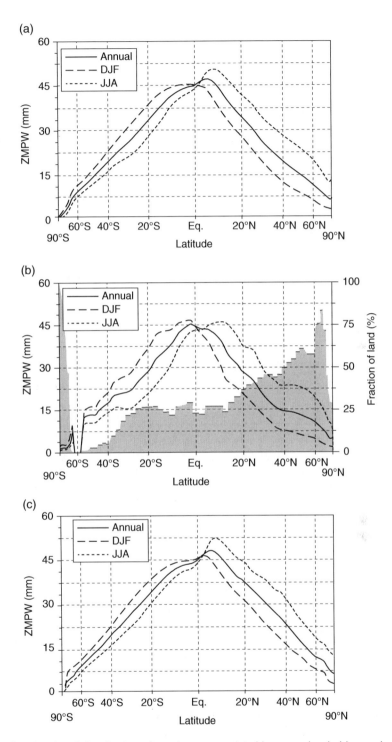

Figure 1.2 Meridional distribution of zonal mean precipitable water (mm): (a) over land and sea, (b) mean over land only, and (c) mean over sea only. Annual mean, December-January-February (DJF) mean, and June-July-August (JJA) mean, for 4 years from 1989 to 1992 (from Oki, 1999)

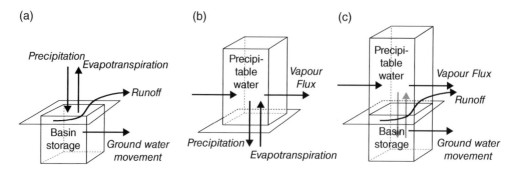

Figure 1.3   Illustrating (a) terrestrial water balance, (b) atmospheric water balance and (c) combined atmospheric–terrestrial water balance (from Oki, 1999)

pascal (Pa), and atmospheric pressure at mean sea level is approximately 10^5 Pa or 1000 hectopascals (hPa). More than 50% of the water vapour is concentrated below the 850 hPa surface, and more than 90% is confined to the layer below 500 hPa (see Peixoto and Oort, 1992).

1.4   Water balance

Water balance at the land surface has generally been estimated using ground observations such as precipitation, water storage in lakes and ground water. This is described by

$$\frac{\partial S}{\partial t} = -\nabla_H \cdot \mathbf{R}_0 - \nabla_H \cdot \mathbf{R}_u - (E - P) \tag{1.1}$$

where S represents the water storage within the area, $\mathbf{R}_0$ is surface runoff, $\mathbf{R}_u$ is the ground water movement, E is evapotranspiration and P is precipitation; ∇_H is horizontal divergence. S includes snow accumulation in addition to soil moisture, ground water and surface water storage, including retention water (see Oki, 1999).

The atmospheric water balance is described by

$$\frac{\partial W}{\partial t} + \frac{\partial W_c}{\partial t} = -\nabla_H \cdot \mathbf{Q} - \nabla_H \cdot \mathbf{Q}_c + (E - P) \tag{1.2}$$

where W represents precipitable water in vapour in a column, W_c is the column storage of liquid and solid water, $\mathbf{Q}$ is the vertically integrated two-dimensional water vapour flux, and $\mathbf{Q}_c$ is the vertically integrated two-dimensional water flux in the liquid and solid phases. Both the land surface and the atmospheric water balances are summarized in Figure 1.3.

1.5   Impact of aerosols on the hydrological cycle

Aerosol particles (liquid and solid particles, usually other than natural water droplets) are an important part of the water cycle as they serve as condensation nuclei for the formation of cloud droplets as well as scattering and absorbing solar radiation.

Ramanathan et al. (2001) have described how this occurs. Atmospheric aerosols range in size from hundredths of a micrometre to many tens of micrometres. They derive from many sources, namely the Earth's surface, oceans, volcanoes and the biosphere. Many aerosols derive from human activity. Large amounts of dust from major deserts, such as the Sahara, produce massive amounts of aerosol. Sea salt produced by bursting air bubbles and winds is the largest natural flux of aerosol mass into the atmosphere, although much of this mass is large particles that are not transported very far. Volcanoes can inject large amounts of particles and gases into the atmosphere, and if they reach the stratosphere they are removed more slowly than if they reside only in the troposphere. The formation of sulphates from sulphur gases, produced from marine phytoplankton, continental biota and the burning of fossil fuels, is an important source of aerosols.

Increases in aerosols generally greater in size than a few hundredths of a micrometre, referred to as cloud condensation nuclei (CCN), increase the droplet number concentration in clouds, which increases cloud extent and leads to global cooling. If the amount of water available for condensation in the cloud is not changed, this means that there will be a greater number of smaller drops, which are less likely to grow to sufficient size to fall out as precipitation (see Chapter 2); thus clouds last longer, again contributing to cooling. The cooling of the surface through increased reflection of solar energy, and the reduced efficiency of clouds in producing precipitation, resulting from increases in aerosol, will weaken the hydrological cycle.

1.6 Coupled models for the hydrological cycle

Studies of the hydrological cycle usually involve feedbacks between atmospheric, ecological and hydrological systems, as well as human society. Often the feedbacks between systems produce unanticipated responses. Hence the coupling of different compartments of the Earth system is a challenge to the numerical modelling community.

Imposing a boundary between components of the modelled hydro-geo-biosphere system is not feasible. Observations of two or more state variables that are coupled cannot be modelled simultaneously without the processes that couple them also being included in the model.

Over the past two decades advances have been made in interfacing different Earth system components within hydrological models (see for example Bronstert et al., 2005). It is apparent that there is a need for hydrology to interact not only with atmospheric sciences but also with soil sciences, geochemistry, biology and even sociology.

Scale issues are important. Processes that dominate system behaviour at small scales may become irrelevant at large scales, and new processes may emerge as being important to the change in scale. In addition, the spatial variability of fluxes such as rainfall, of properties such as hydraulic conductivity, and of state variables such as soil humidity often leads to unexpected system responses. Also land surface and subsurface states, particularly soil moisture, affect the persistence of drier or wetter atmospheric conditions. This is of importance for the occurrence of large floods or droughts. Finally human feedbacks are as important as natural feedbacks.

1.7 Global Energy and Water Cycle Exchanges Project (GEWEX)

GEWEX is an integrated programme of research, observations and science activities that focuses on the atmospheric, terrestrial, radiative, hydrological, and coupled processes and interactions that determine the global and regional hydrological cycle, radiation and energy transitions and their involvement in climate change. It is the core project in the World Climate Research Programme (WCRP) which is concerned with studying the dynamics and thermodynamics of the atmosphere, its interactions with the Earth's surface, and its effects on the global energy and water cycle.

The goal of GEWEX is to reproduce and predict, by means of suitable numerical models, the variations of the global hydrological cycle, its impact on atmospheric and surface dynamics, and variations in regional hydrological processes and water resources and their response to changes in the environment, such as the increase in greenhouse gases. GEWEX will provide an order of magnitude improvement in the ability to model global precipitation and evaporation, as well as accurate assessment of the sensitivity of atmospheric radiation and clouds to climate change.

GEWEX plays a central role in the interaction of WCRP with many international organizations and programmes dealing with climate observations. As part of WCRP's input to the Group on Earth Observations (GEO) Global Earth Observation System of Systems (GEOSS), GEWEX brings its unique expertise in two specific societal benefit areas: climate and water. GEWEX leads in the development of plans for the global data reprocessing effort and observation strategy, and serves as a demonstration project for future climate change understanding, using observational networks in GEOSS. GEWEX supports the Integrated Global Water Cycle Observations (IGWCO) theme.

Clouds cover about 70% of the Earth's surface. The GEWEX Cloud Assessment was initiated in 2005, and provides the first coordinated intercomparison of publicly available global cloud products retrieved from measurements of multispectral imagers, IR sounders and lidar.

1.8 Flooding

Where land not normally covered by water becomes covered by water, a flood occurs. There are many sources of flooding as follows:

- *River (fluvial) flooding*: this occurs when a river cannot cope with the amount of water entering it. Over 12% of the population of the United Kingdom live on fluvial floodplains or areas identified as being subject to the risk of coastal flooding. More than 150 million people and gross domestic product (GDP) of more than $1300 billion are exposed to river flooding each year around the world.
- *Coastal flooding*: weather and tidal conditions can increase sea levels. The frequency and severity of this type of flooding are predicted to increase.
- *Surface water (pluvial) flooding*: this happens when there is heavy rainfall on ground that is already saturated, or on paved areas where drainage is poor.
- *Ground water flooding*: when rainfall causes the water that is naturally stored underground to rise to the surface it can flood low lying areas.
- *Drain and sewer (urban) flooding*: this can occur during heavy rain when drains have become blocked or full.

Jongman et al. (2012) suggested that the potential effects of flooding increased between 1970 and 2010. Three-quarters of those who were affected by river flooding in 2010 lived in Asia. Only 9% lived in Europe, although, after Asia, more people were exposed to river flooding in Europe than anywhere else. The population exposed to river flooding in Europe was predicted to fall between now and 2050. Globally, economic losses from flooding exceeded $19 billion in 2012, and are rising rapidly (Ward et al., 2013).

One major effect of climate change is the increased risk of flooding, which could cause serious loss of life and property in many parts of the world. Recent work by Hirabayashi et al. (2013) uses the output from 11 different climate models to estimate the increased global risk of floods due to climate change. Using a river routing model to calculate future river flows and flooding areas, they were able to predict that floods will become more frequent over 42% of the Earth's surface. Areas including South East Asia, India, eastern Africa and the northern Andes are likely to be most impacted. Floods that used to occur every 100 years are predicted to start occurring every 5–25 years in many areas

In the following chapters we discuss these types of floods and how they are measured, modelled and forecast.

Summary of key points in this chapter

1. The hydrological cycle describes the continuous movement of water above, on and below the surface of the Earth.

2. Models of the biosphere are often referred to as land surface parameterization schemes (LSPs) or soil–vegetation–atmosphere transfer schemes (SVATs).

3. Energy leaves the ocean surface in the form of latent heat in water vapour, and the atmosphere and oceans are strongly coupled by the exchange of energy, water vapour, momentum at their interface, and precipitation.

4. The atmosphere is heated from below by the Sun's energy intercepted by the underlying surface, whereas the oceans are heated from above.

5. The most important processes involved in the hydrological cycle are evaporation, transpiration, condensation and infiltration.

6. The residence times of water in parts of the hydrological cycle may be estimated using the principle of conservation of mass or, for ground water, isotropic techniques.

7. Aerosols are tiny particles in the atmosphere which may enhance the scattering and absorption of solar radiation, produce brighter clouds, and weaken the hydrological cycle.

8. Differential heating by the Sun is the primary cause of the general circulation of the atmosphere, and fluxes of surface latent and sensible heat are different between continental and oceanic surfaces. Clouds transport substantial amounts of water around the atmosphere.

9. Water balance equations describe the movement of water over the land, within the ground, in the atmosphere, and between the atmosphere and the land.

10. Sources of flooding are river (fluvial), coastal, surface water (pluvial), ground water, and drain and sewer (urban) surcharging.

Problems

1. Describe the structure of the hydrological cycle, noting the continuous movement of water.
2. What is the difference between maritime and continental air masses?
3. Briefly describe the processes comprising the hydrological cycle.
4. Describe the residence times of water in the components of the hydrological cycle, and how they may be estimated.
5. Describe the zonal mean precipitation with latitude in summer and winter.
6. Give the equation describing the atmospheric water balance.
7. Outline the nature of aerosols in the atmosphere, and their role in the hydrological cycle.

References

Baumgartner, A. and Reichel, E. (1975) *Die Weltwasserbilanz [World Water Balance]*. Oldenbourg, Munich.

Bronstert, A., Carrera, J., Kabat, P. and Lutkemeier, S. (2005) *Coupled Models for the Hydrological Cycle: Integrating Atmosphere, Biosphere and Pedosphere*. Springer, New York.

Dracos, Th. (1980) *Hydrologie [Hydrology]*. Springer, Wien.

Hirabayashi, Y., Roobavannan, M., Sujan, K., Lisako, K., Dai, Y., Satoshi, W., Hyungjun, K. and Shinjiro, K. (2013) Global flood risk under climate change. *Nat. Clim. Change*, 3, 816–821.

Jongman, B., Ward, P.J. and Aerts, J.C.J.H. (2012) Global exposure to river and coastal flooding: long term trends and changes. *Glob. Environ. Change*, 22(4), 823–835.

Lamontagne, S. and Cook, P.G. (2007) Estimation of hyporheic water residence time *in situ* using ^{222}Rn disequilibrium. *Limnol. Oceanogr. Methods*, 5, 407–416.

Martinec, J. (1985) Time in hydrology. Chapter 9 in *Facets of Hydrology*, vol. II, ed. J.C. Rodda, pp. 249–290. Wiley, Chichester.

Oki, T. (1999) The global water cycle. Chapter 1.2 in *Global Energy and Water Cycles*, ed. K.A. Browning and R.J. Gurney, pp. 10–29. Cambridge University Press, Cambridge.

Peixoto, J.P. and Oort, A.H. (1992) *Physics of Climate*. American Institute of Physics.

Ramanathan, V., Cutzen, P.J., Kiehl, J.T. and Rosenfeld, D. (2001) Aerosols, climate and the hydrological cycle. *Science*, 294(7 December), 2119–2124.

Sellers, P.I., Mintz, Y., Sud, Y.C. and Dalcher, A. (1986) A simple biosphere model (SiB) for use within general circulation models. *J. Atmos. Sci.*, 43, 505–531.

Volker, A. (1970) Water in the world. Public lecture, IAHS Symposium on Representative and Experimental Basins, Wellington, New Zealand.

Ward, P.J., Jongman, B., Sperna Weiland, F., Bouwman, A.A., Van Beek, R., Bierkens, M.F.P., Ligtvoet, W. and Winsemius, H.C. (2013) Assessing flood risk at the global scale: model setup, results, and sensitivity. *Env. Res. Lett.*, 8, 044019.

2

Precipitation

2.1 Introduction

The circulation of water in the hydrological cycle was outlined in the previous chapter. However, water may take a number of different forms in the atmosphere. These forms are collectively termed 'precipitation', which includes rain, drizzle, sleet (partly melted snowflakes, or rain and snow falling together), snow and hail. The intensity and duration of precipitation are extremely variable in most areas of the world.

The source of precipitation is water vapour, which is always present in the atmosphere in varying amounts, although it makes up less than 1% by volume. However, the water vapour in the air must be cooled to allow water to be condensed into cloud droplets. These droplets then grow to form precipitation particles. The mass of water in the atmosphere in both liquid and vapour forms is around 1.3×10^{16} kg, compared with the mass of water in the oceans of around 1.3×10^{21} kg (Nace, 1967). Nevertheless this water is distributed very unevenly, and is transported by the circulation of the atmosphere. In this chapter we will consider the basic thermodynamic processes which lead to the formation of precipitation, and describe the atmospheric systems within which these processes occur. A basic classification of atmospheric systems is shown in Table 2.1.

2.2 Equation of state for a perfect gas

The pressure p(N m^{-2}), molar specific volume V(m^{3}kg^{-1}mol^{-1}), and temperature T (K) of a perfect gas are related by the equation of state,

$$pV = mRT \qquad (2.1)$$

where $m = Nm_p$; N is the total number of molecules each having mass m_p; and R = 8.3144 J K^{-1}mol^{-1} is known as the universal gas constant. This equation reflects the observations of experiments with gases embodied in *Boyle's law*, i.e. pV is a constant at constant temperature; *Charles's law*, i.e. any gas at a given pressure expands uniformly with temperature; and *Avogadro's hypothesis*, i.e. at normal temperature and pressure (273.15 K and 1013.25 mbar) the molecular volume is the same

Hydrometeorology, First Edition. Christopher G. Collier.
© 2016 John Wiley & Sons, Ltd. Published 2016 by John Wiley & Sons, Ltd.
Companion website: www.wiley.com/go/collierhydrometerology

Table 2.1   Atmospheric scale classification

	Synoptic	Mesoscale	Microscale
Period (h)	>48	1–48	<1
Wavelength (km)	>500	20–500	<20

for all gases, namely $22.415 \times 10^3 \, cm^3 kg^{-1} mol^{-1}$. These laws, as well as being true for a perfect gas, are nearly true for atmospheric gases.

Dalton's law states that in a mixture of gases, the total pressure is equal to the sum of the pressures which would be exerted by each gas if it filled the volume under consideration at the same temperature. Hence a mixture of gases, such as the atmosphere, behaves like a single gas provided the mixture gas constant R is given by

$$R = \frac{\sum m_i R_i}{\sum m_i} \tag{2.2}$$

where m_i is the molecular weight of the ith gas and R_i is the corresponding individual gas constant. Equation 2.2 may be evaluated for dry air, giving $R_d = 287 \, J \, K^{-1} \, kg^{-1}$. For atmospheric air, which is usually a mixture of dry air and a variable proportion of water vapour, the gas constant will be a function of the water vapour content.

2.3   Hydrostatic pressure law

Consider a thin rectangular volume of air of cross-sectional area A and depth δz (after Shuttleworth, 2012). This volume of air has a mass $A\delta z \rho_a$, where ρ_a is the density of the moist air. The pressures are p at the lower level z, and $p + \delta p$ at the upper level $z + \delta z$. The forces exerted at the top and bottom of the volume of air are $A(p + \Delta p)$ and Ap respectively, and the additional downward gravitational force is $gA\rho_a \delta z$, where g is the acceleration due to gravity (9.81 m s^{-2}). Therefore

$$AP - A(P + \Delta P) = gA\rho_a \delta z \tag{2.3}$$

Thus

$$\Delta P = -g\rho_a \delta z \tag{2.4}$$

For small δz this gives

$$\frac{\partial P}{\partial z} = -g\rho_a \tag{2.5}$$

2.4   First law of thermodynamics

The first law of thermodynamics states that the energy of a system in a given state relative to a fixed normal state (usually 0 °C, one atmosphere pressure, and at rest in a given position) is equal to the algebraic sum of the mechanical equivalent of all the effects produced outside the system, when it passes in any way from the given state to the normal state. This energy is independent of the manner of the transformation.

The first law is derived from two facts: (i) heat is a form of energy (*Joule's law*); and (ii) energy is conserved. Conservation of energy Q may be expressed by

$$dQ = dU + dW \tag{2.6}$$

where U is the internal energy of the gas, and W is the work done by the gas in moving from one state to another. Generally, equation 2.6 is expressed in terms of unit mass of gas:

$$dq = du + dw \tag{2.7}$$

Consider the work done dw by a unit mass of gas in expanding from V to $V + dV$ by pushing a frictionless, moveable piston of area A. The work done against a pressure p is $pA ds = p dV$. Hence, equation 2.7 may be rewritten as

$$dq = p dV + du \tag{2.8}$$

The internal energy is a function of V and T, that is,

$$du = \left(\frac{\partial U}{\partial T}\right)_V dT + \left(\frac{\partial U}{\partial V}\right)_T dV \tag{2.9}$$

However, for a perfect gas $\left(\partial U / \partial V\right)_T = 0$, and we may define the specific heat at constant volume C_V as $\left(\partial U / \partial T\right)_V$. Equation 2.8 then becomes

$$dq = p dV + C_V dT \tag{2.10}$$

This is sometimes called the *energy equation*. Using equation 2.1,

$$dq = \left(C_V + R\right) dT - V dp \tag{2.11}$$

If we consider a change at constant pressure, then $dp = 0$ and

$$dq = \left(C_V + R\right) dT = C_p dT \tag{2.12}$$

where C_p is the specific heat at constant pressure. Therefore

$$R = C_p - C_V \tag{2.13}$$

In the atmosphere we treat C_p and C_V as constant, with values derived experimentally of $240\,\text{cal kg}^{-1}\text{K}^{-1}\left(10.04 \times 10^6\,\text{erg g}^{-1}\text{K}^{-1}\right)$ and $171\,\text{cal kg}^{-1}\text{K}^{-1}\left(7.17 \times 10^6\,\text{erg g}^{-1}\text{K}^{-1}\right)$ respectively. The value of R is $287\,\text{J kg}^{-1}\text{K}^{-1}\left(2.8704 \times 10^6\,\text{erg g}^{-1}\text{K}^{-1}\right)$, and we define $\gamma = C_p / C_V = 1.4$.

2.5   Atmospheric processes: dry adiabatic lapse rate

The atmosphere receives heat from the Sun or the surface of the Earth (radiation), and from processes involving friction, condensation, evaporation or turbulence. Hence, provided no condensation or evaporation occurs, atmospheric processes may be regarded as *adiabatic*, that is, as involving no exchange of heat between that part of the atmosphere under consideration and its surroundings. Considering equations 2.10, 2.11 and 2.12 for an adiabatic process, $dq = 0$. Therefore

$$p dV + C_V dT = 0$$
$$V dp - C_p dT = 0 \tag{2.14}$$

If the atmospheric process is *reversible* as well as being adiabatic then, using equation 2.1 and integrating,

$$TV^{R/C_V} = \text{constant} \tag{2.15}$$

$$Tp^{-R/C_p} = Tp^{-K} = \text{constant} \tag{2.16}$$

$$pV^{\gamma} = \text{constant} \tag{2.17}$$

where $K = R/C_p$. Equation 2.16 is known as *Poisson's equation*. Taking a reference pressure of 1000 millibars, we may define *potential temperature θ* as follows:

$$\theta = T \left(\frac{1000}{p} \right)^K \tag{2.18}$$

Therefore potential temperature is a constant for a dry adiabatic process, that is, one with no condensation or evaporation. From equation 2.16,

$$\log T - K \log p = \text{constant} \tag{2.19}$$

Differentiating with respect to height Z,

$$\frac{\partial T}{\partial Z} = K \frac{T}{p} \frac{\partial p}{\partial Z} \tag{2.20}$$

It may be shown from the equations of motion for the atmosphere that, to a good approximation,

$$\frac{\partial p}{\partial Z} = -\rho g \tag{2.21}$$

where ρ is the atmospheric density and g is the acceleration due to gravity. This is known as the *hydrostatic equation*. Substitution of equation 2.21 into equation 2.20 gives

$$\frac{\partial T}{\partial Z} = K \frac{T}{p} (-\rho g) \tag{2.22}$$

and using equation 2.1,

$$\frac{\partial T}{\partial Z} = -\frac{g}{C_p} = \gamma_d = -9.76\,°\text{C km}^{-1} \tag{2.23}$$

Therefore if a parcel of air rises dry adiabatically, its temperature will fall at the rate of about 10°C km⁻¹. The quantity γ_d is the *dry adiabatic lapse rate* (DALR).

2.6   Water vapour in the atmosphere

If we compress a sample of air containing water vapour at a temperature greater than 0°C, then the water vapour pressure e will increase until water begins to condense from the sample (Figure 2.1,AA$_1$). From this point the water vapour pressure is constant (A$_1$B), and is referred to as the saturated vapour pressure e_s, which is a function of temperature. Eventually a very large increase of pressure is required to produce a small change in the volume of the liquid (BB$_1$).

If the temperature is 0.01 °C then we find that the horizontal line A$_1$B represents a mixture of water vapour, water and ice. In dealing with unsaturated water vapour, it is sufficiently accurate to treat water vapour as a perfect gas with gas constant R' equal to $R_d/0.622$.

When a change of phase occurs (vapour to liquid; liquid to solid; vapour to solid, known as *sublimation*), heat is released or must be absorbed. The amount of heat required to transform one gram of water into vapour at a constant temperature is defined as the *latent heat of evaporation*, and has a value of 2.50×10^6 J kg^{-1} (at 0 K). The *latent heat of sublimation* (ice to vapour, 2.823×10^6 J kg^{-1} at 0 K) and the *latent heat of fusion* (ice to water, 0.333×10^6 J kg^{-1} at 0 K) are similarly defined. The relationship between the equilibrium pressure e_s and the temperature during a phase change is given by the *Clausius–Clapeyron equation*,

$$\frac{\mathrm{d}e_s}{\mathrm{d}T} = \frac{L_{12}}{T(V_2 - V_1)} \qquad (2.24)$$

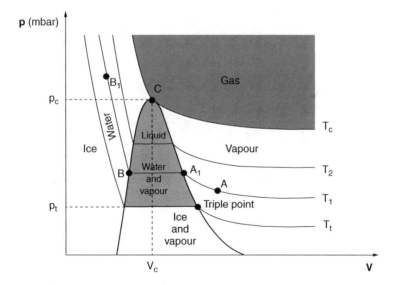

Figure 2.1   Pressure–volume graph for water known as an Amagat–Andrews diagram. Shows phase changes along isotherms in the (p, V) domain. A sample of water vapour is considered at a state corresponding to point A, i.e. at a temperature T_1 and pressure p_1 greater than the triple point temperature T_t and pressure p_t. If the vapour is compressed isothermally, the pressure increases until point A$_1$ is reached when liquid water and water vapour coexist in equilibrium, i.e. some water vapour has condensed to form liquid water (from Tsonis, 2002)

where L is the latent heat, and the subscripts refer to a phase change from a state defined by 1 to that defined by 2. Considering equation 2.21 for the change of phase from ice to water, it may be shown that it takes about 135 atmospheres pressure to lower the melting point of ice to −1 °C. Therefore in the atmosphere the melting point is strictly 0°C. However, water can remain as a liquid at temperatures well below 0°C. In these circumstances a sample of moist air can be supersaturated with respect to ice, but unsaturated with respect to supercooled water. If two surfaces, one of water and one of ice, come into contact with the moist air, then water will evaporate, but ice will grow by deposition of the water vapour. This occurs often in clouds at temperatures below 0°C, and this is important in the formation of precipitation, as we shall see.

2.7   Atmospheric processes: saturated adiabatic lapse rate

For a sample of moist air in which no evaporation or condensation occurs, equation 2.20 may be used. However, when condensation occurs and the resulting water falls out of the sample, the mass of the sample changes and heat is lost with the fallout of the water. This is known as a *pseudo-adiabatic process*. If all the condensed water remains in the sample, then the process is of course reversible. In the atmosphere conditions are usually such that some, but not all, of the condensed water falls out of any sample of moist air.

Using equations 2.1, 2.10 and 2.24, it may be shown that for saturated air which is lifted slightly,

$$\frac{\partial T}{\partial Z} = \gamma_s = \frac{\gamma_d \left(1 + \dfrac{Lx_s}{R_d T}\right)}{1 + \dfrac{L^2 x_s}{R'C_p T^2}} \tag{2.25}$$

where γ_s is the *saturated adiabatic lapse rate* (SALR) and x_s is the *saturated humidity mixing ratio*, that is, the mass of water vapour present in the moist air measured per gram of dry air when the moist air is saturated. Although γ_s varies with temperature and pressure, a typical value in the atmosphere is −5.0 °C km⁻¹.

2.8   Stability and convection in the atmosphere

Moist air can become saturated, and hence produce precipitation, by movement upwards in the atmosphere. Consider a sample, referred to as a parcel, of moist air, for which the pressure in the parcel is the same as that of its environment. Assuming that the parcel can move vertically without disturbing the environment and does not mix with its environment, it can be shown from the equations of motion for the atmosphere that

$$\frac{dw}{dt} = -\frac{1}{\rho}\frac{\partial p}{\partial Z} - g = \frac{(T - T')g}{T'} \tag{2.26}$$

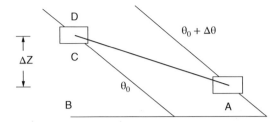

Figure 2.2 Sloping convection: an N–S cross-section with isentropic surfaces (surfaces of constant potential temperature θ) indicated, comparing the potential energy released when a parcel of air moves vertically, horizontally or intermediately

where w is the vertical velocity of the parcel, t is time, T is the temperature of the parcel, and T' is the temperature of the environment. Making the further assumption that the movement is adiabatic,

$$\frac{dw}{dt} = \frac{\left(T_0 - T_0'\right)g}{T_0'} + \frac{(\gamma - \gamma_a)gZ}{T_0'} \tag{2.27}$$

where T_0 and T_0' are the initial temperature of the parcel and the environment respectively, γ is the environmental lapse rate, Z is the vertical coordinate, and γ_a is the appropriate adiabatic lapse rate, being γ_d if the parcel is unsaturated and γ_s if the parcel is saturated. If the temperature is constant in the horizontal then $T_0 = T_0'$ and

$$\frac{dw}{dt} = \frac{(\gamma - \gamma_a)gZ}{T_0'} \tag{2.28}$$

The atmosphere is regarded as stable, neutral or unstable when dw/dt is <0, 0 or >0, respectively. If $\gamma_s < \gamma < \gamma_d$ then the atmosphere is *conditionally unstable*, whereas if $\gamma < \gamma_s$ the atmosphere is *absolutely stable* and if $\gamma > \gamma_d$ the atmosphere is *absolutely unstable*. Hence if moist air is lifted by some means it may become saturated and hence unstable, and may then continue to rise without any external force being applied.

This release of instability in the atmosphere, or convection, is of course much more complex than this; it involves entrainment of environmental air into the rising air, and the effect of descending air on the rising air. The variability in predictability of deep convection has been studied by Done et al. (2012). Indeed, convection is not necessarily a vertical process. Consider the exchange of a parcel of air from A to B in Figure 2.2. The potential energy (energy arising from moving a parcel to a greater height against gravity) of the system does not change, and movement from B to C requires energy to raise the centre of gravity of the parcel upwards, as in the movement of the parcel of air discussed so far.

However, movement from A to D could cause the centre of gravity of the parcel to be lowered, causing energy to be released. A full explanation of this type of convection, known as *sloping* or *baroclinic* convection, is beyond the scope of this book (for a review see for example Hide and Mason, 1975; Emanuel, 1994), but it provides the basic mechanism for the development of large scale atmospheric weather systems which produce most of the precipitation in mid-latitudes. Recent numerical model studies have identified the details of this type of convection (Fantini et al., 2012).

2.9   The growth of precipitation particles

The condensation of water vapour in the atmosphere, brought about by the movement of air upwards, provides water droplets or ice crystals in clouds. Such precipitation particles are denser than the air surrounding them, and therefore they begin to fall at a rate of a few centimetres per second. However, these particles will either evaporate in unsaturated air below the cloud, or be held suspended by vertical currents within the cloud. They will only be able to reach the ground as precipitation if they become large enough to stand evaporative losses and overcome upward air motions.

Jonas (1999) summarizes precipitation microphysics. The droplet concentration is determined by the balance between the rate of supply of vapour by cooling and the rate at which it is removed by deposition. Increases in the supply of vapour, for example with increased vertical air velocity in more vigorous clouds, will result in an increase in the number and density of particles which are activated to become droplets. The concentration of droplets will be reduced by the entrainment of dry air from the environment, and therefore, for example, the droplet concentration in stratocumulus clouds is often higher than in cumulus clouds for similar aerosol particle concentrations, despite the higher updraught in the latter.

The droplet concentration is crucial to the rate of growth of the droplets (see Appendix 2.1) since they compete for the available water. Droplets grow more rapidly by condensation in clouds formed in relatively unpolluted regions than in clouds with high droplet concentrations. The growth rate of droplets by concentration decreases as the radius increases, because the decreasing surface to volume ratio limits the speed with which water can condense. Condensation also tends to produce rather narrow cloud droplet spectra due to the decreasing growth rate as the droplets become larger. Entrainment of dry air into a cloud, which results in partial evaporation of some droplets and the incorporation into the cloud of new nuclei, acts to broaden these narrow spectra.

Over a hundred years ago it was recognized that processes other than condensation were needed to cause the growth of precipitation particles on observed timescales. In 1911 Wegener proposed that rain was formed by the melting of ice particles. The atmosphere normally contains many particles which can act as nuclei for the formation of water droplets. However, there are relatively few particles in the atmosphere which can act as nuclei for the formation of ice crystals except at low temperatures. Consequently clouds rarely become glaciated, i.e. composed mainly of ice crystals rather than water droplets, until their temperature is much less than 0 °C, typically around −15 °C.

In some clouds secondary ice nucleation processes may be active (Mossop, 1978). These can increase the ice particle concentrations significantly. Secondary nucleation is sensitive to temperature and to the cloud droplet spectra. At low temperatures, around −40 °C, nucleation of cloud droplets to form ice crystals will occur even without the presence of a nucleating particle, and therefore at such low temperatures very few clouds contain supercooled drops.

The relatively high saturation mixing ratio over water compared with that over ice, combined with lower concentrations of ice crystals than of water droplets, lead to much more rapid growth of ice crystals by deposition than is the case for droplets. When a cloud of water droplets in air at close to water saturation is cooled and a

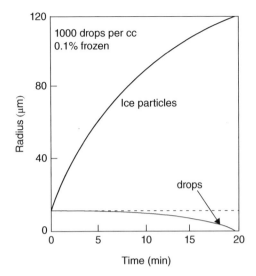

Figure 2.3 Illustrating the growth of ice spheres and decay of water drops in a mixture subject to a constant updraught of 1 m s^{-1}. The air is assumed to be initially saturated with respect to water. The dashed line shows the growth of the droplets in the absence of any ice particles (from Jonas, 1999)

small fraction of the droplets freeze, these ice particles grow very rapidly as shown in Figure 2.3. The rapid growth of the ice crystals at close to water saturation reduces the vapour mixing ratio to a value between those of saturation with respect to ice and water. The result is evaporation of the water droplets.

Ice crystals grow by sublimation when they exist in cloud together with super-cooled water droplets (section 2.6). As water vapour is removed from the air by this process, the air becomes unsaturated with respect to water, so the droplets evaporate. This continues until either all the droplets have been evaporated or the ice crystals become so large that they fall from the cloud. This process takes from 10 to 30 minutes, and as the ice crystals fall they may melt to form rain (droplets with radii $\geq 20\,\mu m$) which can reach the ground. A theory of this process was derived by Bergeron in 1935, and observations confirming the theory were made by Findeisen in 1939. Hence this process of rain formation became known as the *Wegener–Bergeron–Findeisen process*, sometimes shortened to the *Bergeron process*.

This explains the formation of precipitation in mid-latitudes where clouds usually extend well above the 0 °C level; because the cloud particles normally begin as cloud condensation nuclei, it is referred to as the *cold rain process*. However, it is observed that warm clouds with tops below 0 °C also produce rain. Indeed, such clouds occur in mid-latitudes as well as in the tropics. A *warm rain process* is required. It was discovered that cloud particles normally begin to form on *cloud condensation nuclei* (CCN), which consist of partially or completely soluble aerosol particles. As the cloud particles grow by condensation or sublimation on the CCN, they begin to fall and collect other particles. The type of precipitation which is formed by such collisions depends upon the types of cloud particles present. If the cloud contains only water then rain is formed and the process is known as *coalescence*. However, if only ice crystals exist then snow results and the process is called *aggregation*.

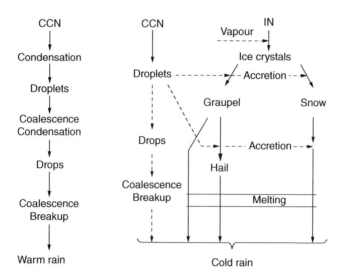

Figure 2.4   The evolution of warm and cold rain starting from cloud condensation nuclei (CCN) and ice nuclei (IN) (from List, 1977)

Owing to the very slow rate of growth of droplets by condensation once they have reached about 20 µm in radius, subsequent growth is mainly by the collision and aggregation process. The collection efficiencies of smaller cloud droplets are relatively low, as is the rate at which droplets encounter one another, owing to the small values of the terminal velocities. However, for larger droplets the process is much more effective, and can lead to the rapid growth of precipitation particles. The distance which large growing droplets fall, relative to the smaller cloud droplets, determines the maximum amount of growth that is possible within a cloud. Turbulent air motions increase the relative path of a fraction of the droplets and this increases the possibility of drizzle formation from shallow clouds such as stratocumulus. Growth by aggregation is also enhanced in clouds with high liquid water contents.

If both water droplets and ice crystals exist then ice and snow pellets (graupel) or hail may form and the process is accretion. Eventually the precipitation particles may reach such a size that they break up, beginning the process again. Detailed descriptions of these processes may be found in Mason (1971). They are sometimes collectively referred to as the *Langmuir process* in recognition of the work of Langmuir in 1948. This process may exist together with the Bergeron process. A summary is given in Figure 2.4.

There have been many reports of ice particle concentrations in convective clouds that are much higher than typical concentrations of ice nuclei (IN) (see for example Mossop et al., 1972; Hobbs and Rangno, 1985; Blyth and Latham, 1993). The large concentrations are true even taking shattering on probe tips into account. The explanation for the large concentrations in most clouds has been found to be the Hallett–Mossop process of splintering during riming (Hallett and Mossop, 1974). Freezing of supercooled raindrops to become instant rimers can be important in short lived clouds (for example Koenig, 1963). Columns of supercooled raindrops have been observed by radar (for example Caylor and Illingworth, 1987; Jameson et al., 1996).

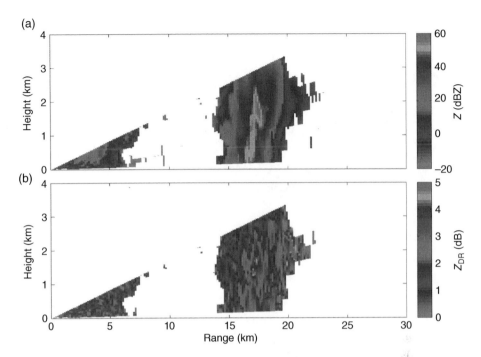

Figure 2.5 Illustrating (a) reflectivity Z (using horizontal polarization of the radar beam, see Chapter 5) and (b) differential reflectivity Z_{DR} (the ratio of horizontal and vertical polarizations of the radar beam, see Chapter 5) reconstructed from measurements made at constant elevation angles by the NCAS mobile X-band dual polarization Doppler radar on 3 August 2013 at 1327 UTC (from Blyth et al., 2015) (*see plate section for colour representation of this figure*)

Figure 2.5 shows radar reflectivity and differential radar reflectivity (Z_{DR}: see Chapter 5) measured in a convective cloud on 3 August 2013 during the COnvective Precipitation Experiment (COPE). It is clear that supercooled raindrops existed in a column in the updraught for a vertical distance of about 0.5 km above the 0 °C level at a height of 2.5 km (i.e. about –3 °C). The reflectivity signal remains strong at altitudes above the 3 km level despite the rapid decrease in Z_{DR}, suggesting that the frozen drops continued to move upwards.

2.10 Precipitation systems

We have seen that water vapour in the atmosphere may be condensed to form clouds when the air is lifted. Cloud particles may then grow by various processes to form precipitation. The atmospheric motions which produce the lifting necessary to trigger these processes are organized on various scales as shown in Table 2.2. Although mesoscale precipitation areas and rain bands are entered in this table as separate phenomena, in fact they occur within larger systems, and represent organization on an intermediate scale of localized convection. Therefore they will be discussed within a description of the larger phenomena. It is not the intention here to provide detailed dynamical and physical descriptions of each system. Comprehensive reviews have

Table 2.2   Scales of precipitation systems (partly after Browning, 1983a)

Precipitation system	Description	Horizontal scale (km)	Vertical velocity (cm s^{-1})
Localized convection	Precipitation from single clouds in the lower atmosphere or from convective generating cells within larger scale systems. They may occupy a large depth of the atmosphere, in which case they are referred to as *thunderstorms*	$10^{0.5}$	10^2
Mesoscale precipitation area (MPA)	Cluster of convective cells	$10^{1.5}$	10
Mesoscale rain band	Convective cells occurring in lines, sometimes almost two-dimensional	Width $10^{1.5}$ Length 10^2	10
Synoptic (or large) scale precipitation systems	Mid-latitude *depressions*	Width 10^2 Length 10^3	1
Tropical storms	Known as *hurricanes* in the southern part of the North Atlantic, *cyclones* in the northern part of the Indian Ocean, *typhoons* in the south-western part of the North Pacific, and *willy-willies* in the north-east of Australia	Radius $10^{1.5}$	10^2

been provided by Houze and Hobbs (1982), Browning (1983a) and Bader et al. (1995). The aim is to note the temporal and spatial distribution of precipitation and the likely intensity extremes which are associated with each system.

2.10.1   Localized convection

As the lower layers of the atmosphere are warmed by the Sun, or cold air passes over the warm sea, the air becomes less dense and rises. If the atmosphere is unstable then vertical motion will occur, giving rise to condensation of water vapour and the formation of cloud. The development of convection may be controlled either by boundary layer characteristics (as discussed by for example Hauck et al., 2011 and Couvreux et al., 2012) or by atmospheric synoptic situations (see for example Clark, 2011, who discussed a case of increasing instability produced by dynamic forcing from ascent associated with an approaching upper level cold pool). The same effect may be produced by air forced to rise over hills. Such clouds grow to a height of several kilometres in 20–40 minutes, producing rain of intensity dependent upon the depth of the cloud.

Equation 2.24 indicates that the vertical velocity will be dependent upon the difference between the environmental lapse rate and the SALR. If this difference is large, then the vertical motion will be large, and a small shower cloud may develop rapidly to a height of around 10 km, becoming a thunderstorm. Electricity is generated by processes which are still not fully understood, producing lightning. Within thunderstorms *hail* may be produced by precipitation particles sweeping up supercooled droplets in an updraught until they are of such a size that they fall out of the updraught. However, the hailstones may be caught in the updraught at a lower level as it is not usually vertical, and therefore they can be recycled several times before falling to the ground. This gives rise to a layered structure to the hail caused by growth in different parts of the cloud. A global climatology of hail has been provided by Hand and Cappelluti (2011).

In recent years a considerable amount of research has been carried out into the structure of thunderstorms. Two main types have been identified: multi-cell storms (Figure 2.6a) and supercell storms (Figure 2.6b). Most storms fall into the multi-cell category. They consist of several cells (clouds) at different stages of development at any one time. New cells may form where the outflow regions from other cells intersect (Purdom, 1976). The air of the downdraught spreads out laterally as shown in Figure 2.7. This has negative buoyancy, and the steady state speed of the outflow, which is a density current, is

$$V = k \sqrt{\left(gd\rho'/\rho_0\right)} \qquad (2.29)$$

where ρ' is the density of the current and ρ_0 is the density of the environment; d is the depth of the density current; and $k = \sqrt{2}$ in the case of the theoretical upper limit in V. Garcia-Carreras et al. (2013) describe the impact of convective cold pool outflows on numerical model biases in the Sahara.

The supercell storms have a structure which consists of a single storm-scale circulation comprising one giant updraught–downdraught pair. Either of these types of storm may produce a *tornado*, a region a few kilometres wide or less of high rotation (winds in excess of 200 mph or 322 km h^{-1}) and low pressure visualized as a funnel cloud. Rainfall may be very heavy, although usually of short duration. Thunderstorms occur in temperate and tropical regions throughout the world, and some examples of extreme storm rainfall totals are given in Table 2.3.

2.10.2   Mesoscale precipitation systems

Several thunderstorms may be grouped together within what Maddox (1980) has referred to as a *mesoscale convective complex* (MCC). The timescale is much longer than one hour, the lifetime of the individual cumulonimbi comprising the system. A typical timescale of an MCC in mid-latitudes is around three hours. Trapp (2013) notes that MCCs are represented as squall lines when near linear. Indeed, it has been suggested that 50–60% of the summer rainfall in the Great Plains and the Midwestern United States is due to MCCs.

The precipitation area is continuous and larger than precipitation from any individual storm, and this has been cited as evidence for an organized circulation system. In the early stages of development the MCC is very convective, and the rainfall is dominated by that produced by the individual cells. However, as the MCC develops, lifting on a larger scale occurs, although the actual mechanism is not yet fully understood. Such systems have also been identified in the United Kingdom (referred to as *mesoscale convective systems*: Browning and Hill, 1984), and elsewhere in Europe, for which sometimes the convective updraught of the associated thunderstorms is fed by air from just above a cold undercurrent (see Marsham et al., 2010; Browning

(a)

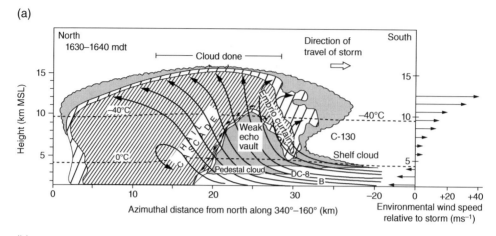

(b)

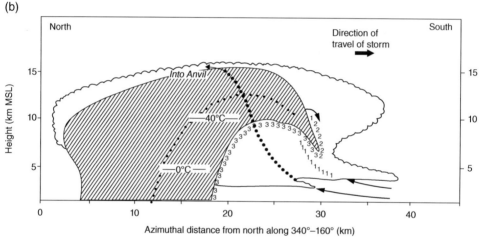

Figure 2.6 (a) Schematic model of a multi-cell hailstorm observed near Raymer, Colorado. It shows a vertical section along the storm's north to south (N–S) direction of travel, through a series of evolving cells. The solid lines are streamlines relative to the moving system. The hail cascade represents the trajectory of a hailstone during its growth from a small particle at cloud base. Lightly stippled shading represents the extent of cloud, and the three darker grades of stippled shading represent radar reflectivities of 35, 45 and 50 dBZ. Environmental winds (m s^{-1}, degrees from north) relative to the storm are shown on the right-hand side of the figure (from Browning et al., 1976). (b) Vertical section corresponding to (a). The radar echo distribution and cloud boundaries are as before. Trajectories 1, 2 and 3 represent the three stages in the growth of large hailstones. The transition from stage 2 to stage 3 corresponds to the re-entry of a hailstone embryo into the main updraught prior to a final up-and-down trajectory during which the hailstone may grow to a large size, especially if it grows close to the boundary of the vault as in the case of the indicated trajectory 3. Other, less favoured hailstones will grow a little farther from the edge of the vault, and will follow the dotted trajectory. Cloud particles growing within the updraught core are carried rapidly up and out into the anvil along trajectory 0 before they can attain precipitation size (from Browning and Foote, 1976)

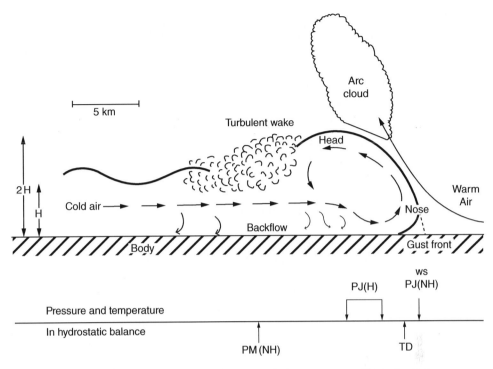

Figure 2.7 Schematic of a convective outflow in vertical cross-section: PM is the surface pressure reaching a local minimum; PJ is the increase or jump due to a dynamic deceleration between the cold and warm air masses; WS is the beginning of the wind shift as the cold air arrives; TD is the temperature break or drop as the cold air passes through; H is the head; and NH is the nose head (from Droegemeier and Wilhelmson, 1987; Trapp, 2013)

Table 2.3 Examples of rainfall totals from intense thunderstorms occurring worldwide

Location	Date	Duration (h)	Rainfall total (mm)
Middle Knoll, Dunsop Valley, Lancashire, UK	8 August 1967	1.5	117
Hampstead, London, UK	14 August 1975	2.5	170
Manchester, UK	6 August 1981	1	83
Boscastle, Devon, UK	16 August 2004	2	75
Ottery St Mary, Devon, UK	30 October 2008	3	200 (plus >200 mm small hail)
Var Department, southern France	15–16 June 2010	12	400
Big Thompson Canyon, Colorado, USA	31 July 1976	4	250
Strongstown, Pennsylvania, USA	20 July 1977	1	71
Rapid City, South Dakota, USA	10 June 1972	4	305
Sydney, Australia	11 November 1976	11 min	100
Kyushu, Japan	27 June 1972	1	100
		10 min	30

et al., 2011). Rainfall totals are similar to those from individual thunderstorms, although large storm totals may accrue. The MCC which caused severe flooding at Johnstown, Pennsylvania, USA on 19–20 July 1977 produced a storm total of over 300 mm in a 9 hour period.

A similar type of system occurs in tropical areas as a cloud cluster (see for example Leary and Houze, 1979), and both systems may develop into squall lines, the configuration of which varies depending upon the environmental wind shear (see Newton and Frankhauser, 1964; Houze and Hobbs, 1982). In all latitudes the squall line tends to travel by a combination of translation and discrete propagation as new cells form on the leading edge of the system.

2.10.3   Mid-latitude depressions

Although thunderstorms produce large amounts of rainfall in mid-latitudes at any one place, they are quite rare. Most of the rainfall originates from low pressure systems termed *depressions*. With the invention of the barometer in the late 17th century it became possible to measure the pressure of the atmosphere frequently, but it was not until the 19th century that particular atmospheric storms were studied using observations made at more than one place. The first theory involved the idea of a *cyclone*, a revolving cylinder of air with a more or less upright axis. No sharp discontinuities or *fronts* were admitted.

This concept held sway until the early 20th century when the idea of fronts within a rotating baroclinic fluid (that is, a fluid with both temperature and density gradients) was developed by what became known as the Bergen School in Norway (Bjerknes and Solberg, 1922). A wave developed on a boundary between warm and cold air, giving rise to a warm front and cold front structure, was considered. Rain from stratiform (layered) cloud is associated with the warm front, and rain from convective cloud and thunderstorms is associated with the cold front. The basis of a dynamical description of the depression is the release of energy in sloping (baroclinic) convection, as outlined in section 2.8.

This model of a depression has formed the basis of practical weather forecasting until recently. However, with the increasing availability of observations from radars and satellites (Bader et al., 1995) it became clear that, in practice, the structure of a depression was much more complex than that depicted by Bjerknes and Solberg (1922) (see Ludlam, 1966). Indeed, many small scale features were evident, and a new model, shown in Figure 2.8, to provide a framework for the development of these features was formulated by Browning (1971), Harrold (1973) and Carlson (1980) (for a review see Browning and Mason, 1980; Houze and Hobbs, 1982). Figure 2.9 shows the lifetime of a depression based upon microwave radiometer data from an open wave through a mature stage to an occluded stage.

In the new model (Figure 2.8) the distribution of rainfall is explained by reference to two main air flows, termed the *warm conveyor belt* (WCB) and the *cold conveyor belt* (CCB). The WCB originates at low levels to the south (in the northern hemisphere) of the depression, and ascends ahead of the cold front, turning southwards above the cold air ahead of the *surface warm front* (SWF). The influence of microphysical processes in the WCB has been discussed by Joos and Wernli (2012). The CCB originates to the north-east of the depression at low levels, ascending

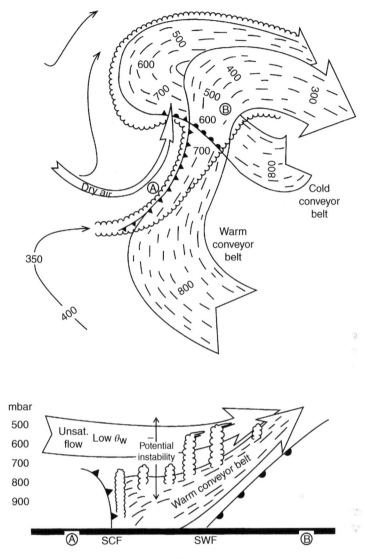

Figure 2.8 Model depicting the main features of the large scale flow that determine the distribution of cloud and precipitation in a mid-latitude depression. The arrows represent flow, the height of which is labelled in millibars. The scalloped line represents the outline of the cloud pattern, and the dashed shading represents the extent of the surface precipitation. SCF is surface cold front, SWF is surface warm front (from Browning and Mason, 1980; after Carlson, 1980)

ahead of the surface warm front beneath the WCB, and then turning to run parallel but below the WCB. Distinct from these conveyor belts, severe extratropical cyclones may produce transient mesoscale jets of air that descend from the tip of the cloud head associated with the cyclone towards the top of the boundary layer. These jets are known as sting-jets (Browning, 2004), which may cause damaging surface winds.

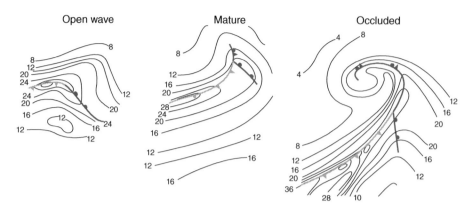

Figure 2.9 Schematic integrated water vapour values covering the lifetime of a depression based upon microwave radiometer data; units are kg m^{-2} (from Katsaros and Brown, 1991 in Bader et al., 1995)

The dry air behind the cold front often overrides the WCB in the region shown in Figure 2.8. This causes the air to become unstable if it is ascending, with the result that convection is released in small cells at mid-levels known as generating cells. These cells tend to occur in clusters and give rise to *mesoscale precipitation areas* (MPAs) which can produce moderate or heavy rainfall. This rainfall is produced by what is known as a *seeder–feeder mechanism* (Bergeron, 1950). Ice particles from the generating cells fall into lower level frontal cloud and act as natural seeding particles, causing a rapid growth of precipitation particles. This mechanism, together with dynamical processes, is important in producing a number of different types of rain band associated with the depression. Figure 2.10 summarizes the types of rain bands which have been identified.

Two types of cold frontal cloud are frequently observed on satellite images. The *classical cold front* appears as one band of mostly cold cloud tops. Alternatively a *split cold front* may be observed which appears as two adjacent bands with distinctly different cloud top temperatures. A cloud band forms as a result of warm moist air ascending through a deep layer of the atmosphere towards the cold side of the front and being undercut by descending cold dry air. The cloud and rain lie on and to the rear of the *surface cold front* (SCF). At the front the air may ascend rapidly through a layer 2–3 km deep, giving a narrow line of heavy precipitation. Ahead of this line is a low level jet (see Browning, 1985). Such bands, referred to as line convection, may extend for hundreds of kilometres, or may be broken into a series of elements separated by gaps. The line elements often move steadily for many hours, and are unaffected by small hills, provided the hills are not similar in height to the width of the convective line. If the hills are of similar dimensions then the convection usually dissipates. It has been found that sometimes small tornadoes occur in the gaps between elements (see Clark, 2013; James and Browning, 1979).

Rainfall totals from mid-latitude depressions are not generally large, although heavy rain may occur from thunderstorms developing on the cold front. Nevertheless, the precipitation often occurs steadily over long periods, particularly, as we shall discuss later, over hilly areas. In such cases, or when a system becomes stationary or slow moving, storm totals of over 100 mm in 24 hours may occur. For example,

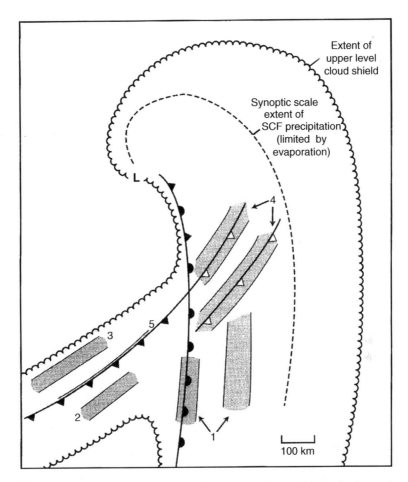

Figure 2.10   Major types of rain bands (stippled) observed in mid-latitude depressions: type 1, warm front bands; type 2, warm sector bands; type 3, cold front bands; type 4, upper level cold surge bands (situated along the leading edge of cold air overrunning the warm front); type 5, line convection (a narrow band or series of line elements along the cold front) (from Browning and Mason, 1980)

211 mm of rain fell over the Rhonda Valley, South Wales, on 10–11 November 1929; 340 mm in 24 hours in Cumbria on 5–6 December 2015; and 229 mm at Richmond, Natal, South Africa on 17–18 May 1959.

One special form of depression occurs in deep polar air over the north-east Atlantic, and is known as a *polar low*. These systems are small and resemble small tropical cyclones. It has been suggested that they arise either from the instability produced by the cold air moving over a relatively warmer ocean (Rasmussen, 1979), or because baroclinic instability occurs in a shallow layer (Harrold and Browning, 1969). More recent work (Føre et al., 2011) has enabled details of the life cycle of polar lows to be specified. Such systems may produce large amounts of snow or rain over short periods. Kolstad (2011) has described a global climatology of favourable conditions for polar lows in the North Atlantic, the North West Pacific and the southern hemisphere.

2.10.4   *Tropical storms*

Tropical cloud clusters were described in section 2.10.2. One favoured area for the formation of these clusters is the Inter-Tropical Convergence Zone (ITCZ) located close to the equator. Here the north-east trade winds of the northern hemisphere converge with the south-east trade winds of the southern hemisphere. A small fraction of clusters contain the necessary circulation structure to enable them to develop into *hurricanes* (*cyclones* or *typhoon*s: see Table 2.2). The upper clouds become circular and concentrated in, usually, several bands around the storm centre or *eye*. Heavy precipitation occurs in the main cloud band known as the *eye wall band* and other bands, although lighter precipitation does occur between them (for a review see Anthes, 1982). An example of the cloud pattern associated with a hurricane is shown in Figure 2.11. Recently it has been highlighted by Smith and Thomsen (2010) that the boundary layer may play an important role in the intensification of these systems. Marsham et al. (2013) describe the role of moist convection in the West African monsoon system.

Hurricanes may be symmetric, having a closed eye wall band, or asymmetric, having an eye wall band which is not closed. In a symmetric system the centre of the wind circulation (anticlockwise in the northern hemisphere, and clockwise in the southern hemisphere) is located in the centre of the circle defined by the eye wall boundary. The eye wall band often contracts as the storm develops. Eventually the eye wall band vanishes, and is replaced by a new eye wall band at a radius of

Figure 2.11   Hurricane Katrina over the Gulf of Mexico at its peak, 28 August 2005; category 5, highest winds 175 mph (280 km h^{-1}), lowest pressure 902 mbar (hPa) (*see plate section for colour representation of this figure*)

Table 2.4   Examples of extreme tropical storm rainfall totals occurring worldwide

Location	Date	Duration (h)	Rainfall total (mm)
Jamaica, West Indies	4–7 October 1963	24	508
Virginia, USA	20 August 1969	8	710
Horian Province, China	5–7 August 1975	72	1605
NW Australia	20 December 1976	6.33	356
Kyushu, Japan	12–13 September 1976	36	1950
Masirah, Arabian Gulf	12–13 June 1977	24	431
Hope, Hong Kong	2–4 August 1979	72	287
Commerson, Réunion, SW Indian Ocean	14–28 January 1980	360 (15 days)	6433
Grand Îlet, Réunion, SW Indian Ocean	14–28 January 1980	1	110
		12	1170
		24	1742
Cratère Commerson, Réunion, SW Indian Ocean Tropical Cyclone Gamede	24–27 February 2007	4 days	4600

50–150 km. This process continues and, each time the eye wall band is replaced, the central pressure of the storm rises and then falls. Asymmetric storms do not appear to follow this cycle. Although hurricanes form over the tropical oceans, the rainfall from them, when they reach land, may be very large indeed, as shown by the examples in Table 2.4.

2.10.5   Orographic effects on precipitation distribution

Orography can affect the distribution of rainfall by forcing air to ascend. The ascent may lead to condensation and rainfall (Figure 2.12a), or may result in convection (see for example Bennett et al., 2011) leading to rainfall (Figure 2.12b). A third process is the enhancement of rain over hills from pre-existing clouds as a result of a natural seeding process (Figure 2.12c). Mechanisms (a) and (c) in Figure 2.12 occur in frontal rainfall situations (section 2.10.3). The natural seeding process was first proposed by Bergeron (1935; 1950). Detailed observations in the United Kingdom (Browning, 1980) have established that this mechanism can substantially enhance rainfall amounts over small hills, as shown in Figure 2.13. Cannon et al. (2012) demonstrated that for tall and wide mountains (such as the southern Andes) precipitation forms efficiently through vapour deposition and collection, even in the absence of embedded convection. However, convection strongly enhances precipitation over short and narrow mountains as for the UK Pennines or the Oregon Coastal Range.

The increase in rainfall over hilly areas is very evident on average annual rainfall maps in mid-latitudes for areas exposed to moist air moving from the sea. Indeed there is almost a relationship between elevation and average annual rainfall amount, although this relationship does vary spatially, as identified by Brunsdon et al. (2001). In the lee of ranges of hills, descending air (or subsidence) may lead to reduced rainfall in areas known as *rain shadows*.

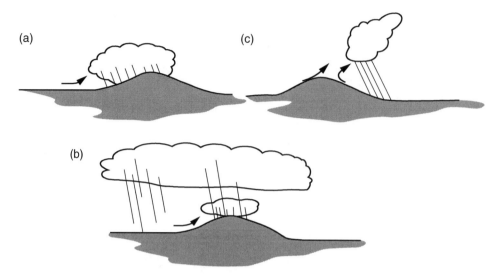

Figure 2.12 Mechanisms of orographic rain generation (see text) (from Smith, 1979)

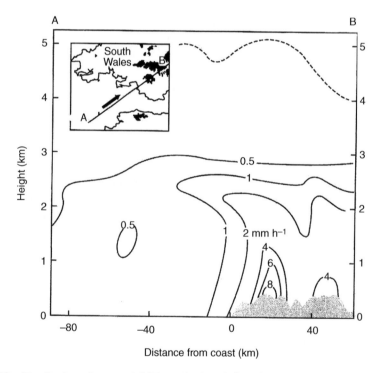

Distance from coast (km)

Figure 2.13 Distribution of mean rainfall intensity (mm h^{-1}) within a vertical section (AB) along the direction of motion of rain areas travelling from the sea over the hills of South Wales, UK, during a 5 hour period of warm sector rain. The inset shows the orientation of the section AB in relation to the coastline and hills (greater than 400 m altitude) (from Browning, 1980)

2.10.6   Topographical effects on precipitation distribution

Precipitation amount may be increased in coastal areas by the difference in the convergence of air caused by increased friction over the land compared to that experienced over the sea. Such an effect is reflected in monthly rainfall totals over the coastal area of The Netherlands (see KNMI, 1972).

Differences between land and sea can cause fronts associated with mid-latitude depressions to become stationary just off-shore, which can lead to very large falls of precipitation in coastal regions, particularly if the land is hilly (see Bosart, 1975, 1981; Browning, 1983b). The sea or land breezes associated with air circulation caused by temperature differences between the land and the sea may organize convection such that precipitation amounts can be very much enhanced. These effects are evident in winter on the Great Lakes in North America (for example Passarelli and Braham, 1981).

2.11   Global atmospheric circulation

In this chapter we have discussed the way in which atmospheric water vapour may be condensed to form precipitation. The average time that a molecule of water is resident in the atmosphere as vapour or within a cloud is 10–12 days (Miller, 1977). During this period the molecule may travel a considerable distance within the atmospheric circulation before being returned to the surface of the Earth as precipitation. Indeed, it is the global circulation of the atmosphere as a whole which dictates the occurrence of the atmospheric systems outlined earlier, and hence the global precipitation distribution.

The climatological distribution of annual rainfall has been summarized by Houze (1981) and Sellers (1965). A more recent visualization is shown in Figure 2.14. The precipitation maxima in mid-latitudes are associated with depressions, showers and thunderstorms. In the tropics the rainfall maximum is associated with cloud clusters and tropical storms. All these features are present most of the time somewhere, as illustrated in the satellite image in Figure 2.15. The water which passes from the atmosphere to the surface of the Earth is replenished by the processes of evaporation and transpiration, which will be discussed in the next chapter.

Appendix 2.1   Growth of a raindrop

Assume T_a is the temperature of the air at a distance a from a raindrop, and T_s is the temperature at the surface of the raindrop. Then

$$T_s - T_a = \frac{D\lambda(\rho_a - \rho_s)}{Mk} \qquad (2.30)$$

where λ is the latent heat of vaporization of water, M is the molecular weight of water, D is the coefficient of diffusion, ρ_a is the vapour density at a great distance from the raindrop, ρ_s is the vapour density at the surface of the raindrop, and k is the thermal

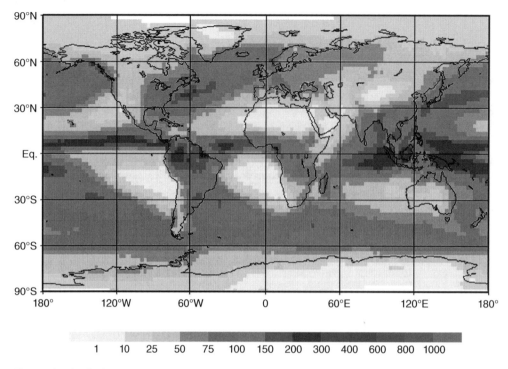

1   10   25   50   75   100   150   200   300   400   600   800   1000

Figure 2.14 Globally averaged annual precipitation 1980–2004 Jan–Dec (mm per month) from the Global Precipitation Climatology Project (GPCP) Version 2 (source: GPCC Visualizer) (*see plate section for colour representation of this figure*)

Figure 2.15 Satellite image in the visible wavelengths taken from visible data gathered by the European geostationary satellite MSG at 1200Z on 5 October 2012. Note the cloud associated with the ITCZ north of the equator. The cloud-free Sahara desert is clearly visible, as are cloud clusters in the tropics and mid-latitude depressions (courtesy of Eumetsat)

conductivity of the air. Taking $\Delta T = T_s - T_a$ and $\rho_s = \rho + (d\rho/dT)\Delta T$, where ρ is the saturation density at temperature T, we may derive (Best, 1957)

$$a_2^2 - a_1^2 = 2C(H-1)t \tag{2.31}$$

where $C = (D\rho Mk)/(Mk + D\lambda(d\rho/dT))$; $100H$ is the relative humidity of the environment; and t is the time for the raindrop to grow from radius a_1 to radius a_2. This formula does not apply when the raindrop is sufficiently small for the effect of dissolved salt to be appreciable. However the formula may be used for the evaporation of a droplet which is not completely saturated, i.e. $H < 1$.

Summary of key points in this chapter

1. The term 'precipitation' includes rain, drizzle, sleet, snow and hail.
2. The source of precipitation is water vapour, present in the atmosphere in varying amounts. When it is cooled, water condenses into cloud droplets.
3. The equation of state for a perfect gas relates pressure, volume and temperature using the molecular weight of the gas and the universal gas constant.
4. The first law of thermodynamics states that the energy of a system in a given state relative to a fixed normal state is equal to the algebraic sum of the mechanical equivalent of all the effects produced outside the system, when it passes in any way from the given state to the normal state.
5. Provided no condensation or evaporation occurs, atmospheric processes may be regarded as adiabatic, that is, as involving no exchange of heat between that part of the atmosphere under consideration and its surroundings.
6. The relationship between the equilibrium pressure and the temperature during a phase change is given by the Clausius–Clapeyron equation.
7. When condensation occurs in a sample of air and the resulting water falls out of the sample, then the mass of the sample changes and heat is lost. This is known as a pseudo-adiabatic process.
8. Moist air can become saturated and precipitation can occur as a result of movement of the air upwards as instability is released or convection occurs. Sloping convection, known also as baroclinic convection, is the basic mechanism for the development of large scale atmospheric weather systems.
9. The production of rain is known as the Wegener–Bergeron–Findeisen process. Cloud particles begin as cloud condensation nuclei above the $0\,°C$ level. This is the cold rain process. Warm clouds, below the $0\,°C$ level, produce rain via a warm rain process. If a cloud contains only water then rain is formed by coalescence, but if ice crystals exist then snow occurs by aggregation.
10. Horizontal scales of precipitation systems range from $10^{0.5}$ km to 10^3 km.

11. If the atmosphere is unstable then vertical air motion occurs, giving rise to condensation of water vapour and the formation of cloud. This gives rise to localized convection, multi-celled thunderstorms and supercells. Intense storms may produce strong outflow regions near the surface which can interact with each other, generating further cells. Several thunderstorms grouped together produce mesoscale convective complexes.

12. Mid-latitude depressions produce cold and warm fronts having banded structures. These may develop into occlusions as they mature. The distribution of rainfall is explained by air flows known as warm and cold conveyor belts.

13. Rainfall may be produced by the seeder–feeder mechanism in which ice particles from generating cells fall into lower level frontal cloud and act as natural seeding particles. This process may cause significant orographic rainfall.

14. Polar lows are produced in deep polar air over the north-east Atlantic. These are small and resemble tropical cyclones, and are produced by cold air moving over a relatively warmer ocean.

15. Tropical cloud clusters may contain the necessary circulation structure to enable them to develop into hurricanes.

16. The average time that a molecule of water is resident in the atmosphere as vapour or within a cloud is 10–12 days. During this period the molecule may travel a considerable distance within the atmospheric circulation.

17. The precipitable maxima in mid-latitudes are associated with depressions, showers and thunderstorms. In the tropics the rainfall maximum is associated with cloud clusters and tropical storms.

Problems

1. The density of water in cgs units is 1 g cm^{-3}. Compute (a) the density in SI (mks) units, (b) the specific volume in m^{3}kg^{-1} and (c) the SI (mks) molar specific volume.

2. (a) At standard conditions, how many kilograms of air are there in a room measuring 10 m × 10 m × 3 m? (b) The mean molecular weight of air is 29; that is, the mass of 1 kg mol of air is 29 kg. What is the weight of the air in pounds?

3. Calculate the vertical velocity of a parcel of air assuming there is no disturbance to the environment. Take the temperature difference between the parcel and the environment as 0.5 K and the environmental temperature at the surface as 280 K.

4. Assuming a constant updraught of 1 m s^{-1}, what is the size of an ice crystal and a raindrop after 10 minutes?

5. Describe the different types of convection.

6. Calculate the speed of outflow from a thunderstorm when the density of the environmental air is 0.00129 g cm^{-3} and the density of the air in the outflow region is 0.00136 g cm^{-3}. The depth of the outflow region is taken as 1 km.

7. Describe the banded structure of rainfall in a mid-latitude depression.

8. What is the vertical gradient of rainfall intensity over hills of height 0.5 km given a strong, moist, low level wind with a speed perpendicular to the hills of strength 20 m s^{-1}?

9. Describe the seeder–feeder mechanism.

10. What is the ITCZ?

References

Anthes, R.A. (1982) Tropical cyclones: their evolution, structure and effects. *Meteorol. Monogr.*, 19, no. 41.

Bader, M.J., Forbes, G.S., Grant, J.R., Lilley, R.B.E. and Waters, A.J. (1995) *Images in Weather Forecasting: A Practical Guide for Interpreting Satellite and Radar Imagery*. Cambridge University Press, Cambridge.

Bennett, L.J., Blyth, A.M., Burton, R.R., Gadian, A.M., Weckworth, T.M., Behrendt, A., Girolamo, P.D., Dorninger, M., Lock, S.-J., Smith, V.H. and Mobbs, S.D. (2011) Initiation of convection over the Black Forest mountains during COPS IOP 15a. *Q. J. R. Meteorol. Soc.*, 137, 176–189.

Bergeron, T. (1935) On the physics of cloud and precipitation. Proceedings of the 5th Assembly of the International Union of Geodesy Geophysics, 1933, vol. 2, pp. 156–161.

Bergeron, T. (1950) Uber den mechanisms der ausgiebigen niederschlage. *Ber. Dtsch. Wetterd.*, 12, 225–232.

Best, A.C. (1957) *Physics in Meteorology*. Pitman, London.

Bjerknes, J. and Solberg, H. (1922) Life cycle of cyclones and the polar front theory of atmospheric circulation. *Geofys. Pub.*, 3(1), 3–18.

Blyth, A.M. and Latham, J. (1993) Development of ice and precipitation in New Mexican summertime cumulus clouds. *Q. J. R. Meteorol. Soc.*, 119, 91–120.

Blyth, A.M., Bennett, L.J. and Collier, C.G. (2015) High-resolution observations of precipitation from cumulonimbus clouds, *Meteorol. Appl.*, 22, 75–89.

Bosart, L.F. (1975) New England coastal frontogenesis. *Q. J. R. Meteorol. Soc.*, 101, 957–978.

Bosart, L.F. (1981) The President's Day snowstorm of 18–19 February 1979: a subsynoptic-scale event. *Mon. Weather Rev.*, 109, 1542–1566.

Browning, K.A. (1971) Radar measurements of air motion near fronts. *Weather*, 26, 320–340.

Browning, K.A. (1980) Structure, mechanism and prediction of orographically enhanced rain in Britain. In *Orographic Effects in Planetary Flows*, ed. R. Hide and P.W. White, GARP Publications Series 23, pp. 85–114. World Meteorological Organization, Geneva.

Browning, K.A. (1983a) Mesoscale structure and mechanisms of frontal precipitation systems. Course on Mesoscale Meteorology, 30 May to 10 June, Pinnarpsbaden, Sweden, Lecture Notes II, No 17, SMHI.

Browning, K.A. (1983b) Air motion and precipitation growth in a major snowstorm. *Q. J. R. Meteorol. Soc.*, 109, 225–242.

Browning, K.A. (1985) Conceptual models of precipitation systems. *Meteorol. Mag.*, 114, 293–319.

Browning, K.A. (2004) The sting at the end of the tail: damaging winds associated with extratropical cyclones. *Q. J. R. Meteorol. Soc.*, 130, 375–399.

Browning, K.A. and Foote, G.B. (1976) Airflow and hail growth in supercell storms and some implications for hail suppression. *Q. J. R. Meteorol. Soc.*, 102, 499–533.

Browning, K.A. and Hill, F.F. (1984) Structure and evolution of a mesoscale convective system near the British Isles. *Q. J. R. Meteorol. Soc.*, 110, 897–913.

Browning, K.A. and Mason, B.J. (1980) Air motion and precipitation growth in frontal systems. *Pageoph., Birkhauser, Basel*, 119, 1–17.

Browning, K.A., Frankhauser, J.C., Chalon, F.P., Eccles, P.J., Strauch, R.C., Merrem, F.H., Musil, D.J., May, E.L. and San, W.R. (1976) Structure of an evolving hailstorm. V: synthesis and implications for hail growth and hail suppression. *Mon. Weather Rev.*, 104, 603–610.

Browning, K.A., Nichol, J.C., Marsham, J.H., Rogberg, P. and Norton, E.G. (2011) Layers of insect echoes near a thunderstorm and implications for the interpretation of radar data in terms of airflow. *Q. J. R. Meteorol. Soc.*, 137, 723–735.

Brunsdon, C., McClatchey, J. and Unwin, D.J. (2001) Spatial variations in the average rainfall–altitude relationship in Great Britain: an approach using geographically weighted regression. *Int. J. Climatol.*, 21, 455–465.

Cannon, D.J., Kirshbaun, D.J. and Gray, S.L. (2012) Under what conditions does embedded convection enhance orographic precipitation? *Q. J. R. Meteorol. Soc.*, 138, 391–406.

Carlson, T.N. (1980) Airflow through mid-latitude cyclones and the comma cloud pattern. *Mon. Weather Rev.*, 108, 1498–1509.

Caylor, I. and Illingworth, A. (1987) Radar observations and modelling of warm rain initiation. *Q. J. R. Meteorol. Soc.*, 113, 1171–1191.

Clark, M.A. (2011) An observational study of the exceptional 'Ottery St Mary' thunderstorm of 30 October 2008. *Meteorol. Appl.*, 18, 137–154.

Clark, M.R. (2013) A provisional climatology of cool-season convective lines in the UK. *Atmos. Res.*, 123, 180–196.

Couvreux, M., Rio, C., Guichad, F., Lothon, M., Bounid, D. and Gougou, A. (2012) Initiation of daytime local convection in a semi-arid region analysed with high resolution simulations and AMP observations. *Q. J. R. Meteorol. Soc.*, 138, 56–71.

Done, J.M., Craig, G.C., Gray, S.L. and Clark, P.A. (2012) Case-to-case variability of predictability of deep convection in a mesoscale model. *Q. J. R. Meteorol. Soc.*, 138(April part A), 638–648.

Droegemeier, K.K. and Wilhelmson, R.B. (1987) Numerical simulation of a thunderstorm outflow dynamics. I: *Outflow sensitivity experiments and turbulence dynamics. J. Atmos. Sci.*, 44, 1180–1210.

Emanuel, K.A. (1994) *Atmospheric Convection*. Oxford University Press, Oxford.

Fantini, M., Malguzzi, P. and Buzzi, A. (2012) Numerical study of slantwise circulations in a strongly sheared prefrontal environment. *Q. J. R. Meteorol. Soc.*, 138(April part A), 585–595.

Findeisen, W. (1939) Zur frage der regentropfenbildung in reinen easserwolken. *Meteorol. Z.*, 56, 365.

Føre, I., Kristjánsson, J.E., Saetra, Ø., Breivik, Ø., Røsting, B. and Shapiro, M. (2011) The full life cycle of a polar low over the Norwegian Sea observed by three research aircraft flights. *Q. J. R. Meteorol. Soc.*, 137, 1659–1673.

Garcia-Carreras, L., Marsham, J.H., Parker, D.J., Bain, C.L., Milton, S., Saci, A., Salah-Ferroudj, M., Ouchene, B. and Washington, R. (2013) The impact of convective cold pool outflows on model biases in the Sahara. *Geophys. Res. Lett.*, 40, 1–6.

Hallett, J. and Mossop, S. (1974) Production of secondary ice crystals during riming process. *Nature*, 249, 26–28.

Hand, W.H. and Cappelluti, G. (2011) A global hail climatology using the UK Met Office convection diagnosis procedure (CDP) and model analyses. *Meteorol. Appl.*, 18, 446–458.

Harrold, T.W. (1973) Mechanisms influencing the distribution of precipitation within baroclinic disturbances. *Q. J. R. Meteorol. Soc.*, 99, 232–251.

Harrold, T.W. and Browning, K.A. (1969) The polar low as a baroclinic disturbance. *Q. J. R. Meteorol. Soc.*, 95, 719–730.

Hauck, C., Barthlott, C., Krauss, L. and Kalchoff, N. (2011) Soil moisture variability and its influence on convective precipitation over complex terrain. *Q. J. R. Meteorol. Soc.*, 137, 42–56.

Hide, R. and Mason, B.J. (1975) Sloping convection in a rotating fluid. *Adv. Phys.*, 24(1), 47–100.

Hobbs, P. and Rangno, A. (1985) Ice particle concentrations in clouds. *J. Atmos. Sci.*, 42, 2523–2549.

Houze, R.A. (1981) Structure of atmospheric precipitation systems: a global survey. *Radio Sci.*, 16(5), 671–689.

Houze, R.A. and Hobbs, P.V. (1982) Organisation and structure of precipitating cloud systems. *Adv. Geophys.*, 24, 225–315.

James, P.K. and Browning, K.A. (1979) Mesoscale structure of line convection at surface cold fronts. *Q. J. R. Meteorol. Soc.*, 105, 371–382.

Jameson, A., Murphy, M. and Krider, E. (1996) Multiple-parameter radar observations of isolated Florida thunderstorms during the onset of electrification. *J. Appl. Meteorol.*, 35, 343–354.

Jonas, P. (1999) Precipitating cloud systems. *Chapter 3.4 in Global Energy and Water Cycles*, ed. K.A. Browning and R.J. Gurney, pp. 109–123. Cambridge University Press, Cambridge.

Joos, H. and Wernli, H. (2012) Influence of microphysical processes on the potential vorticity development in a warm conveyor belt: a case study with limited-area model COSMO. *Q. J. R. Meteorol. Soc.*, 138, 407–418.

Katsaros, K.B. and Brown, R.A. (1991) Legacy of the Seasat mission for studies of the atmosphere and air–sea–ice interactions. *Bull. Am. Meteorol. Soc.*, 72, 967–981.

KNMI (1972) *Klimaatatlas van Nederland.* Koninklijk Nederlands Meteorologisch Instituut. Staatsuitgeverij, 's-Gravenhage.

Koenig, L. (1963) The glaciating behaviour of small cumulonimbus clouds. *J. Atmos. Sci.*, 20, 29–47.

Kolstad, E.W. (2011) A global climatology of favourable conditions for polar lows. *Q. J. R. Meteorol. Soc.*, 137, 1749–1761.

Langmuir, I. (1948) The production of rain by a chain reaction in cumulus clouds at temperatures above freezing. *J. Meteorol.*, 5, 175–192.

Leary, C.A. and Houze, R.A. (1979) The structure and evolution of convection in a tropical cloud cluster. *J. Atmos. Sci.*, 36, 437–457.

List, R. (1977) The formation of rain. *Trans. R. Soc. Canada IV*, XV, 333–334.

Ludlam, F.H. (1966) The cyclone problem: a history of models of the cyclonic storm. Inaugural Lecture, 8 November, Imperial College, University of London.

Maddox, R.A. (1980) Mesoscale convective complexes. *Bull. Am. Meteorol. Soc.*, 61, 1374–1387.

Marsham, J.H., Browning, K.A., Nichol, J.C., Parker, D.J., Norton, E.G., Blyth, A.M., Corsmeier, U. and Perry, F.M. (2010) Multi-sensor observations of a wave beneath an impacting rear-inflow jet in an elevated mesoscale convective system. *Q. J. R. Meteorol. Soc.*, 136, 1788–1812.

Marsham, J.H., Dixon, N.S., Garcia-Carreras, L., Lister, G.M.S., Parker, D., Knippertz, P. and Birch, C.E. (2013) The role of moist convection in the West African monsoon system: insights from continental-scale convection-permitting simulations. *Geophys. Res. Lett.*, 40, 1843–1849.

Mason, B.J. (1971) *The Physics of Clouds.* Clarendon, Oxford.

Miller, D.H. (1977) *Water at the Surface of the Earth: An Introduction to Ecosystem Hydrodynamics.* Academic Press, New York.

Mossop, S. (1978) The influence of drop size distribution on the production of secondary ice particles during graupel growth. *Q. J. R. Meteorol. Soc.*, 104, 323–330.

Mossop, S., Cottis, R. and Bartlett, B. (1972) Ice crystal concentrations in cumulus and stratocumulus clouds. *Q. J. R. Meteorol. Soc.*, 98, 105–123.

Nace, R.L. (1967) Water resources: a global problem with local roots. *Environ. Sci. Technol.*, 1, 550–560.

Newton, C.W. and Frankhauser, J.C. (1964) On the movements of convective storms with emphasis on size discrimination in relation to water-budget requirements. *J. Appl. Meteorol.*, 3, 651–688.

Passarelli, R.E. and Braham, R.R. (1981) The role of the winter land breeze in the formation of Great Lake snow storms. *Bull. Am. Meteorol. Soc.*, 62(4), 482–491.

Purdom, J. (1976) Some uses of high resolution GOES imagery in mesoscale forecasting of convection and its behaviour. *Mon. Weather Rev.*, 104, 1494–1483.

Rasmussen, E. (1979) The polar low as an extratropical CISK disturbance. *Q. J. R. Meteorol. Soc.*, 105, 531–549.

Sellers, W.D. (1965) *Physical Climatology*. University of Chicago Press, Chicago.

Smith, R.B. (1979) The influence of mountains on the atmosphere. *Adv. Geophys.*, 21, 87–230.

Smith, R.K. and Thomsen, G.L. (2010) Dependence of tropical-cyclone intensification on the boundary-layer representation in a numerical model. *Q. J. R. Meteorol. Soc.*, 16, 167–1685.

Shuttleworth, W.J. (2012) *Terrestrial Hydrometeorology*. Wiley-Blackwell, Chichester.

Trapp, A.J. (2013) *Mesoscale-Convective Processes in the Atmosphere*. Cambridge University Press, Cambridge.

Tsonis, A.A. (2002) *An Introduction to Atmospheric Thermodynamics*. Cambridge University Press, Cambridge.

Wegener, A. (1911) *Thermodynamik der Atmosphäre*. Barth, Leipzig.

3

Evaporation and Transpiration

3.1 Introduction

Evaporation is the result of a balance between externally applied atmospheric demand and the availability of water at an evaporating surface. For evaporation to occur it is necessary to have (i) a supply of water; (ii) a source of heat such as direct solar energy R_c, sensible heat H, heat from the ground G, or stored heat from the water R_s; and (iii) a gradient of concentration $e_s - e_d$, where e_s is the saturated vapour pressure at temperature T and e_d is the vapour pressure for dry air. The physics of evaporation is the same regardless of the evaporating surface, although different surfaces, such as open water, bare soils and vegetation covers, impose different controls on the processes (for a review see for example Ward and Robinson, 2000).

Sometimes total evaporation is referred to as *evapotranspiration*; this is the combination of *evaporation* from soil and water surfaces and *transpiration* (evaporation of water from plant tissue) from plants, which is closely linked to photosynthesis (see Ward and Robinson, 2000). An important simplification has been the development of the concept of *potential evaporation*, determined by atmospheric demand, as distinct from the evaporation that actually takes place. The calculated potential rate may then be reduced to give an estimate of the actual evaporation according to the soil water content and a model whereby inadequate soil water supply reduces evaporation below the atmospheric demand. The potential evaporation concept assumes that there is at all times sufficient water to supply the requirements of the transpiring vegetation cover.

3.2 Modelling potential evaporation based upon observations

Thornthwaite (1948) derived a formula for estimating potential evapotranspiration based upon lysimeter and catchment observations in the central and eastern United States. Monthly potential evaporation E_p(cm) is calculated as an exponential function of air temperature:

$$E_p = \frac{16t}{I} \tag{3.1}$$

Hydrometeorology, First Edition. Christopher G. Collier.
© 2016 John Wiley & Sons, Ltd. Published 2016 by John Wiley & Sons, Ltd.
Companion website: www.wiley.com/go/collierhydrometerology

where t (°C) is the mean monthly temperature and I is the annual heat index, the sum of 12 individual monthly indices i, where

$$i = (t/5)^{1.514} \tag{3.2}$$

The only data requirements are mean air temperature and hours of daylight to adjust for the unequal day lengths in different months.

The formula seems to work well in the temperate, continental climate of North America where it was derived, and where temperature and radiation are strongly correlated. However, elsewhere it is less successful. It may produce severe underestimation of evaporation rates in dry climates (Monteith, 1985).

3.3   Aerodynamic approach

Dalton (1801) provided a means of quantifying evaporation rates using an aerodynamic approach. This led to the energy balance approach at the evaporating surface (see for example Wiesner, 1970; Calder, 1990). Later this approach was combined with the aerodynamic approach, and treatments of the availability of water at vegetative surfaces.

Early work led to the equation for evaporation E, sometimes referred to as the *Dalton equation*:

$$E = (e_0 - e)f(u) \tag{3.3}$$

where e_0 is the vapour pressure at the surface, e is the vapour pressure of the air, and $f(u)$ is some function of wind speed. Following the formulation of Calder (1990), the *aerodynamic function* is expressed as a resistance r_a to the transfer of water vapour down the vapour concentration gradient that exists between the evaporating surface and the atmosphere, as shown in Figure 3.1. The evaporation rate is expressed as the equivalent rate of latent heat λE, where λ is the latent heat of vaporization of water, so that

$$\lambda E = \frac{\lambda \rho_a \varepsilon}{r_a} \left(\frac{e_0}{p - e_0} - \frac{e}{p - e} \right) \tag{3.4}$$

where ρ_a is the density of dry air, ε is the ratio of the molecular weights of water and air, and p is the total air pressure. The total air pressure p is usually much larger than the terms $(1-\varepsilon)e$ and $(1-\varepsilon)e_0$; then equation 3.4 can be expressed as

$$\lambda E = \frac{\rho C_p (e_0 - e)}{r_a \gamma} \tag{3.5}$$

where C_p is the specific heat of air and $\gamma = C_p p / \lambda \varepsilon$ is known as the *psychrometric constant*, which has a value of 0.066 kPa °C^{-1} at a temperature of 20 °C and a pressure of 100 kPa (we discuss this further in section 3.6). Note that γ is not a constant,

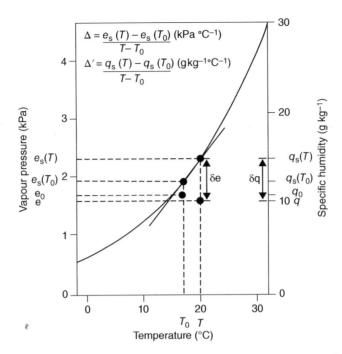

Figure 3.1 Relationship between saturated vapour pressure (or saturated specific humidity) and temperature, illustrating what is referred to as the *del approximation* (see section 3.5) (from Calder, 1990)

although it is referred to as such, because it varies with atmospheric pressure and temperature.

Specific humidity q may be defined as the mass of water vapour contained per unit mass of moist air (kg kg^{-1}; g kg^{-1}); then

$$q = \frac{\varepsilon e}{p - (1 - \varepsilon) e} \qquad (3.6)$$

Equation 3.3 may then be rewritten as (Calder, 1990)

$$\lambda E = \frac{\lambda \rho (q_0 - q)}{r_a} \qquad (3.7)$$

where ρ is the density of moist air and q_0 and q are the specific humidities of the surface and the air respectively. The sensible heat H may be given as a similar relationship,

$$H = \frac{\rho C_p (T_0 - T)}{r_a} \qquad (3.8)$$

where T_0 and T are the surface and air temperatures respectively.

3.4 Energy balance

The energy balance approach to estimating evaporation involves the process of vapour transfer, the factors in which are the following (Wiesner, 1970):
- R_c: the incoming radiation from the surface, which may be reflected short wave or long wave
- $R_c(1-\text{alb})$: the incoming radiation into a surface of albedo (alb)
- R_b: the outgoing radiation from the surface, which may be reflected short wave or long wave
- H: the sensible heat transfer from air to surface or in the opposite direction
- LE: the heat used in converting liquid to vapour, where L is the latent heat and E is the evaporation
- G: the heat flux into the ground or vegetation or in the opposite direction
- R_s: heat stored in the water
- R_p: heat converted to chemical energy in the process of photosynthesis
- R_i: heat moved into the air or out of the system by water inflow or outflow
- R_n: the net radiation received by the surface, where $R_n = R_c(1-\text{alb}) - R_b$.
  Hence at the evaporating surface, the conservation of energy gives

$$R_n = H + LE + G + R_s + R_p + R_i \tag{3.9}$$

Neglecting the storage terms, which are not usually significant over short periods, gives

$$R_n = H + LE + G \tag{3.10}$$

Combining the aerodynamic and energy balance methods of computing lake evaporation is outlined in Appendix 3.1.

3.5 The Penman equation

In practice it is not usually possible to solve either the aerodynamic or the energy balance equations as the surface temperature, the humidity terms and the sensible heat term are generally unknown. However, Penman (1948) provided a solution using knowledge of the change with temperature of the saturated vapour pressure of water. In Figure 3.1,

$$\Delta = \frac{e_s(T) - e_s(T_0)}{T - T_0} \tag{3.11}$$

where e_s represents the saturated vapour pressure at the temperature T or T_0. If the surface of the vegetation is wet, the humidity at the surface expressed as a vapour pressure is given by

$$e_0 = e_s(T_0) \tag{3.12}$$

Hence eliminating surface temperature, surface humidity and sensible heat from equations 3.7, 3.8, 3.11 and 3.12 (Calder, 1990) gives

$$\lambda E = \frac{\Delta H + \dfrac{\rho C_p \left(e_s(T) - e \right)}{r_a}}{\Delta + \gamma} \tag{3.13}$$

A similar equation can be derived in terms of specific humidity.

Subsequently several developments of the Penman equation have been derived to allow for the empirical aerodynamic and net radiation terms. Thom and Oliver (1977) suggested two additional parameters m and n to take account of different roughness and surface factors, giving

$$E_T = \frac{\Delta Q_n + m\gamma E_a}{\Delta + \gamma (1 + n)} \tag{3.14}$$

where E_T is the potential transpiration, E_a is the aerodynamic evaporation rate, Q_n is the evaporation equivalent of the available energy, and typical values are $m = 2.5$ and $n = 1.4$. An alternative equation was developed by Priestley and Taylor (1972) for use over land sites:

$$\lambda E = \alpha \frac{\Delta}{\Delta + \gamma} (R_n - s) \tag{3.15}$$

where α is a constant fraction expressing the advective term of the equilibrium rate. A typical value for short crops was found to be 1.26.

Monteith (1985) introduced another parameter r_s into equation 3.13 to deal with the resistance that crops provide against the transfer of water from within their internal organs (Figure 3.2). The result is the *Penman–Monteith equation*,

$$\lambda E = \frac{\Delta H + \dfrac{\rho C_p \delta q}{r_a}}{\Delta + \gamma \left(1 + \dfrac{r_s}{r_a} \right)} \tag{3.16}$$

This form of the equation is that most generally used by hydrometeorologists. When vegetation surfaces are covered by precipitation, the surface resistance term in this equation tends to zero. For aerodynamically rough vegetation such as forest ($r_a \approx 3.5\,\mathrm{s\,m^{-1}}$) the aerodynamic term is often much larger than the radiation term and high rates of evaporation may occur (see for example Calder, 1979). When the surface is covered by snow or ice, the latent heat of fusion must be taken into account.

3.6 Sensible and water vapour fluxes

Eddy diffusion theory (flux–gradient theory) provides a model for the transport of momentum, heat and water vapour between the surface and the atmosphere (see for example Mak, 2011). Transfer coefficients for the various entities may be calculated.

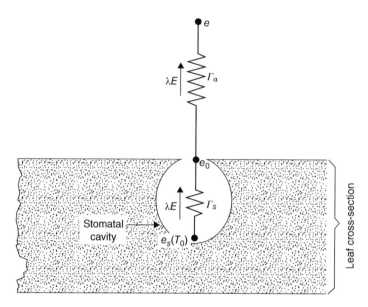

Figure 3.2   Schematic diagram illustrating how the latent heat flux λE is driven by humidity gradients between the inside of the stomatal cavity $e_s(T_0)$ and the leaf surface e_0 and the bulk atmosphere e against the stomatal and aerodynamic transfer resistances (from Calder, 1990)

Bowen (1926) showed that heat and water transfer, assuming laminar flow, are proportional: hence the *Bowen ratio*

$$\beta = \frac{H}{LE} \tag{3.17}$$

This means that there is a constant division of the available energy between the transfer of heat and water vapour. The Bowen ratio may be given by

$$\beta = \frac{\rho C_p K_{\mathrm{H}} \dfrac{\partial T}{\partial z}}{L K_{\mathrm{W}} \dfrac{\partial q}{\partial z}} \tag{3.18}$$

where K_{H} is the transfer coefficient of heat, K_{W} is the transfer coefficient of water vapour, C_p is the specific heat at constant pressure, and the psychrometric constant is $\gamma = C_p p/0.622\lambda = 0.0677\,\mathrm{kPa}\,{}^{\circ}\mathrm{C}^{-1}$, where p is the total air pressure and λ is the latent heat of vaporization of water at $20\,{}^{\circ}\mathrm{C}$ and a pressure of 101.2 kPa. Δq and ΔT are the specific humidity and temperature differences measured over the same height interval. However, in convective conditions Bowen's ratio and the ratio of K_{H} and K_{W} can vary considerably.

3.7   Evaporation of water from wet vegetation surfaces: the interception process

Early studies of the evaporation of water from wet vegetative surfaces, known as *transpiration*, attempted to relate interception losses to rainfall amounts through simple empirical relationships. Rutter et al. (1971; 1975) considered the water balance of stored precipitation on the canopy and related it to rates of precipitation. Inputs to canopy storage are assumed equal to the precipitation rate minus a constant *free throughfall* fraction p. Drainage from the canopy is uniquely related to the canopy storage via two drainage parameters b and k, and evaporation is assumed proportional to the canopy storage for storage less than the canopy capacity S. Following the formulation of Calder (1990),

$$\frac{dc}{dt} = Q - k\left(e^{bc} - 1\right)$$

(3.19)

where

$$Q = (1-p)R - E^{PM}\frac{c}{S} \quad \text{for } c < S$$

(3.20)

$$Q = (1-p)R - E^{PM} \quad \text{for } c \geq S$$

(3.21)

and where c is the canopy storage (mm), R is the precipitation rate (mm min^{-1}), b is a drainage parameter (mm^{-1}), k is a second drainage parameter (mm min^{-1}), p is the free throughfall fraction (dimensionless), S is the canopy storage capacity (mm), Q is the rate of precipitation plus evaporation (mm min^{-1}), and E^{PM} is the Penman–Monteith evaporation rate in wet conditions, calculated from equation 3.13 (mm min^{-1}).

An alternative approach based upon a stochastic model of the processes involved was proposed by Calder (1987). The surface area of a tree is considered to be composed of elemental areas, each of which has the same probability of being struck by raindrops. If the elemental areas can retain an average number of drops then the excess number of drops will be shed by drainage.

3.8   Measuring evaporation and transpiration

If *soil moisture content* (SMC) over wide areas can be monitored accurately on a regular basis, then evapotranspiration can be deduced. A number of methods of observing soil moisture content are currently used, including neutron probes and Penman–Monteith methods based upon meteorological observations which we discuss later in this chapter. Fox et al. (2000) discuss the use of data from *synthetic aperture radar* (SAR) to measure soil moisture in a naturally vegetated upland area using the surface water and radiation balance equations.

Figure 3.3 shows both soil moisture content measurements made using a capacitance probe, a grass weighing lysimeter and the satellite SAR, and the retrieved cumulative evapotranspiration (ET) totals from a capacitance probe and a grass lysimeter. Estimates of evaporation and evapotranspiration may be made using

(a) (b)

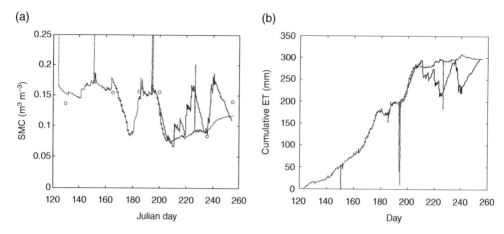

Figure 3.3 (a) Comparison of soil moisture content as measured by a capacitance probe (fainter line), lysimeter (darker line) and SAR (circles). (b) Retrieved cumulative evapotranspiration totals from a capacitance probe (fainter line) and a lysimeter (darker line) (from Fox et al., 2000)

tracers such as deuterium implanted in trees and plants (Calder et al., 1986), or more usually using evaporation pans and weighing devices.

A *time domain reflectometry* (TDR) instrument, model 6050X1 Trase System, was used in the study described by Kong et al. (2011). The TDR instrument can instantaneously measure volumetric water content of soils due to the great difference in the dielectric constant of water from the other constituents in the soil (Topp et al., 1980). While the standard probe of the TDR instrument is 15 cm, two 8 cm mini-probes were used on some occasions in order to make measurements more comparable with the estimates from satellite images and the MOSES model (see later) in the near-surface layer. *In situ* soil moisture data from different measurement depths (8 cm against 15 cm) could cause ±3.5% error for the individual point; however the error for the field mean is approximately ±1% which can be neglected.

3.9   Water circulation in the soil–plant–atmosphere continuum

Soil, plants and atmosphere constitute a physical entity from the viewpoint of water transfer. Philip (1966) was the first to use the term *soil–plant–atmosphere continuum* (SPAC). Water potential Ψ in the SPAC describes the effect of different linking forces, such as osmosis, capillarity and imbibition, which exist between water molecules and the component parts of the soil or plant. It may be defined as the work which it is necessary to furnish to a unit mass of water at one point in the system to transfer it to a reference state corresponding to that of free water at the same temperature at a reference point (e.g. soil level) (Guyot, 1998). The potential defined in this way has a negative value as it is necessary to do a certain amount of work to overcome the forces restraining the water. The work which must be done is expressed in $J\,m^{-3}$ which is equivalent to a pressure expressed in pascals. Hence water potentials are generally expressed in Pa, kPa or $MPa\left(1\,MPa = 10^{6}\,Pa\right)$.

It is possible to show that water potential can also be expressed as a function of the relative humidity of air in equilibrium with bound water at the same temperature.

Hence

$$\Psi = \frac{RT}{V_w^0} \ln\left(\frac{e}{e_s}\right) \tag{3.22}$$

where R is the ideal gas constant, T is the temperature in kelvins, e is the water vapour pressure in equilibrium with bound water, e_s is the saturation vapour pressure of water at the same temperature, and $V_w^0 = 18 \times 10^{-6} \, m^3 mol^{-1}$ is the molar volume of water.

Figure 3.4 shows schematically how the fall in water potential occurs in the course of the SPAC. Curve 1 corresponds to optimal water supply conditions. The water potential of the soil, and therefore of the external root surfaces, will have a value close to zero, corresponding to its maximal value. In leaves, water potential is weaker, but still greater than that which would bring on withering. Therefore the stomata can open fully.

In the situation represented by curve 2, the soil potential is still fairly high ($\Psi_s = -0.5\,MP_a$), but the water flux in the plant is such that water potentials drop sharply. At the level of the roots, the potential becomes such that values triggering closure of the stomata are approached. These conditions are similar to those met during a warm sunny day, even for a well watered crop. It only needs a small increase in evapotranspiration for the stomata to be closed, leading to a reduction in transpiration, and the gas exchanges of photosynthesis. Curves 3 and 4 correspond to conditions of insufficient water supply and relatively dry soil. In curve 3 climatic demands are moderate, and the water potential of roots and leaves is relatively low, which would lead to a partial closure of the stomata. In curve 4 the plant is withered up and the stomata are completely closed, leading to the onset of wilting point. The shaded

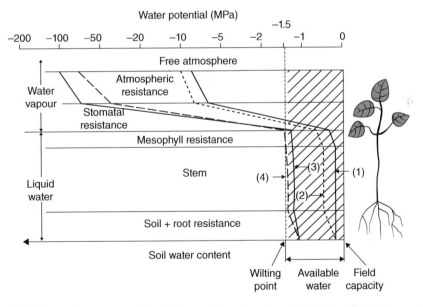

Figure 3.4 Schematic representation of the variation of water potential in different sectors of the SPAC: (1) moist soil, low evaporation rate; (2) moist soil, high evaporation rate; (3) dry soil, low evaporation rate; (4) dry soil, high evaporation rate (from Cruiziat, 1991; Guyot, 1998)

part of Figure 3.4 illustrates the range of variation in water potential of the plant compared with that of the atmosphere.

3.10   Water circulation and transpiration

Water circulates through a plant from the soil and into the atmosphere by means of a difference in water potential along the way, and there is hydraulic continuity of soil, plant and atmosphere. Apart from certain periods such as the onset of spring, when rising of the sap can play a very significant role, the driving force of water circulation is leaf transpiration.

Leaf transpiration is a function of the energy received by the leaves, and the state of the stomata. Transpiration causes water to be extracted from the soil through the plant against gravity and resistances to circulation such as frictional and viscous forces. Guyot (1998) describes these processes in more detail. Transpiration occurs mainly at the surface of substomatal cavity walls; below these and down to the ground, water is found as a liquid. It propagates through tissues, partly through a complex capillary network and partly through cell walls. Stomatal transpiration represents about 90% of the total over 24 hours.

When open, stomata make up only 1–2% of the total leaf surface, being only 5–30 µm wide. Generally a leaf with stomata fully open evaporates an amount of water comparable to that of a water covered surface of the same dimensions. Note that 99% of the passage of water through the plant is usually in liquid phase.

Modelling soil moisture wetness is described in Appendix 3.2.

3.11   Water flux in plants

A conservative water flux is the condition in which absorption in a plant is equal to transpiration from the plant. This cannot be attained during plant growth, as the increase in size of the plant results in the water content remaining constant to maintain water potential. However, if withering occurs during the day the fluxes can be conservative. In these conditions an analogy with Ohm's law may be valid, and the plant likened, where water circulation is concerned, to resistances in series, as shown in Figure 3.5a.

If V is the volume of water contained in a plant and Ψ its potential, then the water capacity C_w may be defined as

$$C_w = \frac{dV}{d\Psi} \tag{3.23}$$

Water capacity is analogous to thermal capacity, or to the capacity of a condenser. It varies with potential, and the contribution of different water capacities in plant sections allows an assessment of the difference which may exist between absorption of water and transpiration. This is a non-conservative flux (Figure 3.5b) which is much nearer reality.

Figure 3.6 shows the phase difference between transpiration and absorption in plants. In a single day the amount of water transpired by a plant is approximately equal to that absorbed by the roots (Klepper, 1990). Therefore the flux of water through the plant is conservative.

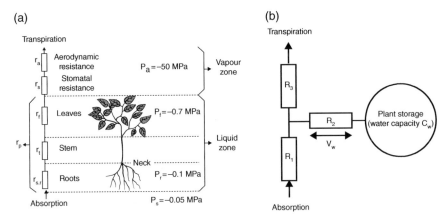

Figure 3.5 (a) Schematic representation of a plant in ohmic form for a conservative flux (from Cruiziat and Tyree, 1990; Guyot, 1998); (b) schematic representation of a plant in ohmic form for a non-conservative flux (from Guyot, 1998)

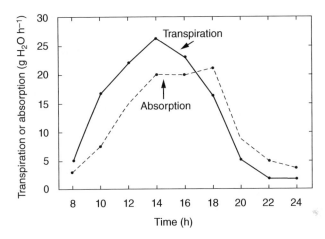

Figure 3.6 Illustrating the phase difference between transpiration and absorption in plants (from Guyot, 1998)

3.12   Modelling land surface temperatures and fluxes

Numerous parameterizations have been published for modelling land surface temperatures and fluxes, often with the inclusion of a canopy model, as listed by Rooney and Claxton (2006) (e.g. Deardorff, 1978; Sellers et al., 1986; Choudhury and Monteith, 1988; Mintz and Walker, 1993; Bosilovich and Sun, 1995; and the review by Avissar and Verstraete, 1990). In the United Kingdom the first operational use of equation 3.16 was the Meteorological Office Rainfall and Evaporation Calculation System (MORECS), which has been described by Thompson et al. (1981). The net short wave radiation R_{NS} at the surface was derived from

$$R_{NS} = (1-\alpha)R_C \tag{3.24}$$

Table 3.1   MORECS albedos (from Thompson et al., 1981)

Surface type	Albedo	Surface type	Albedo
Grass	0.25	Upland	0.25
Cereals	0.25	Riparian	0.25
(wheat, barley)		Bare rock	0.12
Potatoes	0.25	Conifers	0.12
(early and maincrop)		Impervious urban	0.10
Sugar beet	0.25	Water	0.05
Orchards	0.25	Bare soil (dry)	0.10, 0.20, 0.30[*]
Deciduous trees	0.17	Bare soil (wet)	0.05, 0.10, 0.15[*]

[*] Values refer respectively to soils of high, medium and low available water capacity without developed crop cover and originate from field experiments.

where α is the albedo of the surface and R_C is the downward daily total of short wave radiation at the surface evaluated using the procedure described by Cowley (1978). Values of the albedo used for different surfaces are given in Table 3.1, and refer to a fully developed crop cover. An incomplete cover was allowed for by an empirical correction based upon a leaf area index. Estimates of incoming long wave radiation are made using equations of radiative transfer, and the technique described by Brutsaert (1975). Heat storage in soil, available water capacity, and aerodynamic resistance allowing for crop height are also derived in MORECS. Hence, values of daily potential evapotranspiration, actual evaporation, soil moisture and hydrologically effective rainfall for 40 km × 40 km squares are derived.

A replacement system known as the Met Office Surface Exchange Scheme (MOSES) was developed in the 1990s and introduced operationally around 2000 (Cox et al., 1998; 1999; Essery et al., 2001) to model the land surface sensible and latent heat fluxes. This system incorporated a canopy model and a subsurface moisture and temperature model, and has been implemented in the Nimrod now-casting system as the MOSES-PDM (probability distributed model) (Moore, 1985; Blyth, 2002; Smith et al., 2006).

Rooney and Claxton (2006) describe MOSES, in which heterogeneous surfaces may be treated using a tiled representation. The model is one-dimensional, with explicit fields of subsoil temperature and moisture. After the model has been initialized with the settings of a particular location, it can then be forced with meteorological data to produce estimates of the surface heat fluxes and temperature. The data used to force the model are surface downward short wave irradiance, surface downward long wave irradiance, rainfall rate, snowfall rate, air temperature, westerly wind component, southerly wind component, surface pressure and specific humidity. Generally these forcing data are supplied by another model, although actual observations can be used. MOSES assumes that the wind, temperature and humidity inputs are those from the first model level, i.e. $z_1 = 10$ m.

Bulk aerodynamic formulae are used in MOSES to calculate the surface heat fluxes from the mean surface and atmospheric (first model level) values of temperature and humidity. These involve an exchange coefficient formulation based on the momentum and scalar roughness lengths z_{0M} and z_{0S} respectively, von Karman's constant k, and the bulk Richardson number R_{iB}. The surface exchange coefficient C_H is obtained as $C_H = f_H C_{Hn}$, where

$$C_{Hn} = k^2 \left[\ln\left(\frac{z_1 + z_{0M}}{z_{0M}}\right) \ln\left(\frac{z_1 + z_{0M}}{z_{0S}}\right) \right]^{-1} \tag{3.25}$$

$$f_H = \begin{cases} \left(1 + 10 R_{iB}/P_r\right)^{-1} & \text{for } R_{iB} \geq 0 \, (\text{stable}) \\ 1 - 10 R_{iB} \left[1 + 10 C_{Hn} \sqrt{(-R_{iB})/f_z}\right]^{-1} & \text{for } R_{iB} < 0 \, (\text{unstable}) \end{cases} \tag{3.26}$$

$$f_z = 0.25 \left(\frac{z_{0M}}{z_1 + z_{0M}}\right)^{0.5} \tag{3.27}$$

$$P_r = \ln\left(\frac{z_1 + z_{0M}}{z_{0M}}\right) \left[\ln\left(\frac{z_1 + z_{0M}}{z_{0S}}\right)\right]^{-1} \tag{3.28}$$

The default value of scalar roughness is $z_{0S} = 0.1 z_{0M}$.

The sensible heat flux is calculated using a purely aerodynamic scheme. The latent heat flux comprises aerodynamic evaporation from saturated surfaces, e.g. lakes, wet vegetation canopies and snow, as well as transpiration by plants and evaporation from bare soil controlled by a surface conductance g^*. The calculation of g^* uses a plant model taking wind, pressure, humidity, soil moisture, surface temperature and short wave radiation as input (see Cox et al., 1998).

A comparison between MOSES-PDM and MORECS has been reported by Hough (2003). Figure 3.7 shows a flow chart for MOSES and MORECS. In the original version of MOSES surface runoff is generated if the rainfall intensity is greater than the infiltration capacity of the soil. If any soil layer is supersaturated, the water is routed to the layer below or to drainage if it is the bottom layer. In the PDM enhancement to MOSES, surface runoff is generated assuming that the water holding capacity of the soil has a specified unresolved (statistical) distribution within the area covered by the grid square, i.e. parts of a grid square will be saturated and others will not be saturated.

In spring and early summer it was found that MOSES soil moisture deficit (SMD) tended not to increase as rapidly as MORECS, so that by early June MOSES SMD was rather less than MORECS. However in the very dry summers of 1975–6 the MOSES SMD continued to increase whereas the MORECS SMD remained fixed at a maximum value. It was concluded that MORECS has limitations and MOSES-PDM attempts to correct these limitations.

In recent years a Met Office Joint UK Land Environment System (JULES) has evolved from the MOSES system. This is packaged as a community land surface model that can be used both offline and as part of the Met Office numerical weather prediction and climate Unified Modelling System. This allows studies to be made of

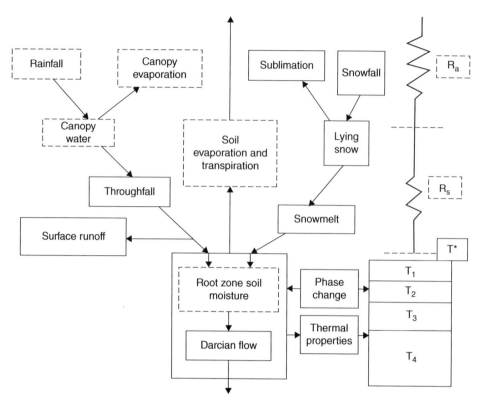

Figure 3.7   Flow chart for MOSES and MORECS. All boxes are part of MOSES; those with a dashed border show the structure of MORECS. The arrows show the flows of water. T_1 etc. are the mean temperatures of the four soil layers in MOSES, and T^* is a 'skin' surface temperature which is calculated for each MOSES land use. R_s is the canopy resistance to moisture flow, which has fixed monthly values in MORECS but is interactive in MOSES. R_a is the resistance to vapour flow from the canopy to the level where temperature and humidity are measured. The box labelled 'phase change' refers to the physical processes when ice changes to water and back again. The box showing 'Darcian flow' refers to flow for which the specific discharge is related to the hydraulic gradient (from Hough, 2003)

the impact of land surface processes on weather prediction and climate. For example Taylor et al. (2012) note that rain occurs more often over drier soils.

3.13   Soil–vegetation–atmosphere transfer schemes

Shuttleworth (2012) lists the requirements in a soil–vegetation–atmosphere transfer (SVAT) scheme as follows:

1. *Basic requirements in meteorological models*
   - Momentum absorbed from the atmosphere by the land surface.
   - Proportion of incoming solar radiation captured by the land surface.
   - Outgoing long wave radiation (calculated from area average land surface temperature).

- Effective area-average surface temperature of the soil–vegetation–atmosphere interface.
- Area-average fraction of surface energy leaving as latent heat (with the remainder leaving as sensible heat).
- Area-average of energy entering or leaving storage in the soil–vegetation–atmosphere interface.

2. *Required in hydrometeorological models to better estimate area-average latent heat and to describe the hydrological impacts of weather and climate*
   - Area-average partitioning of surface water into evapotranspiration, soil moisture, surface runoff, interflow and base flow.

3. *Required in meteorological models to describe indirect effect of land surfaces on climate through their contribution to changes in atmospheric composition*
   - Area-average exchange of carbon dioxide (and possibly other trace gases).

Over the last 40 years or so the complexity and realism of the SVATs have increased significantly (Shuttleworth, 2012). A schematic diagram of the SVATs used in early studies of the effect of land surfaces on weather and climate is shown in Figure 3.8. A form of potential evaporation rate was assumed and calculated using the Penman–Monteith equation. Zero surface resistance, and aerodynamic resistance equal to that for momentum transfer in neutral conditions with an appropriate roughness length, were assumed. A water balance is carried out of the water level in a hypothetical 'bucket' located at the land surface, filled with precipitation, and emptied by evaporation when the water level d' stored in the bucket exceeds a critical d_{max} and also by runoff.

Improvements by Deardoff (Deardorff, 1978) included the representation of soil heat fluxes; Dickinson et al. (1986) and Sellers et al. (1986) also improved the representation of vegetation controls on evaporation. Further improvements of the representation of hydrological processes (see for example Liang et al., 1994; Schaake et al., 1996) are shown in Figure 3.9. Yet further improvements by for example Bonan (1995) and Sellers et al. (1996) led to a representation of the carbon cycle in SVATs. Work to develop SVATs further continues (see Shuttleworth, 2012).

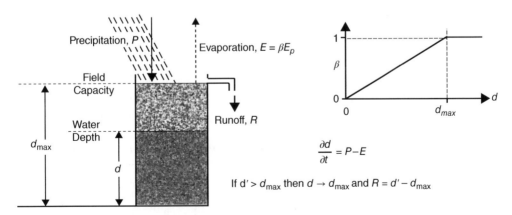

Figure 3.8   Schematic diagram of early SVATs (after Budyko, 1956 from Shuttleworth, 2012)

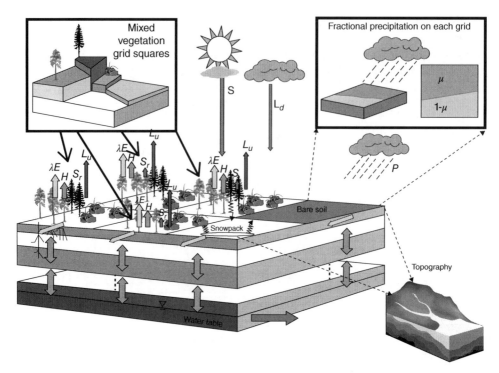

Figure 3.9 Schematic diagram of SVATs with improved representation of hydrologic processes: S_r is canopy capacity (saturation of canopy), H is sensible heat, L is latent heat, P is precipitation, S is incoming solar radiation, μ is fraction of precipitation in each grid square (from Shuttleworth, 2012)

3.14 Estimation of large scale evapotranspiration and total water storage in a river basin

Following a simplified version of the atmospheric water balance from Oki (1999) available over short timescales such as months or days,

$$E = \frac{\partial W}{\partial t} + \nabla_H \cdot \mathbf{Q} + P \tag{3.29}$$

where E is evapotranspiration, W is precipitable water, $\mathbf{Q}$ is the vertically integrated two-dimensional water vapour flux, ∇_H is the horizontal divergence and P is precipitation. The region over which the evapotranspiration is estimated is not limited to a river basin.

The water storage including soil moisture, ground water, surface water and snow accumulation may be estimated from

$$\frac{\partial S}{\partial t} = -\frac{\partial W}{\partial t} - \nabla_H \cdot \mathbf{Q} - \nabla_H \cdot \mathbf{R}_0 \tag{3.30}$$

This indicates that the change of water storage on land can be estimated from atmospheric and runoff data. An initial value is required to obtain the absolute value of storage. This is useful in estimating the seasonal change of total water storage in large river basins.

Appendix 3.1   Combination of aerodynamic and energy balance methods of computing lake evaporation

The combination of aerodynamic and energy balance equations (WMO, 2008) is

$$E_i = \frac{R_n \Delta + E_a \gamma}{\Delta + \gamma} \tag{3.31}$$

where E_i is the estimated evaporation from a free water surface; $\Delta = (e_s - e_{sz})/(T_s - T_z)$ is the slope of the saturation vapour pressure curve at any temperature θ_a, tabulated as γ/Δ versus T_z (Brutsaert, 1982); R_n is the net radiation; γ is the constant in the wet and dry bulb psychrometer equation; the direct evaporation of the air is $E_a = k(e_s - e_a)$, where k is a coefficient depending upon the wind velocity, atmospheric pressure and other factors; and e_s and e_a are saturation vapour pressures corresponding to the water surface temperature and the vapour pressure of the air respectively.

The psychrometric constant γ is the same constant as for the Bowen ratio (equation 3.18), having a value at 1000 hPa pressure of 0.61. The net radiation R_n (MJ m^{-2}day) can be estimated from

$$R_n = \left(0.25 + 0.5\frac{n}{N}\right)S_0 - \left(0.9\frac{n}{N} + 0.1\right)\left(0.34 - 0.14\sqrt{e_d}\right)\sigma T^4 \tag{3.32}$$

where n/N is the ratio of actual to possible hours of sunshine, S_0 is the extraterrestrial radiation (MJ m^{-2}day), e_d is the actual vapour pressure of the air in mm of mercury, σ is the Stefan–Boltzmann constant expressed in equivalent evaporation (mm day^{-1}), and T is the mean air temperature expressed in kelvins. Lake evaporation computed for short periods by this method would be applicable only to very shallow lakes with little or no advection of energy to the lake. For deep lakes and conditions of significant advection due to inflow and outflow, it is necessary to correct the computed lake evaporation for net advected energy and change in energy storage.

Appendix 3.2   Modelling soil moisture wetness

Soil moisture content may be defined and modelled using several soil layers (see for example Sellers et al., 1986). Consider here just two soil layers, 1 and 2, following Laval (1997); then the equations for soil moisture wetness w_1 are (Manabe, 1969)

$$\theta_s D_1 \frac{\partial w_1}{\partial t} = P_r - E_1 - Q_{12} - R_s \tag{3.33}$$

$$\theta_s D_2 \frac{\partial w_2}{\partial t} = E_2 + Q_{12} - Q_2 \tag{3.34}$$

where $w_i = \theta_i / \theta_s$, θ_i is volumetric soil moisture, θ_s is the soil saturation value, D_1 and D_2 are the soil layer depths, P_r is the precipitation rate reaching the soil (including snowmelt), E_1 and E_2 are the transpiration of water resulting from its transfer from layer 1 or 2 through plants, Q_{12} is diffusion and/or drainage between the two layers, Q_2 is the subsurface drainage, and R_s is the surface runoff.

The water diffusion between layers depends upon the humidity gradient in the soil, given by *Darcy's law*,

$$Q = K \left(\frac{\partial \Psi}{\partial z} + 1 \right) \tag{3.35}$$

where the first term represents capillary forces and the second term represents the gravity force. K and Ψ are the unsaturated hydraulic conductivity and the soil moisture potential, which are related to the saturated values as follows:

$$K = K_s w^{2B+3} \tag{3.36}$$

$$\Psi = \Psi_s w^{-B} \tag{3.37}$$

where B is a coefficient which represents the soil type.

Further discussion of Darcy's law is given in section 4.7.

Summary of key points in this chapter

1. Evaporation is the result of a balance between externally applied atmospheric demand and the availability of water at an evaporating surface. Total evaporation is referred to as evapotranspiration, a combination of evaporation from soil and water surfaces plus transpiration (evaporation from plant tissue) from plants. Potential evaporation is determined by atmospheric demand as distinct from the evaporation that actually takes place.

2. Evaporation rate is expressed as the equivalent rate of latent heat λE, so that

$$\lambda E = \frac{\lambda \rho_a \varepsilon}{r_a} \left(\frac{e_0}{p - e_0} - \frac{e}{p - e} \right)$$

where ρ_a is the density of dry air, ε is the ratio of the molecular weights of water and air, and p is the total air pressure; r_a is a resistance to the transfer of water vapour down the vapour concentration gradient that exists between the evaporating surface and the atmosphere. The sensible heat is given by $H = \rho C_p (T_0 - T)/r_a$ where T_0 and T are the surface and air temperatures respectively.

3. At an evaporating surface, conserving energy gives

$$R_n = H + LE + G + R_s + R_p + R_i$$

where R_n is the net radiation, H is the sensible heat, LE is the heat used in converting liquid to vapour, G is the heat flux into the ground, R_s is the heat stored in the water, R_p is the heat converted to chemical energy, and R_i is the heat moved into the air or out of the system by water inflow or outflow.

4. The Penman–Monteith equation uses knowledge of the change of temperature of the saturated vapour pressure of water over vegetation surfaces to estimate the evaporation rate.

5. The Bowen ratio is the sensible heat divided by the latent heat.

6. Soil moisture content may be measured using a capacitance probe, a grass weighing lysimeter or the satellite synthetic aperture radar (SAR). The time domain reflectometer (TDR) measures the instantaneous volumetric water content of soils due to the great difference in the dielectric constant of water from the other constituents in soil.

7. Soil, plants and atmosphere constitute a physical entity from the viewpoint of water transfer, referred to as the soil–plant–atmosphere continuum (SPAC). Water potential can be expressed as a function of the relative humidity of air in equilibrium with bound water at the same temperature. Leaf transpiration is a function of energy received by leaves and the state of the stomata.

8. Water circulates through a plant from the soil and into the atmosphere by means of a difference in water potential along the way, and the hydraulic continuity of soil, plant and atmosphere. A conservative water flux is the condition in which absorption in a plant is equal to transpiration from the plant.

9. A soil–vegetation–atmosphere transfer (SVAT) scheme comprises representations of hydrologic processes including potential evaporation rate, momentum transfer, water balance, soil heat fluxes and vegetation controls on evaporation.

Problems

1. Define transpiration and potential evaporation.

2. Estimate the monthly potential evaporation (in cm) for the temperate, continental climate of North America.

3. Estimate evaporation rate assuming the specific heat of air is $C_p = 1.0 \times 10^3 \, \mathrm{J \, kg^{-1} K^{-1}}$ at a temperature of 20 °C, with air pressure $= 100 \, \mathrm{kPa}$, vapour pressure $= 2.3 \, \mathrm{kPa}$, air density $= 0.00129 \, \mathrm{g \, cm^{-3}}$ and aerodynamic function for a rough surface $= 3.5 \, \mathrm{s \, m^{-1}}$.

4. Draw a diagram illustrating how latent heat flux is driven by humidity gradients between the inside of a stomatal cavity, the leaf surface and the bulk atmosphere.

5. Write down the Penman–Monteith equation.

6. Describe mathematically how the rate of change of the canopy storage of water may be estimated.

7. What is a SPAC?

8. Illustrate the flux of water in plants in ohmic form.

9. Give a formula for the net short wave radiation.

10. Draw a flowchart illustrating the flow of water via land surface fluxes, throughfall from a canopy, runoff and movement into the soil.

11. Estimate large scale evapotranspiration and water storage in a river basin over days, assessing a precipitable water amount and vertically integrated two-dimensional water vapour flux.

References

Avissar, R. and Verstraete, M.M. (1990) The representation of continental surface processes in atmospheric models. *Rev. Geophys.*, 28, 35–52.

Blyth, E.M. (2002) Modelling soil moisture for a grassland and a woodland site in south-east England. *Hydrol. Earth Syst. Sci.*, 6, 39–47.

Bonan, G.B. (1995) Land–atmosphere CO_2 exchange simulated by a land surface process model coupled to an atmospheric general circulation model. *J. Geophys. Res.*, 100(D2), 2817–2831.

Bosilovich, M.G. and Sun, W.-Y. (1995) Formulation and verification of a long-surface parameterization for atmospheric models. *Boundary-Layer Meteorol.*, 73, 321–341.

Bowen, I.S. (1926) The ratio of heat losses by conduction and by evaporation from any water surface. *Phys. Rev.*, 27, 779–789.

Brutsaert, W. (1975) On a derivable formula for long wave radiation from clear skies. *Water Resour. Res.*, 11, 742–744.

Brutsaert, W. (1982) *Evaporation into the Atmosphere*. Reidel, Dordrecht.

Budyko, M.I. (1956) *The Heat Balance of the Earth's Surface* [in Russian]. Gidrometeoizdat, Leningrad.

Calder, I.R. (1979) Do trees use more water than grass? *Water Serv.*, 83, 11–14.

Calder, I.R. (1987) A stochastic model of rainfall interception. *J. Hydrol.*, 89, 65–72.

Calder, I.R. (1990) *Evaporation in the Uplands*. Wiley, Chichester.

Calder, I.R., Narayanswamy, M.N., Scrinivasalu, N.V., Daling, W.G. and Lardner, A.J. (1986) Investigation into the use of deuterium as a tracer for measuring transpiration from eucalypts. *J. Hydrol.*, 84, 335–351.

Choudhury, B.J. and Monteith, J.L. (1988) A four-layer model for the heat budget of homogeneous land surfaces. *Q. J. R. Meteorol. Soc.*, 114, 373–398.

Cowley, J.P. (1978) The distribution over Great Britain of global solar irradiation on a horizontal surface. *Meteorol. Mag.*, 107, 317–373.

Cox, P.M., Huntingford, C. and Harding, R.J. (1998) A canopy conductance and photosynthesis model for use in a GCM land surface scheme. *J. Hydrol.*, 212–213, 79–94.

Cox, P.M., Betts, R.A., Bunton, C.B., Essery, R.L.H., Rowntree, P.R. and Smith, J. (1999) The impact of new land surface physics on the GCM simulation of climate and climate sensitivity. *Clim. Dyn.*, 15, 183–203.

Cruiziat, P. (1991) L'eau et les cultures. *Tech. Agric.*, 1165–6, 1–26.

Cruiziat, P. and Tyree, M.T. (1990) La montée de la sève dans les arbres. *La Recherche*, 220, 406–414.

Dalton, J. (1801) On the constitution of mixed gases; on the force of steam or vapour from water and other liquids in different temperatures both in a Torricellian Vacuum and in the air; on evaporation; and on the expansion of gases by heat. *Mem. Lit. Phil. Soc.*, 5, 366–535.

Deardorff, J.W. (1978) Efficient prediction of ground surface temperature and moisture with inclusion of a layer of vegetation. *J. Geophys. Res.*, 83, 1889–1903.

Dickinson, R.E., Henderson-Sellers, A., Kennedy, P.J. and Wilson, M.E. (1986) Biosphere–atmosphere transfer scheme (BATS) for the NCAR Community Climate Model. NCAR Technical Note TN-275+STR.

Essery, R., Best, M. and Cox, P. (2001) MOSES 2.2 technical documentation. Hadley Centre Technical Note 30, Met Office, Exeter. www.metoffice.com/research/hadleycentre/pubs/HCTN/index.html.

Fox, N.I., Saich, P. and Collier, C.G. (2000) Estimating the surface water and radiation balance in an upland area from space. *Int. J. Remote Sens.*, 21, 2985–3002.

Guyot, G. (1998) *Physics of the Environment and Climate*. Wiley-Praxis, Chichester.

Hough, M. (2003) An historical comparison between the Met Office Surface Exchange Scheme Probability Distributed Model (MOSES-PDM) and the Met Office Rainfall and Evaporation Calculation System (MORECS). Note. Environment Agency/Met Office.

Klepper, B. (1990) Root growth and water uptake. In *Irrigation of Agricultural Crops*, ed. B.A. Stewart and D.R. Neilsen, pp. 281–322. Agronomy Monograph 30, ASA-CSSA-SSSA, Madison, WI.

Kong, X., Darling, S. and Smith, R. (2011) Soil moisture modelling and validation at an agricultural site in Norfolk using the Met Office Surface Exchange Scheme (MOSES). *Meteorol. Appl.*, 18, 18–27.

Laval, K. (1997) Hydrological processes in GCMs. In *Land Surface Processes in Hydrology: Trials and Tribulations of Modeling and Measuring*, ed. S. Sorooshian, H.V. Gupta and S.C. Rodda, pp. 45–61. Springer, Berlin.

Liang, X., Lettenmaier, D.P., Wood, E.F. and Burges, S.J. (1994) A simple hydrologically based model of land surface water and energy fluxes for general circulation models. *J. Geophys. Res.*, 99(D7), 14415–14428.

Mak, M. (2011) Atmospheric Dynamics. Cambridge University Press, Cambridge.

Manabe, S. (1969) Climate and the ocean circulation: the atmospheric circulation and the hydrology of the Earth's surface. *Mon. Weather Rev.*, 97, 739–774.

Mintz, Y. and Walker, G.K. (1993) Global fields of soil moisture and land-surface evapotranspiration derived from observed precipitation and surface air temperature. *J. Appl. Meteorol.*, 32, 1305–1333.

Monteith, J.L. (1985) Evaporation from land surfaces: progress in analysis and prediction since 1948. In *Advances in Evapotranspiration*, pp. 4–12. American Society of Agricultural Engineering.

Moore, R.J. (1985) The probability-distributed principle and runoff production at point and basin scales. *Hydrol. Sci.*, 30, 273–297.

Oki, T. (1999) The global water cycle. Chapter 1.2 in *Global Energy and Water Cycles*, ed. K.A. Browning and R.J. Gurney, pp. 10–29. Cambridge University Press, Cambridge.

Penman, H.L. (1948) Natural evaporation from open water, bare soil and grass. *Proc. R. Soc. Lond. A*, 193, 120–145.

Philip, J.R. (1966) Plant water relations: some physical aspects. *Ann. Rev. Plant Physiol.*, 17, 245–268.

Priestley, C.H.B. and Taylor, R.J. (1972) On the assessment of surface heat flux and evaporation using large-scale parameters. *Mon. Weather Rev.*, 100, 81–92.

Rooney, G.G. and Claxton, B.M. (2006) Comparison of the Met Office's surface exchange scheme, MOSES, against field observations *Q. J. R. Meteorol. Soc.*, 132, 425–446.

Rutter, A.J., Kershaw, K.A., Robins, P.C. and Morton, A.J. (1971) A predictive model of rainfall interception in forests. I: Derivation of the model from observations in a plantation of Corsican pine. *Agric. Meteorol.*, 9, 367–384.

Rutter, A.J., Morton, A.J. and Robins, P.C. (1975) A predictive model of rainfall interception in forests. II: Generalization of the model and comparison with observations in some coniferous and hardwood stands. *J. Appl. Ecol.*, 12, 367–380.

Schaake, J.C., Koren, V.I., Duan, Q.Y., Mitchell, K. and Chen, F. (1996) Simple water balance model for estimating runoff at different spatial and temporal scales. *J. Geophys. Res.*, 101(D3), 7461–7475.

Sellers, P.J., Mintz, Y., Sud, Y.C. and Dalcher, A. (1986) A simple biosphere model (SiB) for use within general circulation models. *J. Atmos. Sci.*, 43, 505–531.

Sellers, P.J., Randell, P.A., Collatz, C.J., Berry, J.A., Field, C.B., Dazlich, D.A., Zhang, C., Collello, G. and Bounoua, L. (1996) A revised land surface parameterization (SiB2) for atmospheric GCMS. I: Model formulation. *J. Clim.*, 9, 676–705.

Shuttleworth, W.J. (2012) *Terrestrial Hydrometeorology*. Wiley, Chichester.

Smith, R.N.B., Blyth, E.M., Finch, J.W., Goodchild, S., Hall, R.L. and Madry, S. (2006) Soil state and surface hydrology diagnosis based on MOSES in the Met Office Nimrod nowcasting system. *Meteorol. Appl.*, 13, 89–109.

Taylor, C.M., De Jeu, R.A.M., Guichard, F., Harris, P.P. and Wouter, A.D. (2012) Afternoon rain more likely over drier soils. *Nature*, 489, 423–426.

Thom, A.S. and Oliver, H.R. (1977) On Penman's equation for estimating regional evaporation. *Q. J. R. Meteorol. Soc.*, 103, 345–357.

Thompson, N., Barrie, I.A. and Ayles, M. (1981) The Meteorological Office Rainfall and Evaporation Calculation System: MORECS. Hydrological Memorandum 45, Meteorological Office, Bracknell.

Thornwaite, C.W. (1948) An approach towards a rational classification of climate. *Geogr. Rev.*, 38, 55–94.

Topp, G.C., Davis, J.L. and Annan, A.P. (1980) Electromagnetic determination of soil water content: measurement in coaxial transmission lines. *Water Resour. Res.*, 16, 574–582.

Ward, R.C. and Robinson, M. (2000) *Principles of Hydrology*, 4th edn. McGraw-Hill, London.

Wiesner, C.J. (1970) *Hydrometeorology*. Chapman and Hall, London.

WMO (2008) *Guide to Hydrological Practices. Vol. I: Hydrology: From Measurement to Hydrological Information*, WMO 168, 6th edn. World Meteorological Organization, Geneva.

4

Snow and Ice

4.1 Introduction

The processes involved in the generation of precipitation, which includes snow, were discussed in Chapter 2. However, snowfall has a very large impact on the activities of humans, particularly on transport and in snowmelt flooding, and therefore it merits more detailed discussion. Indeed in many areas of the world snowfall is a primary factor in hydrological forecasting. For example in the United States it has been estimated (Castruccio et al., 1980) that a forecast improvement of 6% attained by the use of operational satellite measurement of snow covered areas resulted in an annual benefit to agriculture of $28 million, and to hydroelectricity production of $10 million.

4.2 Basic processes

4.2.1 Formation of snow

As warm moist air ascends, water vapour begins to condense to form cloud. When the cloud temperatures drop below freezing, conditions are suitable for the formation of snow. However, several processes are involved which govern the different types of snow that may be produced. Figure 4.1 summarizes these processes. At about –5 °C the aerosol nuclei (size 0.01 to 1 μm), which are always present in the atmosphere, form small crystals (diameters less than 75 μm) by ice nucleation. These crystals have simple shapes, but may continue to grow by sublimation (Chapter 2) to form snow crystals which often have very complex shapes. A number of snow crystals together form a snowflake. Sometimes snow crystals pass through parts of the cloud which have many cloud droplets (size 10–40 μm), and therefore *riming* (droplets freeze when they come into contact with the crystals) occurs if the crystals are larger than about 300 μm. This occurs at temperatures from –5 to –20 °C. If riming continues for a significant time then snow pellets (*graupel*) may be formed.

Hydrometeorology, First Edition. Christopher G. Collier.
© 2016 John Wiley & Sons, Ltd. Published 2016 by John Wiley & Sons, Ltd.
Companion website: www.wiley.com/go/collierhydrometerology

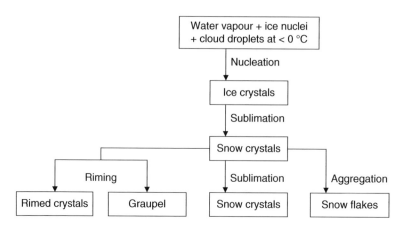

Figure 4.1   Flow diagram of the formation of different types of snow (after Gray and Male, 1981)

At temperatures appropriate for the formation of snow, a cloud may only be slightly supersaturated with respect to water, but 10–20% supersaturated with respect to ice. Hence there is a transfer of water vapour from the cloud droplets to the ice crystals, which consequently grow. This is the Bergeron process, discussed in Chapter 2. The shape of an ice crystal depends upon the temperature at which it grows, but its rate of growth and secondary crystal features depend upon the degree of supersaturation. The range of shapes is: 0 to –4 °C, plates; –4 to –10 °C, prism-like crystals, scrolls, sheaths and needles; –10 to –20 °C, thick plates, dendrites and sector plates; and –20 to –35 °C, sheaths and hollow columns. Figure 4.2 shows examples of the structure of snowflakes. The rate of increase with time t of a mass m of a crystal through diffusion of water vapour onto its surface is

$$\frac{dm}{dt} = 4\pi CDFA\left(\rho_{\infty} - \rho_{0}\right) \tag{4.1}$$

where C is a shape factor; D is the diffusivity of water vapour in air; F is a ventilation factor depending on the relative motion of the crystal with respect to the air; A is a function of crystal size; ρ_{∞} is the vapour density (mass of water vapour per unit volume of moist air) at a large distance from the crystal; and ρ_{0} is the vapour density at the surface of the crystal.

The mass growth rate due to riming depends upon the fall speed of the ice crystals relative to the cloud droplets, and the efficiency with which the droplets freeze and remain attached to the crystals. Hence

$$\frac{dm}{dt} = \pi r^{2} abwW \tag{4.2}$$

where r is the radius of the ice crystals; a is the adhesion efficiency; b is the collision efficiency (ab is between 0.1 and 1); w is the fall speed of the ice crystals relative to the droplets (less than 5 m s^{-1}); and W is the liquid water content of the cloud.

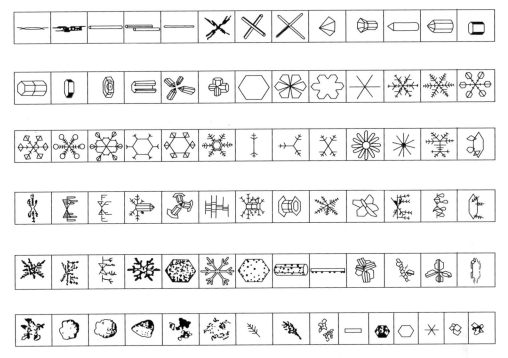

Figure 4.2   Snowflake structures (from Gray and Male, 1981)

4.2.2   Formation of snow cover and its effects on the atmosphere

Several conditions have been found to be necessary though not sufficient for the occurrence of significant snowfall:

1. sufficient moisture and aerosol nuclei for the formation and growth of ice crystals
2. sufficient depth of cloud to permit snow crystal growth
3. temperatures below 0 °C in most of the layer through which the snow falls
4. sufficient moisture and aerosol nuclei to replace losses caused by precipitation.

Operational weather forecasts tend to use the temperature of the lower part of the atmosphere from 500 hPa to 1000 hPa, known as the thickness temperature, as an indicator of likely snowfall. However, the other factors are very important, and currently numerical prediction models cannot represent the processes involved accurately enough to provide completely reliable guidance. Fulkes (1935) indicated the rate of precipitation obtained from adiabatically ascending air for a 100 m layer with a vertical velocity of 1 m s^{-1} (Figure 4.3). Such information is useful, but the vertical velocity of the air is determined by the nature of the atmospheric system (see for example Browning and Mason, 1981), orographic effects (Browning, 1983) and topographic effects (Lavoie, 1972). Hence the estimation of the likely snowfall on particular occasions is a very difficult problem, although the recent developments of high resolution numerical models hold out much promise (see Chapter 8).

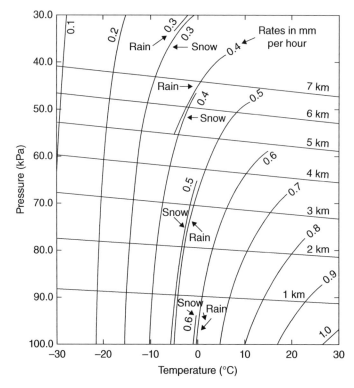

Figure 4.3 Rates of precipitation from adiabatically ascending air for a 100 m layer with a vertical velocity of 1 m s⁻¹ (Fulkes, 1935)

Heavy falls of snow are not always associated with high latitudes, although clearly the lower temperatures experienced in these regions are conducive to snowfall for much of the year. Other areas where very cold air occasionally crosses relatively warm stretches of water are also likely to be subjected to heavy snow. Hence, in winter, snow cover can persist for some time in mid-latitudes. This persistence is prolonged if the snow cover is over land at high latitudes. The Rocky Mountains, the Alps and the Himalayas all retain snow cover at the highest elevations all year round. Even the Scottish Highlands, only about 1 km high, lose their snow cover for just one or two months each year.

Both land and sea (the Arctic) areas are covered by snow. The areas involved vary with the seasons, with night and day, and with the day-to-day weather over particular areas. It is not surprising to discover that snow cover is an important factor in determining climate and climatic change. Snow cover influences the atmospheric circulation by interacting with, and affecting, overlying air masses, which results in either the amplification or the stabilization of circulation anomalies that often cause the weather (Rango, 1985).

Snow strongly reflects visible and near infrared light, that is it has a high *albedo*. Consider Table 4.1, which compares the albedo of snow with that of other natural surfaces. The albedo of snow varies with the age of the snow, but is considerably higher than the albedos of most natural surfaces. Therefore because of the seasonal changes in the extent of snow cover, the albedo of the surface of the Earth varies from

Table 4.1 Percentage of incident short wave radiation reflected by some surfaces

Surface	Albedo (%)	Surface	Albedo (%)
Fresh snow	80–95	Bare fields	15–25
Old snow	50–60	Field crops	3–25
Sea ice	50–70	Barley	23
Open water	5–30	Sugar beet	26
Clouds	1–80	Grass	20–25
Coniferous forest	15	Tundra	15–20
Deciduous forest	18	New concrete	55
Desert	24–30		
Desert sand	40	Average over all Earth	30–35

season to season. Since snow also has a high thermal emissivity, a low thermal conductivity, and a low water vapour pressure, the energy balance within the atmosphere will change as the seasons change. As melting occurs, the water vapour pressure and the thermal conductivity of the snow increase, and the latent heat of the snow must be taken into account in the energy balance.

Since the snow cover at higher latitudes persists for much of the year, less solar energy is absorbed at these latitudes, causing a profound effect upon the meridional flux of energy. Higher latitudes act as source regions for cold air masses which move to mid-latitudes, resulting in the atmospheric systems discussed in Chapter 2 (Walsh, 1984). Deviations from the regular seasonal changes in snow cover may be one factor in triggering climatic change (Chapter 13).

4.2.3   Formation of ice

As the surface of an area of water cools, the cooler, denser water near the surface sinks, and is replaced by less dense water from below. For temperatures less than about 4 °C, the less dense, cooler water at the surface begins to freeze even though the water below it is relatively warmer.

The initial orientation of the ice crystals is random, but as the ice thickens some crystals grow more quickly downwards than others. In sea ice more horizontally oriented crystals are favoured, whereas for ice on lakes this is not the case (see for example Hobbs, 1974). The water below the ice continues to lose heat by conduction through the ice sheet, albeit much more slowly, and as a consequence the ice thickens downwards. If snowfall is low then the layer of ice may be over 3 m thick and is known as *black ice*.

As snow accumulates on the ice its density increases with depth, and the existing black ice is depressed below the original water level. The temperature within the lower layers of the snow cover may rise slightly, so that sublimation (Chapter 2) occurs, and a layer of *white ice* is formed on the black ice, separated by a layer in which some melting occurs due to the flooding caused by the snow cover depression. This area of *slush* may also be caused by warmer water moving upwards by capillary action through cracks in the black ice, but it quickly refreezes.

Like a snow cover, the ice on lakes delays runoff from a catchment area, and therefore may have a significant impact upon hydrological forecasting. Melting will require appropriate modelling, which is discussed later in this chapter. Extensive sea ice affects the weather on a much larger scale in the same way that extensive snowfields influence atmospheric systems.

The snow that makes up ice sheets and glaciers undergoes a sequence of conversion stages: snowfall; settling snow; *nivation*, thaw–freeze alternation leading to the conversion of snow to ice crystals; *firn* or névé, pressure melting; *sintering*, continued fusion and squeezing out of air; and glacier ice. Models of glaciers have been constructed (see for example Mernild et al., 2006).

4.3   Characteristics of snow cover

The characteristics of snow cover are determined by atmospheric conditions and the condition of the land surface. The rates of precipitation, deposition and condensation, together with the magnitudes of the turbulent transfer of heat and moisture, radiative exchange and air movement, all contribute to the form of the snow cover. In addition the features of the land may modify the atmospheric conditions, and therefore the snow cover and the amount of moisture retained by the land will also have their effects. The longevity of the snow cover is as important as the amount of snow because it determines the maintenance of high albedo versus low albedo surfaces. Figure 4.4 shows qualitatively the patterns of snow cover from high temperate latitudes to the high Arctic (Rouse, 1993).

Snow depth will be increased as temperatures fall with increasing land elevation. Indeed snow cover may not be established below particular elevations dependent on season and type of weather (see for example Minnich, 1986). Wet snow, which is not easily blown by the wind, often occurs when air temperatures are near 0 °C, as may occur when air flows over large bodies of water (Passarelli and Braham, 1981). Terrain slope and aspect also have their effects. There are large differences between snowfall on the windward and lee sides of hills, and over steep and gentle slopes. In the temperate zone, snow accumulates quickly to a maximum in midwinter and then melts rapidly. In the high Arctic it collects rapidly in early winter with a second phase of accumulation in late winter, but there is little or no accumulation for a long period in midwinter.

The roughness of the land surface causes turbulent flow in the lowest layers of the atmosphere. This turbulence influences the snow cover distribution and properties, mainly density. The density increases as wind causes drifting. The effects of forest canopies on snow accumulation can be very important in mid-latitude and high latitude areas. Andreadis et al. (2009) used a mass and energy balance model to investigate processes in forestry environments using data from a Boreal Ecosystem–Atmosphere Study (BOREAS) site in Canada. Ellis et al. (2010) reviewed studies of the simulation of snow accumulation and melt in needle leaf forests. Table 4.2 shows the range of densities measured in snow cover. Condensation and melting also affect density.

The rate of transport of snow depends upon the wind speed, the nature of the terrain, and the type of climatic region. In the Rocky Mountains Martinelli (1973) measured a transport flux rate of 136 tonne m^{-1}, whereas along the Arctic coast Mikhel' et al. (1969) measured a rate of 907 tonne m^{-1}. The snow will be deposited where the wind speed decreases markedly, often over rough ground. The result will be large drifts if, upwind of the deposition area, there is a long fetch covered with loose

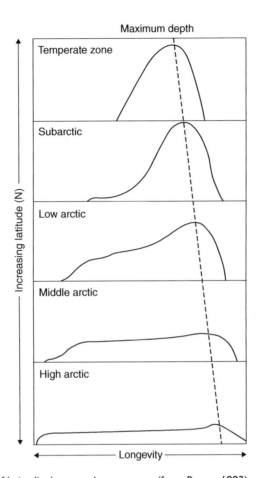

Figure 4.4   Patterns of latitudinal seasonal snow cover (from Rouse, 1993)

Table 4.2   Densities of snow cover (from Gray and Male, 1981)

Snow type	Density (kg m^{-3})
Wild snow	10 to 30
Ordinary new snow immediately after falling in still air	50 to 65
Settling snow	70 to 90
Very slightly toughened by wind immediately after falling	63 to 80
Average wind-toughened snow	280
Hard wind slab	350
New firn*	400 to 550
Advanced firn snow	550 to 650
Thawing firn snow	600 to 700

*Firn snow consolidated partly into ice (after Seligman, 1962).

snow, and a high wind speed has been sustained for a considerable time. The presence of forests often causes drifting on the upwind edge provided the forest is reasonably dense. Since the forests also modify atmospheric conditions and flow, they also change the snow cover within them relative to that in non-forested areas, sometimes in quite complex and poorly understood ways, which has led to modelling studies as mentioned previously.

Melting caused by changing atmospheric conditions will clearly modify the snow cover significantly, and this is discussed in the next section. However, before major melting begins, the characteristics of the snow will be modified by radiative fluxes. The amount of solar (short wave) radiation absorbed by the snow cover, and therefore available for melting, is governed by the albedo of the snow, as discussed in section 4.2.2. In cloudy conditions or at night, long wave radiation is lost from the snow cover. Whilst this is usually not significant in the melting process, Olyphant (1986) has demonstrated that the surrounding rock walls in alpine snowfields cause energy reflection which may reduce this loss of long wave radiation by about 50%, and so reduce the rate of melting. Snow fields on Ben Nevis in Scotland, which are sheltered by steep slopes, may occasionally persist throughout the year – one of the very few locations in the United Kingdom where this may occur. Observational and numerical studies of the maritime snow cover in the Antarctic suggest that snow on top of sea ice plays a major role in shaping the seasonal growth and decay of the ice pack in the Southern Ocean. Fichefet and Morales Maqueda (1999) used a coupled snow/sea-ice/upper-ocean model to investigate the response of the Antarctic sea ice cover to snowfall.

4.4   Glaciers

Alpine glaciers, also known as mountain glaciers or cirque glaciers, form on the crests and slopes of mountains, and they may fill a valley, in which case they may be called a valley glacier. A glacier mass balance is derived from the annual accumulation, ablation and change in mass of ice held in storage. If the difference between accumulation and ablation is positive there is a net gain of snow, ice and water to the glacier. A negative balance indicates a net loss.

Glacier accumulation results from snowfall, avalanche, snow drifting onto the glacier and superimposed ice. Compacted snow, known as *firn*, has a density that exceeds 500 kg m^{-3}, and turns into glacier ice when most of the gaps between snow crystals are filled (Woo, 1993).

Glacier runoff responds to the energy budget that causes the snow, firn and glacier ice to melt. Note that rain falling on the glacier generates much less runoff. Glacier runoff is low during cool years and higher in warm years. Runoff occurs on the glacier surface (supraglacial), within the glacier (englacial) or along the base of the glacier (subglacial). Runoff with Arctic glaciers ceases as winter starts and begins again with the melt season.

At the edges of glaciers where the ice meets land, dams may occur. Small lakes may build up from runoff of melt water and may then drain rapidly as the ice dam is breached by overflow or subglacial drainage. Large dams may release large volumes of water in a short time producing large floods.

4.5   Sea ice

The freezing point of water decreases from 0 °C for fresh water to –2 °C for water of 35% salinity. Also the temperature of maximum density decreases from 4 °C for fresh water to –1.3 °C at 24.7% salinity (see for example Barry, 1993). When sea water freezes a skin of pure ice crystals forms on the surface. This is followed by ice which aggregates into slush ice with brine being trapped in pockets, giving young ice with a salinity of 12–15%. The rate of thickening depends upon the surface temperature and thickness of the ice or snow.

4.6   Permafrost

Permafrost or *cryotic soil* is soil at or below the freezing point of water (0 °C) for two or more years. Most permafrost is located in high latitudes (in and around the Arctic and Antarctic regions), but alpine permafrost may exist at high altitudes in much lower latitudes. Ground ice is not always present, as may be the case with non-porous bedrock, but it frequently occurs and may be in amounts exceeding the potential hydraulic saturation of the ground material. Permafrost accounts for 0.022% of total water on Earth and exists in 24% of exposed land in the northern hemisphere.

Consider the situation where the mean annual surface temperature falls below 0 °C so that the depth of winter freezing will exceed the depth of summer thaw. Then a layer of permafrost would grow downward from the base of the seasonal frost, getting thicker each winter. However, the depth is limited by the heat escaping from the Earth's interior due to the temperature increasing with depth by about 30 °C per kilometre. Therefore the base of the permafrost reaches an equilibrium depth (see for example Smith, 1993). The ground temperature profile in a permafrost region is shown in Figure 4.5. Changes in the thermal regime of the ground that may lead to the degradation or formation of permafrost can result from climate change or changes in vegetation or soil.

In Arctic environments, limited infiltration in permafrost areas, low topographic gradient, and large amounts of water available after snowmelt give rise to large areas of wetlands or peat lands. These wetlands form a large reservoir of carbon that may be released when temperature and evaporation rise and a thaw occurs (French, 1996). As wetlands are usually frozen at the end of winter, flow during snowmelt is restricted (Woo, 1986). Likewise the recharge and movement of ground water is limited.

4.7   The physics of melting and water movement through snow

The total melt water which is produced from a snow cover is controlled by the energy exchanges at the upper and lower snow surfaces. When the surface air temperature is at 0 °C melting will begin, but not until the temperature is at 0 °C throughout the depth of the snow will substantial melt water be produced. In practice the air temperature should be at least 5 °C to initiate melting. At this point the snow crystals are coated with a thin film of water; indeed there are usually small pockets of water in cracks. If further solar energy passes into the snow cover, melt water will begin

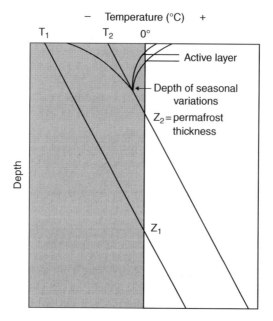

Figure 4.5 Ground temperature profile in a permafrost region; T_1 and T_2 represent different mean surface temperature conditions (from Smith, 1993)

to drain to the ground at speeds from 2 to 60 cm min^{-1} (see later), and more than 20% of the snow cover may become water. The energy balance (see for example Zeinivand and DeSmedt, 2009) is

$$E_{\mathrm{m}} = E_{\mathrm{sn}} + E_{\mathrm{ln}} + Q_{\mathrm{h}} + Q_{\mathrm{e}} + Q_{\mathrm{g}} + Q_{\mathrm{p}} - \frac{\mathrm{d}u}{\mathrm{d}t} \tag{4.3}$$

where E_{m} is the energy flux available for melting; E_{sn} is the net short wave (0.2–100 μm) radiation flux absorbed by the snow; E_{ln} is the net long wave radiation flux at the snow–air interface; Q_{h} is the sensible heat flux from the air at the snow–air interface; Q_{e} is the flux of the latent heat at the snow–air interface; Q_{g} is the flux of heat from the snow–ground interface by conduction; Q_{p} is the flux of heat from rain; and $\mathrm{d}u/\mathrm{d}t$ is the rate of change of internal (stored) energy per unit area of snow cover.

Although rain may penetrate the snow cover quite substantially, the depth of penetration being affected by afforestation (Berris and Haar, 1987), melt water is most usually produced at the snow–air interface. The daily amount of melt is

$$M = \frac{E_{\mathrm{m}}}{L\rho F} \tag{4.4}$$

where E_{m} is snowmelt water equivalent (cm day^{-1}), ρ is the density of water (1000 kg m^{-3}), L is the latent heat of fusion (333.5 kJ kg^{-1}), and F is the fraction of ice in a unit mass of wet snow, typically 0.95–0.97. Estimates of catchment snowmelt inflow rates are discussed in Appendix 4.1.

At the top of the atmosphere the solar radiation is equal to the solar constant (1.365 kW m^{-2}), which varies a few percent with the annual changes in the distance between the Sun and the Earth. However, the short wave radiation reaching the surface of the Earth comprises direct sunlight and radiation scattered within the atmosphere, which depend upon the transmissivity of the atmosphere, the slope of the surface and the time of day relative to solar noon (see for example Kontratyev, 1972). As described earlier, a large portion of the short wave radiation reaching the surface of the snow will be reflected, the actual amount depending upon the amount of melting which has taken place, as shown in Figure 4.6.

If rain falls on the snow, it may freeze, releasing latent heat of fusion, or it may remain liquid, in which case

$$Q_p = \rho C_p (T_r - T_s) D_r / 1000 \tag{4.5}$$

where Q_p is the flux of heat from rain; ρ is the density of water; C_p is the specific heat of water at constant pressure; T_r is the temperature of the rain, usually taken as the air temperature; T_s is the temperature of the snow; and D_r is the depth of rain.

When the melt water reaches the ground it is absorbed, but if the infiltration rate is exceeded some water will flow on the surface, forming a saturated layer moving at speeds from 10 to 60 cm min^{-1}. Eventually channels may form and the speed of travel increases rapidly. If the snow cover is unsaturated then the volume of water per unit area and time moving vertically, q_z, may be estimated from *Darcy's law* as follows:

$$q_z = \frac{k}{\mu} \frac{\partial p}{\partial z} + \rho g \tag{4.6}$$

where k is the permeability of snow, μ is the viscosity of water, ρ is the density of water, g is the acceleration due to gravity, z is the vertical direction, and $\partial p / \partial z$ is the

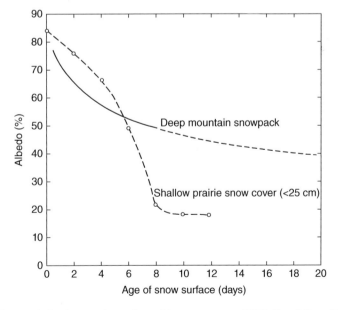

Figure 4.6   Temporal albedo variations of a melting snow cover (O'Neill and Gray, 1973)

capillary pressure gradient. Similarly, the volume of water per unit area and time moving along a shallow slope in the x direction, q_x, is given by

$$q_x = -\frac{k}{\mu}\frac{\partial p_1}{\partial x} - \rho g \theta \qquad (4.7)$$

where θ is the angle of slope and p_1 is liquid pressure.

4.8   Water equivalent of snow

The water content of snowfall may be estimated from measurements of the depth of snow and the density of snow. The long term average of snow density, often referred to as the 'general rule', is around $0.1 \times 10^3 \, kg \, m^{-3}$, although densities may vary as shown in Table 4.2.

Sevruk (1985) has attempted to derive corrections to the general rule for (i) the seasonal and regional pattern, (ii) the effects of altitude and site exposure, (iii) the effects of temperature, and (iv) wind speed. This work has led to the presentation of nomograms such as that shown in Figure 4.7 in which the snowfall density is related to air temperature, wind speed and exposure. In addition a variation with altitude occurs.

Jackson (1977a) found that a correction of 8.8 mm water equivalent per 100 m elevation was appropriate in SE Scotland and NE England. This is supported by Galeati et al. (1986) in Italy, who measured a variation of about 10 mm per 100 m.

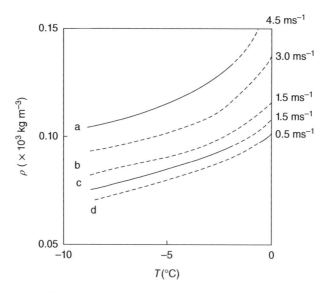

Figure 4.7 Nomogram for the assessment of snowfall density ρ. Average wind speed u at the 2 m level on days with snowfall: (a) exposed locations, $u > 4 \, m \, s^{-1}$; (b) open locations, $u = 1.5$ to $4 \, m \, s^{-1}$; (c) sheltered locations, $u < 1.5 \, m \, s^{-1}$; (d) completely sheltered locations, $u < 0.5 \, m \, s^{-1}$ (after Sevruk, 1985)

However consistent changes with altitude are often masked by the effects of exposure and wind (Sevruk, 1985). In some countries enough snowfall measurements are available to work out what water equivalents of lying snow might accumulate at a particular place. An example from the United Kingdom of the five-year return water equivalent of lying snow reduced to sea level is shown in Figure 4.8.

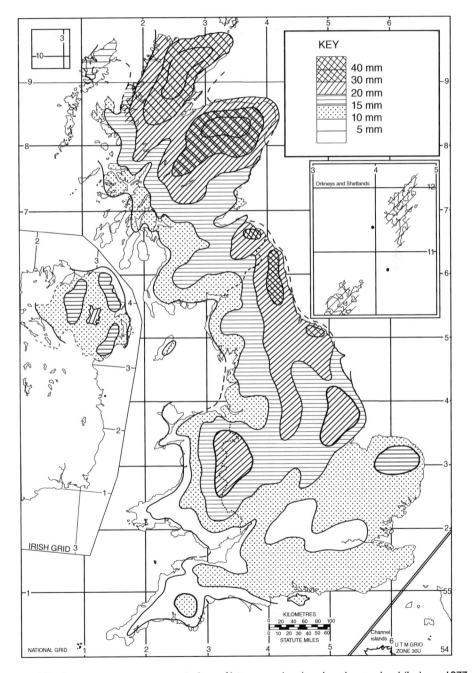

Figure 4.8 Five-year return water equivalent of lying snow (mm), reduced to sea level (Jackson, 1977c)

4.9 Modelling snowmelt and stream flow

In many areas of the world the river runoff from melting snow is a significant factor in water management, and may contribute significantly to flooding. During certain times of the year, water from snowmelt can be responsible for almost all of the stream flow in a river. The runoff from snowmelt varies not only by season, but also by year. The lack of water stored as snowpack in the winter can affect the availability of water for the rest of the year.

Various procedures have been developed to forecast the contributions of snowmelt to stream flow. The most important factors which govern estimates of the amount and timing of melt release to a river are:

1. the energy available to melt the snow over a given area
2. the areal extent of the melting snow cover
3. the effects of storage on the movement of the melt quantities as they move from the surface of the snow to the river.

Snowmelt models are generally of four types (i) 'complex' mass energy balance models (e.g. Flerchinger and Saxton, 1989); (ii) temperature index models (e.g. Maidment, 1993); (iii) hybrid temperature index models (e.g. Kustas et al., 1994); and (iv) 'simple' mass and energy models (e.g. Marks et al., 1998; Brooks and Boll, 2005).

Most operational forecasting procedures are based upon the use of air temperature as the index of the energy available for melting the snow. Unfortunately there is no universally applicable temperature index, each index depending upon the characteristics of the river basin for which it has been developed. The most common expression used is

$$M = M_f \left(T_i - T_b \right) \tag{4.8}$$

where M is the melt produced (cm water per unit time); M_f is a melt factor (cm per °C per unit time); T_i is the index air temperature (e.g. maximum or mean daily temperature); and T_b is the base temperature (usually 0°C). The expression may be interpreted as a flow chart such as that used by Archer (1983) in NE England, shown in Figure 4.9. Peña and Nazarala (1984) describe a similar procedure used in Chile. For periods of rain, M_f may be adjusted as follows:

$$M_f \left(\text{rain} \right) = M_f + 0.00126P \tag{4.9}$$

where M_f (rain) is a melt factor for rain and P is rainfall (mm).

Values of M_f vary between 1.3 and 3.7 mm °C^{-1}day^{-1} (see for example Anderson, 1973), and must be evaluated from a complete water balance of the river basin concerned. However, where it can be assumed that the soil moisture deficit and other storage terms can be either neglected or satisfied during the initial melt period, and that the loss due to infiltration is small, the value of the melt factor approaches the snowmelt runoff factor. In Canada, Erickson et al. (1978) found that the cumulative

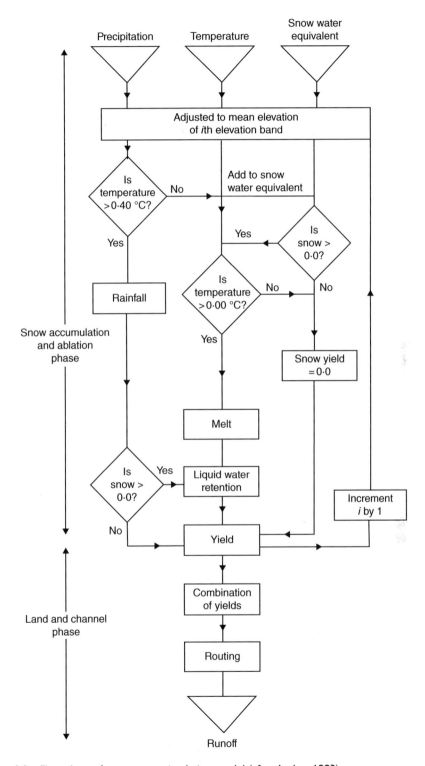

Figure 4.9 Flow chart of a computer simulation model (after Archer, 1983)

daily runoff per unit snow covered area is related to the number of degree-days (days with $T>0\,°C$): that is,

$$\sum_{i=1}^{n}\left(\frac{Q_i}{a}\right)=M_q\sum_{i=1}^{n}\left(\frac{A_i}{a}\right)T_i \qquad (4.10)$$

where Q_i is the mean daily runoff on the ith day (mm), a is the drainage area of the river basin (m²), M_q is the snowmelt runoff coefficient (mm °C⁻¹day⁻¹), A_i is the mean area of the river basin that is snow covered, T_i is the mean daily temperature ($> 0\,°C$) on the ith day (°C) and n is the number of days with runoff.

Similar results have been found to apply in Switzerland (Hall and Martinec, 1985), in Scotland (Ferguson, 1984), in Chile (Peña and Nazarala, 1984) and other places. Hence the result seems to have wide applicability, and Dey et al. (1983) have shown how satellite data can be used to estimate seasonal snowmelt runoff from the Himalayas in Pakistan, an area where alternative data sources are not available. An example is shown in Figure 4.10. Values of M_q have been found to depend upon land use, and slightly upon the type of terrain: for example, $M_q = 0.248\,\text{mm}\,°C^{-1}\text{day}^{-1}$ for stubble, 0.255 for pasture and 0.820 for summer fallow. Values ranging from 1.7 to 4.57 mm °C⁻¹day⁻¹ have been measured in forested river basins.

Archer (1981) (see also Koren and Bel'chikov, 1983) found that the rate of snow depth depletion does not give a direct indication either of melt or of yield. Nevertheless, the total area of snow cover has been found to be a good indicator for runoff forecasts. Hall and Martinec (1985) describe different types of depletion curves and their behaviour in varying hydrometeorological conditions. Each day's snowmelt volume is converted to areal shrinkage of the snowpack using

$$V_i = A_i D_i M_i \qquad (4.11)$$

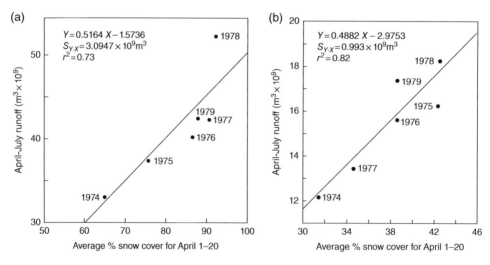

Figure 4.10   Satellite-derived snow cover estimates versus measured runoff for (a) the Indus River above Besham, Pakistan, 1974–9; and (b) the Kabul River above Nowshera, Pakistan, 1974–9. Regression values for the lines are shown for each graph, where y is runoff, S is snow cover and r is the correlation coefficient (after Dey et al., 1983)

where V_i is the melt volume on the ith day (m³), A_i is the snowpack area (km²) and D_i is the degree-day index (°C⁻¹day⁻¹).

The shortage of A_i measurements depends upon the details of how the snowpack melts. Ferguson (1984), assuming that the snowpack is of the same water equivalent W everywhere and melts entirely at the edges, used the equation

$$A_{i+1} = A_i - \frac{V_i}{W} \tag{4.12}$$

However, it was recognized that snowpacks are not always uniform, and if it is assumed that the melt rate decreases linearly to zero at the snowiest point, and W is replaced by a constant mean value $\overline{W}$, then the depletion equation 4.12 becomes

$$A_{i+1} = \left(A_i^2 - \frac{V_i A_0}{W_0} \right)^{1/2} \tag{4.13}$$

where W_0 is the initial mean water equivalent depth. Buttle and McDonnell (1987) have found that models which assume that melt occurs predominantly at the edges of the snowpack are best suited to conditions of discontinuous snow cover. However, models which assume that the snowpack water equivalent varies spatially are best suited to continuous snow cover.

The total time required for snowmelt water to reach a particular point will depend upon the sum of the times required for it to move through the snowpack to the surface of the ground, move along the ground and enter the river. These delays in the melt water entering the river will be important in forecasting the peak flow, particularly in small river basins. The most common procedure for estimating this time delay is to keep a continuous estimate of the heat storage or internal energy changes in the snow. When the snowpack is isothermal at 0°C, melt water will move to the surface of the ground. This situation may be monitored by measuring the cold content W_c of the snowpack,

$$W_c = \frac{\rho d T_s}{160} \tag{4.14}$$

where ρ is the average density of the snowpack (kg m⁻³), d is the depth of the snow (m) and T_s is the average absolute temperature of the snowpack below 0 °C. W_c is the heat required per unit area to raise the temperature of the snowpack to 0°C, and is usually expressed as the amount of liquid water in millimetres which, when freezing, will warm the snowpack to 0°C by releasing latent heat of fusion.

Anderson (1973) derived the following empirical formula for estimating the lag time of water in the snowpack:

$$LAG = 5.33 \left[1.0 - \exp(-0.03W/EXCESS) \right] \tag{4.15}$$

where LAG is the lag time (h), W is the water equivalent of the solid portion of the snowpack (mm), and $EXCESS$ is the excess liquid water available for runoff generated in a 6 hour period (mm). The outflow is attenuated, this attenuation being estimated from

$$PACKRO = (S+I)/\left[0.5\exp\left(-220.4I/W^{1.3}\right)+1\right] \tag{4.16}$$

where $PACKRO$ is the snow cover outflow for a 6 hour period (mm), S is the amount of excess water in the snowpack at the beginning of the period (mm), and I is the amount of lagged inflow for the 6 hour period (mm).

An alternative using the analogue of a flood wave moving through reservoir storage (a routing equation approach) has been proposed by Eggleston et al. (1971). In fact this is the same approach as that subsequently used to route the river flow with its snowmelt contribution. The basic routing equation used is

$$\frac{ds}{dt} = I(x,y,t) - O(x,y,t) \tag{4.17}$$

where ds/dt is rate of change in storage within the part of the river considered, $I(x, y, t)$ is the inflow rate at a point as a function of (x, y, t), and $O(x, y, t)$ is the outflow rate at a point as a function of (x, y, t). This will be discussed further in Chapter 9.

4.10   Snow avalanches

An avalanche is a rapid flow of snow down a hill or mountainside. Although avalanches can occur on any slope given the right conditions, certain times of the year and certain locations are more dangerous than others. Wintertime, particularly from December to April, is when most avalanches tend to happen. There are two common types of avalanche. The first is a *surface avalanche*, which occurs when a layer of snow slides over another layer of snow that has different properties. One example is when a layer of dry loosely packed snow slides over a dense layer of wet snow; a second example is shown in Figure 4.11. An avalanche that occurs in dry snow at below freezing temperatures can be composed of either loose snow or slabs. The vast majority of avalanche fatalities are caused by dry slab avalanches. The second type of avalanche is known as a *full-depth avalanche*, which occurs when an entire snow cover, from the earth to the surface, slides over the ground.

Snow avalanches are most likely to occur after a fresh snowfall adds a new layer to a snowpack (National Geographic Education: education.nationalgeographic.org/encyclopedia/avalanche/). If new snow occurs during a storm, the snowpack may become overloaded, setting off an avalanche. Earthquakes can set off avalanches, but much smaller vibrations can trigger them as well: a single skier can cause enough vibrations to set off an avalanche. Currently, scientists are not able to predict with certainty when and where avalanches will happen. However, they can estimate hazard levels by checking on the snowpack, temperature and wind conditions as follows:

1. Snowstorm and wind direction: heavy snowstorms are more likely to cause avalanches. The 24 hours after a storm are considered to be the most critical. Wind normally blows from one side of the slope of a mountain to another side. While blowing up, it will scour snow off the surface which can overhang a mountain.

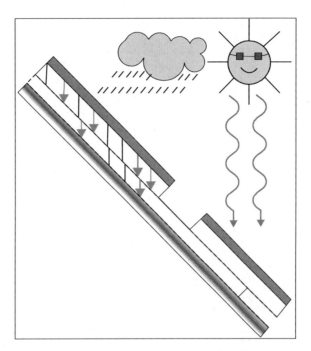

Figure 4.11   Rain and melt close to the surface form water in the snow cover; when water accumulates at a boundary between layers the strength of the snow is reduced and avalanches can occur (from the Swiss Federal Institute for Forest, Snow and Landscape Research, Davos)

2. Heavy snowfall may deposit snow in unstable areas and put pressure on the snowpack. Precipitation during the summer months is the leading cause of wet snow avalanches.

4.11   Worldwide distribution and extremes of snow cover

Figure 4.12 shows the world distribution of snow cover, differentiating between areas with 100% snow cover and areas with 20% or less snow cover. The depth of snow north of 50°N and in the polar regions varies between 40 and 80 cm, but in high latitude mountainous areas, depths over 200 cm are common. The mean annual snow depth in parts of the Rocky Mountains exceeds 280 cm. The Antarctic is the southern polar region, specifically the region around the Earth's South Pole, and is opposite the Arctic region around the North Pole. The Antarctic comprises the continent of Antarctica and the ice shelves, waters and islands in the Southern Ocean. It covers some 20% of the southern hemisphere, of which 5.5% (14 million km²) is the surface area of the continent itself. The Arctic is an ice covered ocean, surrounded by treeless, frozen ground (Figure 4.13). Greenland is the world's largest island. More than 80% of the island is covered by an ice cap (Figure 4.13), 4 km thick in places. Recent studies have raised fears that global warming (see Chapter 13) is causing Greenland's ice cover to melt increasingly fast, and that this could have serious implications for future sea levels and ocean currents.

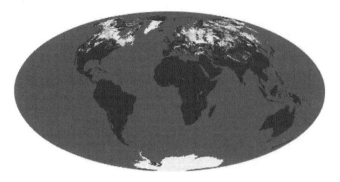

Figure 4.12 World distribution of snow cover, February 2000: *grey areas*, less than 20% of the land area covered; *white areas*, 100% (derived from the MODIS instrument on the TERRA satellite, courtesy NASA)

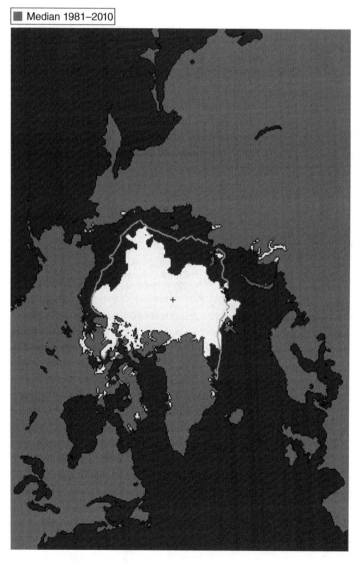

Figure 4.13 Arctic sea ice extent for 17 September 2014; the grey line shows the 1981 to 2010 median extent for that day; the black cross indicates the geographic North Pole (after National Snow and Ice Data Center, Boulder, CO, USA)

Table 4.3 Snowfall accumulations (cm) in specified durations for a frequency of occurrence per year and specified return periods at Birmingham (Elmdon) and (in brackets) Eskdalemuir (Dumfries and Galloway) (from Jackson, 1977b)

Duration (h)	Frequency (per year)		Return period (years)		
	2	1	2	5	10
1	1.9 (3.2)	3.4 (4.3)	4.6 (5.3)	6.6 (6.6)	8.5 (8.5)
3	3.2 (5.1)	4.8 (6.8)	6.5 (8.3)	9.2 (9.9)	12.7 (12.8)
6	3.6 (6.3)	5.7 (8.3)	7.5 (10.4)	10.5 (12.5)	15.4 (16.2)
24	3.9 (8.5)	7.3 (11.2)	9.5 (14.7)	13.6 (19.0)	20.8 (23.8)

Although lowland areas in mid-latitudes generally have much less snow than the mountain areas, local topography can generate large snowfall. For example, south-east of Lake Huron in Canada, moisture from the lake results in more than 250 cm of snow per year. Even areas with much smaller average snow depths are sometimes prone to exceptional falls. Table 4.3 shows the snowfall accumulations in specified durations and with specified return periods at Birmingham (Elmdon) and Eskdalemuir (Dumfries and Galloway) in the United Kingdom. The table shows, for example, that for a return period of once in 10 years it is likely that a snowfall of over 20 cm in 24 hours will occur at either location.

Dickson and Posey (1967) and Hardy (1974) provide maps and statistics of snow cover probabilities for the northern hemisphere including the United Kingdom, which show how the variability in mid-latitudes changes with the seasons. For structural design purposes it is important to know the probability of getting a particular rate of precipitation (snowfall). Although rain may occur with air temperatures less than 2.5 °C, to a good approximation these data can be taken as indicative of the likely rates of snowfall or sleet.

Appendix 4.1   Estimates of catchment snowmelt inflow rates

The daily snowmelt inflow to a catchment is given by

$$Q_n = \frac{m}{(1-\alpha_0)} f_1(M, I_f) f_2(M, \alpha_0) \tag{4.18}$$

where m is the daily melt; α_0 is the initial relative amount of melt water retained by snow, which is equal to the upper limit of water retention capacity of the snow cover; $f_2(M, \alpha_0)$ is a function expressing the relative area of the snowmelt inflow related to the accumulated melt M and the initial water retention of snow; and $f_1(M, I_f)$ is a runoff coefficient as a function of the accumulated melt and the index of the infiltration capacity of the catchment I_f. The simplest technique for estimating the snowmelt inflow area assumes uniform melt and constant initial water retention of snow over the whole catchment. Therefore the function $f_2(M, \alpha_0)$ can be found from the difference between the two further functions,

$$f_2(M, \alpha_0) = f_3(M) - f_4(M) \tag{4.19}$$

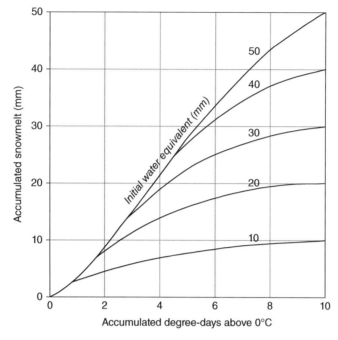

Figure 4.14 Temperature–snowmelt relationship for different values of initial water equivalent for a lowland catchment (from WMO, 1994)

where $f_4(M)$ is the area of the catchment from which the snow has melted and $f_3(M)$ is an integral function of the relative area on which snow becomes saturated,

$$f_3(M) = f_4\left(\frac{M}{\alpha_0}\right) \qquad (4.20)$$

This may be used to derive graphical relationships between accumulated degree-days and accumulated snowmelt for different mean water equivalents of snow cover, as shown in Figure 4.14.

Summary of key points in this chapter

1. World snowfall is a primary factor in hydrological forecasting, and in the United States a forecast improvement of 6% was attained by the use of operational satellite measurement of snow covered areas.
2. Several processes are involved in governing the different types of snow which may be produced, including nucleation and sublimation of ice and snow crystals, the aggregation of snowflakes, and riming to form crystals and graupel (snow pellets).

3. The rate of increase with time t of a mass m of a crystal through diffusion of water vapour onto its surface is $dm/dt = 4\pi CDFA / (\rho_\infty - \rho_0)$, where C is a shape factor, D is the diffusivity of water vapour, F is a ventilation factor depending on the relative motion of the crystal with respect to the air, A is a function of crystal size, ρ_∞ is the vapour density at a large distance from the crystal, and ρ_0 is the vapour density at the surface of the crystal.

4. Conditions necessary, but not sufficient, for the occurrence of significant snowfall: (i) sufficient moisture and aerosol nuclei for the formation and growth of ice crystals; (ii) sufficient depth of cloud; (iii) temperature below 0°C in most of the layer through which the snow falls; and (iv) sufficient moisture and aerosol nuclei to replace losses caused by precipitation.

5. Snow cover is an important factor in determining climate and climate change.

6. Snow strongly reflects visible and near infrared light, that is, it has a high albedo. It also has a high thermal emissivity, a low thermal conductivity and a low water vapour pressure.

7. For temperatures less than about 4°C, less dense, cooler water at the surface begins to freeze even though the water below it is relatively warmer.

8. For sea ice more horizontally oriented crystals are favoured, whereas for ice on lakes this is not so. In both cases the ice thickens downwards.

9. Like snow cover, the ice on lakes delays runoff from a catchment area.

10. The snow that makes up ice sheets and glaciers undergoes a sequence of conversion stages including nivation (thaw–freeze alternation leading to the conversion of snow to ice crystals), firn or névé (pressure melting) and sintering (continued fusion and squeezing out of air).

11. The characteristics of snow cover are determined by atmospheric conditions and the condition of the land surface; atmospheric turbulence influences the snow cover distribution and properties, mainly density.

12. Glaciers form on the crests and slopes of mountains and may fill a valley. Their mass balance is derived from the annual accumulation, ablation and change in mass of ice held in storage.

13. The freezing point of water decreases from 0°C for fresh water to –2°C for water of 35% salinity.

14. Permafrost is soil at or below the freezing point of water (0°C) for two or more years.

15. The total melt water which is produced from a snow cover is controlled by the energy exchanges at the upper and lower snow surfaces. The daily amount of melt is $M = E_m/L\rho F$, where E_m is the snowmelt water equivalent, ρ is the density of water, L is the latent heat of fusion, and F is the fraction of ice in a unit mass of wet snow.

16. The water content of snowfall may be estimated from measurements of the depth of snow and the density of snow.

17. Most operational forecasting procedures are based upon the use of air temperature as the index of energy available for melting the snow. The rate of

snow depth depletion does not give a direct indication either of melt or of yield, but the total area of snow cover has been found to be a good indicator for runoff forecasts.

18. The total time for snowmelt water to reach a particular point will depend upon the sum of the times required for it to move through the snowpack to the surface of the ground.

19. There are two common types of avalanche: a surface avalanche, which occurs when a layer of snow slides over another layer of snow that has different properties; and a full-depth avalanche, which occurs when an entire snow cover, from the earth to the surface, slides over the ground.

20. The depth of snow north of 50°N and in the polar regions varies between 40 and 80 cm.

21. The Antarctic covers some 20% of the southern hemisphere, of which 5.5% is the surface area of the continent itself.

22. Local topography can generate large snowfall, e.g. south-east of Lake Huron in Canada. Here moisture from the lake results in more than 250 cm of snow per year.

Problems

1. Draw a flow diagram of the formation of different types of snow.

2. Calculate the mass growth of an ice crystal through the diffusion of water vapour onto its surface, assuming the collision efficiency is 0.5 and the fall speed of the ice crystals is 1 m s^{-1}. Specify the radius of the ice crystals and the adhesion efficiency used.

3. What are the conditions necessary for the occurrence of significant snowfall?

4. List the albedo for fresh snow, open water, clouds, desert sand, field crops, grass and concrete.

5. Describe how ice forms.

6. Give five examples of the density of snow cover.

7. Describe how glaciers are formed.

8. Draw a diagram illustrating the ground temperature profile in a permafrost region.

9. Calculate the daily amount of snowmelt in centimetres per day, assuming the density of water is 1000 kg m^{-3}, the latent heat of fusion is 333.5 kJ kg^{-1} and the fraction of ice in a unit mass of wet snow is 0.96.

10. Give the formula for Darcy's law.

11. What is the correction for water equivalent as a function of elevation in northern England and southern Scotland?

12. Give a common expression for the snow melt produced in cm of water in unit time, and an expression for the conversion of each day's snow melt volume to areal shrinkage of a snow pack.

13. What is the 'cold content' of a snow pack, and how might it be calculated?

14. Describe how avalanches occur.

15. Give examples of daily heavy snowfall in two different parts of the world.

References

Anderson, E.A. (1973) National Weather Service river forecast system: snow accumulation and ablation model. NOAA Technical Memorandum NWS HYDRO-17, US Department of Commerce, Washington, DC.

Andreadis, K.M., Storek, P. and Lettenmaier, D.P. (2009) Modeling snow accumulation and ablation processes in forested environments. *Water Resour. Res.*, 45, W05429. doi 10.1029/2008WR007042.

Archer, D.R. (1981) Severe snowmelt runoff in north-east England and its implications. *Lond. Inst. Civ. Eng. Proc. Part 2*, 71(12), 1049–1060.

Archer, D.R. (1983) Computer modelling of snowmelt flood runoff in north-east England. *Lond. Inst. Civ. Eng. Proc. Part 2*, 75(6), 155–173.

Barry, R.G. (1993) Canada's cold seas. Chapter 2 in *Canada's Cold Environments*, ed. H.M. French and C. Slaymaker, pp. 29–61. McGill–Queen's University Press, Montreal.

Berris, S.N. and Haar, R.D. (1987) Comparative snow accumulation and melt during rainfall in forested and clear-cut plots in the Western Cascades of Oregon. *Water Resour. Res.*, 23, 135–142.

Brooks, E.S. and Boll, J. (2005) A simple GIS-based snow accumulation and melt model. Proceedings of the Western Snow Conference, 11 April, Great Falls, MT, pp. 123–129.

Browning, K.A. (1983) Air motion and precipitation growth in a major snowstorm. *Q. J. R. Meteorol. Soc.*, 108, 225–242.

Browning, K.A. and Mason, B.J. (1981) Air motion and precipitation growth in frontal systems. *Pure Appl. Geophys. Basle*, 119, 577–593.

Buttle, J.M. and McDonnell, J.J. (1987) Modelling the areal depletion of snowcover in a forested catchment. *J. Hydrol.*, 90, 43–60.

Castruccio, P.A., Loats, H.L. Jr, Lloyd, D. and Newman, A.B. (1980) Cost/benefit analysis. Workshop on Operational Applications of Satellite Snowcover Observations, Sparks, NV, 1979, NASA Conference Publication 2116, pp. 239–254.

Dey, B., Goswami, D.C. and Rango, A. (1983) Utilization of satellite snow-cover observations for seasonal stream flow estimates in the Western Himalayas. *Nord. Hydrol.*, 257–266.

Dickson, R.R. and Posey, J. (1967) Maps of snow cover probability for the Northern Hemisphere. *Mon. Weather Rev.*, 95, 347–355.

Eggleston, K.O., Israelson, E.K. and Riley, J.P. (1971) Hybrid computer simulation of the accumulation and melt processes in a snowpack. Report PRWG65-1, Utah State University, Logan, UT.

Ellis, C.R., Pomeroy, J.W., Brown, T. and MacDonald, J. (2010) Simulation of snow accumulation and melt in needleleaf forest environments. *Hydrol. Earth Syst. Sci. Discuss.*, 7, 1033–1072.

Erickson, D.E.L., Lin, W. and Steppuhn, H. (1978) Indices for estimating prairie runoff from snowmelt. Seventh Symposium Water Standards Institute, Applied Prairie Hydrology, Saskatchewan (source: Gray and Male, 1981).

Ferguson, R.J. (1984) Magnitude and modelling of snowmelt runoff in the Cairngorm Mountains, Scotland. *Hydrol. Sci. J.* 29(1), 3/1984, 49–62.

Fichefet, T. and Morales Maqueda, M.A. (1999) Modelling the influence of snow accumulation and snow-ice formation on the seasonal cycle of the Antarctic sea-ice cover. *Clim. Dyn.*, 15, 251–268.

Flerchinger, G.N. and Saxton, K.E. (1989) Simultaneous heat and water model of a freezing snow–residue–soil system. 1: Theory and development. *Trans. Am. Soc. Agric. Eng.*, 32, 565–571.

French, H.M. (1996) *The Periglacial Environment*. Addison Wesley Longman, Harlow.

Fulkes, J.R. (1935) Rate of precipitation from adiabatically ascending air. *Mon. Weather Rev.*, 63, 291–294.

Galeati, G., Rossi, G., Pini, G. and Zilli, G. (1986) Optimization of a snow network by multivariate statistical analysis. *Hydrol. Sci. J.*, 31(1), 3/1986, 93–108.

Gray, D.M. and Male, D.H. (eds) (1981) *Handbook of Snow: Principles, Processes, Management and Use*. Pergamon, Toronto.

Hall, D.K. and Martinec, J. (1985) *Remote Sensing of Ice and Snow*. Chapman and Hall, London.

Hardy, R.N. (1974) Rates of fall of snow, mixed precipitation and freezing rain. London Meteorological Office, Investigations Division Technical Note 9.

Harr, R.D. (1986) Effects of clear cutting on rain-on-snow runoff in Western Oregon: a new look at old studies. *Water Resour. Res.*, 22(7), 1095–1100.

Hobbs, P.V. (1974) *Ice Physics*. Clarendon, Oxford.

Jackson, M.C. (1977a) The water equivalent of lying snow. *J. Inst. Water Eng. Sci.*, 31(1), 54–56.

Jackson, M.C. (1977b) The occurrence of falling snow over the United Kingdom. *Meteorol. Mag.*, 106, 26–38.

Jackson, M.C. (1977c) A method of estimating the probability of occurrence of snow water equivalents in the United Kingdom [Méthode pour l'évaluation de la probabilité de l'occurrence des équivalents neige-eau au Royaume Uni]. *Hydrol. Sci. Bull.*, 22(1), 127–142.

Kondratyev, K.Ya. (1972) Radiation processes in the atmosphere. Second IMO Lecture, WMO, Geneva, no. 309.

Koren, V.I. and Bel'chikov, V.A. (1983) The effect of the amount of detail in initial data on the accuracy of model calculation of snow–rain runoff. *Nauch.-Issled. Tsent. SSR, Trudy*, vyp. 265. Gidromet, Leningrad [English translation UK Met Office Library].

Kustas, K.P., Rango, A. and Uijlenhoet, A. (1994) A simple energy budget algorithm for the snowmelt runoff model. *Water Resour. Res.*, 30, 1515–1527.

Lavoie, R. (1972) A mesoscale numerical model and lake-effect storms. *J. Atmos. Sci.*, 29, 1025–1040.

Maidment, D.R. (ed.) (1993) *Handbook of Hydrology*. McGraw-Hill, New York.

Marks, D., Kimball, J., Tingey, D. and Link, T. (1998) The sensitivity of snowmelt processes to climate conditions and forest cover during rain-on-snow: a case study of the 1996 Pacific Northwest flood. *Hydrol. Proc.*, 12, 1569–1587.

Martinelli, M. Jr (1973) Snow fences for influencing snow accumulation. Proceedings Banff Symposium on the Role of Snow and Ice in Hydrology, September 1972, UNESCO-WMO-IAHS, Geneva–Budapest–Paris, vol. 2, pp. 1394–1398.

Mernild, S.H., Liston, G.E., Hasholt, B. and Knudsen, N.T. (2006) Snow distribution and melt modeling for Mittivakkat Glacier, Ammassalik Island, Southeast Greenland. *J. Hydrometeorol.*, 7, 808–824.

Mikhil', V.M., Rudneva, A.V. and Lipouskaya, V.I. (1969) Snowfall and snow transport during snowstorms over the USSR. [English translation 1971, Israel Program for Scientific Translations 5909.]

Minnich, R.A. (1986) Snow levels and amounts in the mountains of southern California. *J. Hydrol.*, 89, 37–58.

Olyphant, G.A. (1986) Long wave radiation in mountainous areas and its influence on the energy balance of Alpine snowfields. *Water Resour. Res.*, 22(1), 62–66.

O'Neill, A.D.J. and Gray, D.M. (1973) Spatial and temporal variations of the albedo of prairie snowpack. Proceedings Banff Symposium on the Role of Snow and Ice in Hydrology, September 1972, UNESCO-WMO-IAHS, Geneva–Budapest–Paris, vol. 1, pp. 176–186.

Passarelli, R.E. Jr and Braham, R.R. (1981) The role of the winter land breeze in the formation of Great Lake snow storms. *Bull. Am. Meteorol. Soc.*, 62(4), 482–491.

Peña, H.T. and Nazarala, B.G. (1984) Short term snowmelt flow forecast system. Proceedings of the WMO Technical Conference on Microprocessors in Operational Hydrology, Geneva, 4–5 September, pp. 155–161. Reidel, Dordrecht.

Rango, A. (1985) An international perspective on large-scale snow studies. *Hydrol. Sci. J.*, 30(2), 6/1985, 225–238.

Rouse, W.R. (1993) Northern climates. Chapter 3 in *Canada's Cold Environments*, ed. H.M. French and O. Slaymaker, pp. 65–92. McGill–Queens University Press, Montreal.

Seligman, G. (1962) *Snow Structure and Ski Fields* (reprint of 1936 edn). Adam, Brussels.

Sevruk, B. (1985) Conversion of snowfall depths to water equivalents in the Swiss Alps. WMO Workshop on the Correction of Precipitation Measurements, 1–3 April, Zurich, pp. 81–88.

Smith, M.W. (1993) Climatic change and permafrost. Chapter 12 in *Canada's Cold Environments*, ed. H.M. French and O. Slaymaker, pp. 291–311. McGill–Queen's University Press, Montreal.

Walsh, J.E. (1984) Snow cover and atmospheric variability. *Am. Sci.*, 72(1), 50–57.

WMO (1994) Snow melt runoff analysis. Chapter 31 in *Guide to Hydrological Practices: Data Acquisition and Processing, Analysis, Forecasting and Other Applications*, pp. 431–443. World Meteorological Organization, Geneva.

Woo, M.-K. (1986) Permafrost hydrology. *Atmos. Ocean*, 24, 201–234.

Woo, M.-K. (1993) Northern hydrology. Chapter 5 in *Canada's Cold Environments*, ed. H.M. French and C. Slaymaker, pp. 117–142. McGill–Queen's University Press, Montreal.

Zeinivand, H. and DeSmedt, F. (2009) Simulation of snow covers area by a physical based model. *World Acad. Sci. Eng. Tech.*, 55, 469–474.

5

Measurements and Instrumentation

5.1   Measurement, resolution, precision and accuracy

The appropriate requirement for resolution, frequency of measurement, accuracy and precision of observations will depend upon the application to which they are to be put. The precision with which measurements should be made refers to their reproducibility in space, time and quantity. Precision is expressed quantitatively as the standard deviation of results from repeated trials under identical conditions. A measurement may be quantitatively accurate but imprecise, and vice versa (Figure 5.1). This depends upon the interrelationship between resolution, frequency and numerical accuracy. A measurement from a raingauge is spatially precise, but may be imprecise in time and quantity. A radar measurement may be quantitatively imprecise, but precise in time and space.

5.2   Point measurements of precipitation

5.2.1   Raingauge types

Non-recording gauges are used by most hydrological and meteorological services. They consist of open receptacles with vertical sides, i.e. right cylinders. Various sizes of orifice and height are used in different countries, and therefore measurements are not easily comparable. The depth of precipitation caught by the gauge is measured by means of a graduated flask or 'dipstick'. However in recent years recording gauges have been more commonly used. Three types are in operation. In the *weighing type*, the weight of a receiving can plus the precipitation accumulating in it is recorded continuously. In the *float type*, rainfall is passed into a float chamber containing a light float. As the water level rises, the vertical movement of the float is transmitted to a pen on a chart. In the *tipping-bucket type*, a light metal container is divided into two compartments and is balanced in unstable equilibrium above a horizontal axis. After a predetermined amount of rain the bucket tips over a relay contact to produce a record of the number of tips. Each tip may be 0.5 mm, 0.2 mm or 0.1 mm. Other types of gauge may use electronic or optical mechanisms and be capable of rapid response (see for example Norbury, 1974).

Hydrometeorology, First Edition. Christopher G. Collier.
© 2016 John Wiley & Sons, Ltd. Published 2016 by John Wiley & Sons, Ltd.
Companion website: www.wiley.com/go/collierhydrometerology

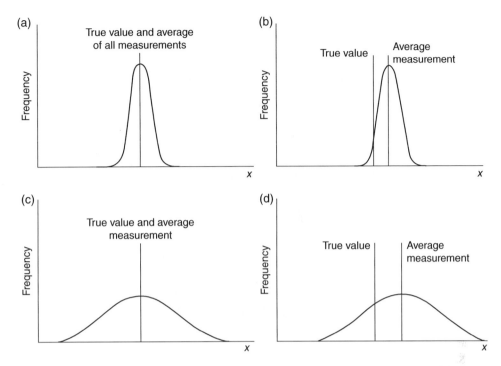

Figure 5.1 Precision and accuracy: (a) accurate and precise, (b) not accurate but precise, (c) accurate but not precise, (d) not accurate and not precise (from Chatfield, 1983)

A *disdrometer* is an instrument used to measure the drop size distribution and velocity of falling hydrometeors. Some disdrometers can distinguish between rain, graupel and hail. The best known type is the Joss–Waldvogel disdrometer (Joss and Waldvogel, 1967) (see Figure 5.2). This is an impact device in which the diameter of a raindrop is given by a relationship involving the mechanical momentum of an impacting drop.

The 2D-video-disdrometer (2DVD) (Schonhuber et al., 2007) measurement principle is based on two high speed line scan cameras, watching particles falling through a $25 \times 25 \, \text{cm}^2$ inlet of the housing. The common area of the crossed optical paths is approximately $10 \times 10 \, \text{cm}^2$, constituting the virtual measurement area. The imaging grid resolution for raindrops is finer than 0.2 mm, in both horizontal and vertical directions.

A sophisticated optical disdrometer has been developed by several groups. A particularly good example has been produced by the University of Iowa, USA. Two beams of laser light, vertically spaced by 1 cm, are produced by two 670 nm laser beams passing through a collecting lens and culminating lens, respectively. The two beams of laser light then pass through a convex lens located 20 cm from the lasers, which focuses the light on a photodetector. A computer reads and stores the voltages at 10 kHz. The velocity, diameter, shape and drop size distribution of raindrops are extracted from the voltage measurements. The system is simple, low cost, and requires no calibration.

Strangeways (2004) points out that precipitation amount is not measured adequately for many precise applications. He describes the field testing over 3 years of the design for an aerodynamic rain collector proposed by Folland (1988) (Figure 5.3).

Figure 5.2 The Joss–Waldvogel disdrometer at the Karlsruhe Institute of Technology; the measurement surface has an area of 48 cm² (copyright Karlsruhe IT)

Although shown to have aerodynamic advantage over a standard 5 inch gauge, the new rain collector was found to suffer from outsplash in heavy rain. Its performance was investigated in turbulent, real-world conditions rather than in the controlled and simplified environment of a wind tunnel or mathematical model as in the past. To do this, video records were made using thread tracers to indicate the path of the wind, giving new insight into the complex flow of natural wind around and within rain-gauges. A new design resulted, and 2 years of field testing have shown that the new gauge has good aerodynamic and evaporative characteristics and minimal outsplash, offering the potential for improved precipitation measurement. Further discussion of errors in raingauge measurements is contained in section 5.2.3.

5.2.2   Measuring snow and hail

The depth of lying fresh snow is measured on open ground with a graduated ruler or scale. A mean of several vertical measurements should be made in places where there is no drifting snow. The depth of snow may also be measured in a fixed container after the snow has been levelled without compressing it. Further details are given in Collier (2000). The water equivalent of a snowfall is the amount of liquid precipitation contained in that snowfall using one of the following methods (Sevruk, 1992):

1. Weighing or melting.
2. Using raingauges: snow collected in a non-recording raingauge should be melted immediately.

Figure 5.3 The Folland-shaped gauge with the door open showing the collecting bottle (from Strangeways, 2004)

3. Radioisotope gauges: radioactive gamma sources are used to measure water equivalent from the attenuation of the gamma radiation signal which is proportional to snow depth and the water equivalent. The measuring element consists of two vertical tubes of the same length at a distance of 0.5 to 0.7 m from each other. One tube contains the source, the other the detector.
4. Natural gamma radiation surveys: based upon attenuation by snow of the gamma radiation emitted from natural radioactive elements in the ground.

Snowfall density, on average, decreases with decreasing temperature. Also the density decreases with decreasing wind speed, and these problems may lead to inaccuracies in the measurements. A decrease in the errors can be achieved by taking the mean of several measurements at each separate point. The necessary number of measurements to ensure that the desired relative accuracy of the mean values will be achieved is determined from

$$N = V_x / V_a \tag{5.1}$$

where N is the number of measurements required to obtain the prespecified accuracy, V_x is the variance of the errors, and V_a is the square of the acceptable standard error of the estimate of the mean. This also applies to rainfall measurements.

A melting snow cover typically contains from 2% to 5% by weight of liquid water, although sometimes as much as 10%. Absorbed solar radiation results in melting, but the reflectivity of a snow surface ranges from 90% for newly fallen snow to around 40% for old snow.

Hail is measured directly by observers from the indentations made on hail pads. These pads are usually about 2.5 cm thick by 0.5 m × 0.5 m. Their material is wrapped in aluminium foil. The hail size is correlated with the foil indentations. Usually a very dense network of hail pads is necessary to provide reliable estimates of the hail swath; for example, 96 hail pads may be distributed over an area of about 260 km².

5.2.3   Errors in measurement

The use of can-type raingauges may be subject to appreciable systematic error (3% to −30% or more), particularly if they are exposed above ground level (WMO, 1982). The components of the systematic error are listed in Table 5.1. The model for corrections is

$$P_k = kP_c = k\left(P_g + \Delta P_1 + \Delta P_2 + \Delta P_3 + \Delta P_4 - \Delta P_5\right) + P_r \qquad (5.2)$$

Table 5.1   Main components of the systematic error in precipitation measurements and their meteorological and instrumental factors, listed in order of general importance (after WMO, 1982)

Symbol	Component of error	Magnitude	Meteorological factors	Instrumental factors
k	Loss due to wind field deformation above the gauge orifice	2–10%	Wind speed at the gauge rim during precipitation and the structure of precipitation	The shape, orifice area and depth of both gauge rim and collector
$\Delta P_1 + \Delta P_2$	Losses from wetting on internal walls of the collector and in the container when it is emptied	10–50%	Frequency, type and amount of precipitation, the drying time of the gauge and the frequency of emptying the container	The same as above, plus the material colour and age of both gauge collector and container
ΔP_3	Loss due to the evaporation from the container	2–10%	Type of precipitation, saturation deficit and wind speed at the level of the gauge rim during the interval between the end of precipitation and its measurement	The orifice area and the isolation of the container, the colour and, in some cases, the age of the container, or the type of funnel (rigid or removable)
ΔP_4	Outsplash and insplash	0–4%	Rainfall intensity and wind speed	The shape and depth of the gauge collector and the kind of gauge installation
ΔP_5	Blowing and drifting snow	1%	Intensity and duration of snowstorm, wind speed and the state of snow cover	The shape, orifice area and depth of both gauge rim and collector

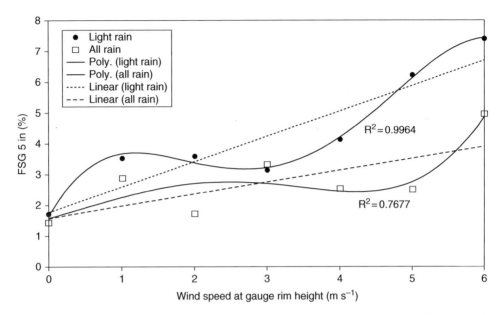

Figure 5.4 The percentage difference in catch between the Folland-shaped gauge (FSG) and the 5 inch gauge for increasing wind speed (from Strangeways, 2004)

where P_k is the corrected amount of precipitation; P_g is the gauge catch, i.e. the gauge measured precipitation; k is a conversion factor due to wind field deformation; P_c is the amount of precipitation caught by the gauge; ΔP_1 to ΔP_5 are corrections for various components of systematic error according to Table 5.1; and P_r are random observational and instrumental errors.

The conversion factor k and the corrections ΔP_1 to ΔP_5 for a particular gauge can be estimated experimentally by comparisons between field measurements and laboratory tests. The meteorological factors can be estimated using standard meteorological observations at the gauge site or nearby.

Figure 5.4 was produced from field data and shows the difference between the catch of the Folland-shaped gauge (FSG) and a 5 inch gauge. In light (low intensity) rain there is a clear aerodynamic gain by the FSG over the 5 inch gauge (upper linear trend line). The difference in catch increases from 1.8% to around 6.7% as the wind speed increases from calm to strong (6 m s^{-1}), an increase of 4.9%. The line does not go through zero because evaporation has not been completely compensated.

Strangeways (2004) noted that the low walls around a gauge to minimize impacts of wind flow, and the shallow-angled funnel of the Folland-shaped gauge, could result in outsplash in heavy rain, and this does occur as confirmed by Folland (1988). Outsplash is not a simple process in which drops 'bounce' out of the collector like balls, although this does occur with hail. The drops break up on impact with a wetted surface (but not with a dry one) into many hundreds of small drops (as a spray), some of them leaving the collector and others falling back or striking the sides.

5.3   Areal measurements of precipitation using raingauge networks

Shaw (1994) summarizes the ways point raingauge measurements may be used to estimate areal rainfall, and the reader is referred to this text. The simplest method is to sum gauge values for a selected duration and divide the total by the number of gauges. This is satisfactory provided the gauges are uniformly spaced and the area does not vary greatly in topography. A more sophisticated technique is the use of Thiessen polygons. Here individual gauge values are first weighted by the fractions of the catchment area represented by each gauge. A more accurate method involves the use of isohyetals, although this does require a skilled analyst. Finally the rainfall surface may be represented by a multi-quadratic surface derived from the network of gauges implemented via a computer program. This is discussed in more detail in Chapter 7. The combination of dissimilar estimates is outlined in Appendix 5.1.

5.4   Radar measurements of rainfall

5.4.1   Basics

Over the last 30 years or so, weather radar networks have become an integral part of operational meteorological observing systems. Measurements of rainfall using radar are used qualitatively by weather forecasters. However, there remains a reluctance by operational hydrologists to use these data quantitatively, although improvements in algorithm development and the advent of dual polarization radar are now changing this situation. In this section we review the current accuracy achieved and new approaches to radar measurements, drawing particularly upon the reviews of Collier (2000) and Illingworth (2004).

The radar beam rotates about a vertical axis, and pulses of radar energy are transmitted. The backscattered energy from precipitation particles in volumes of size dependent on the beam width and pulse size above the ground (Figure 5.5) is received by the radar. The relationship for the power received at the radar antenna is the sum of the contributions from the individual scatterers (hydrometeors), and is represented by the radar equation (see Probert-Jones, 1962; Collier, 1996). For a radar wavelength λ and a spherical raindrop with diameter D, a backscattering cross-section $\sigma_b(D)$ and a total attenuating cross-section $\sigma_a(D)$ are proportional to D^6/λ^4, assuming Rayleigh scattering (small drops). This is the justification for introducing a physical parameter called the radar reflectivity factor Z, defined as

$$Z = \int_0^{D_{max}} N(D)D^6 dD \tag{5.3}$$

where $N(D)$ is the drop size distribution (DSD) within the resolution cell (Z in $mm^6 m^{-3}$, D in mm, $N(D)$ in $mm^{-1} m^{-3}$). In the absence of attenuation along the radar path, and as long as Rayleigh theory is appropriate, the backscattered power is proportional to Z. However, when the ratio $\pi D/\lambda$ becomes larger than 0.1, that is for very

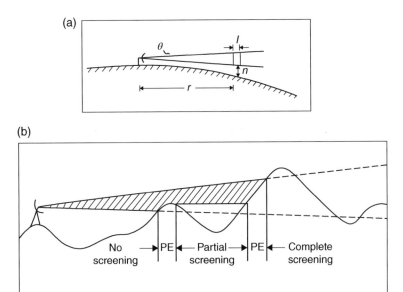

Figure 5.5   (a) The geometry of the radar beam relative to the curvature of the Earth. A radar pulse having a length l is located at a range r from the radar. The height of the base of the beam at range r having an angular width of θ is h. (b) Permanent echoes (PE) and screening caused by hills close to the radar site are also shown (from Browning and Collier, 1989)

large drops and hail, Mie theory should be used. To take account of this effect, an equivalent radar reflectivity factor Z_e is used which is the same as Z for light rain.

If liquid precipitation uniformly fills the pulse volume, the average power returned from precipitation at range r is proportional to Z/r^2, where the radar reflectivity factor Z is given by the sum of the precipitation particle diameters raised to the sixth power. Z is related to the rate of rainfall R by the empirical relationship

$$Z = AR^B \tag{5.4}$$

where A and B depend on the rainfall. Values often used are $A = 200$ and $B = 1.6$ (see section 5.4.3). An example of reflectivity measurements is shown in Figure 5.6.

5.4.2   *Errors in radar measurements*

Some problems in making measurements with radar arise from the characteristics of the radar and the radar site. The maintenance of a stable system is extremely important. Joss and Waldvogel (1990) state that it is possible to keep the cumulative error of transmitted power, antenna parameters, noise figure, amplification in the receiver chain and analogue-to-digital conversion to better than 2 dB, corresponding to a 36% error in precipitation rate.

Generally a low horizon (0.5° or less) is required for surface precipitation measurements. However the consequence of this is that the main radar beam or its side lobes may intersect the ground, as shown in Figure 5.5b. This causes strong

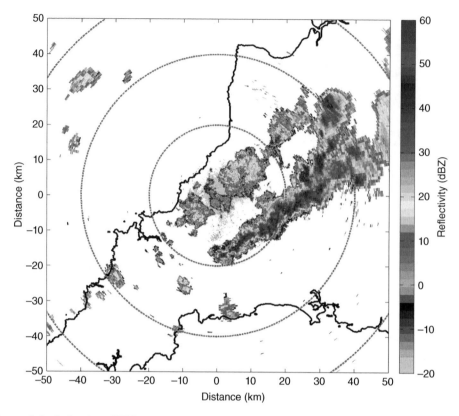

Figure 5.6   Reflectivity (dBZ) measured by the NCAS mobile radar at a constant elevation angle of 2.5° on 3 August 2013 during the COnvective Precipitation Experiment (COPE) (from Blyth et al., 2015)

persistent ground echoes (ground clutter) and screening (occultation), causing partial loss of the radar beam. A ground clutter map collected on a day when there is no precipitation may enable ground clutter to be avoided and interpolation across it to be achieved. Likewise provided about 60% of the beam remains beyond areas of screening it may be possible to apply corrections. However, anomalous refraction of the radar beam caused by warm, dry air overlaying cooler, moist air may cause additional ground echoes known as anomalous propagation echoes (anaprop).

As the radar beam passes through precipitation it may be attenuated. Use of a long wavelength such as 10 cm (S-band) avoids such a problem, although for the same antenna size the beam width at 10 cm is twice as large as at 5 cm (C-band), causing reduced spatial resolution at long range. Also, as pointed out by Ulaby et al. (1981), precipitation echoes observed at 10 cm are 12 dB weaker than those observed at 5 cm for the same output power. A 3 cm radar (X-band) is even more effective than a 5 cm radar having the same beam width, but will suffer more attenuation. However, the use of X-band radars has become possible with the development of procedures for attenuation correction based upon the application of polarization parameters (see section 5.4.5). Radars often have radomes to protect them from strong winds and the effects of snow and ice. However when the radome gets wet on radars with wavelengths of 5 cm or less, additional attenuation may

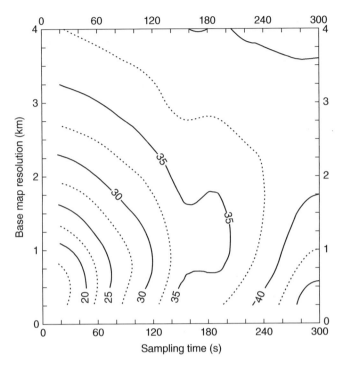

Figure 5.7 Absolute error (%) in 5 minute accumulations as a function of the resolution of the reflectivity maps and sampling intervals (from Fabry et al., 1994)

result. As we shall discuss later, the application of dual polarization technology mitigates the impact of attenuation to a large degree and therefore makes shorter wavelength radars more attractive.

In addition to the problems due to the hardware and siting, further problems arise from the characteristics of the precipitation itself. Averaging radar signals over areas of non-uniform precipitation and samples fluctuate, and significant variations within the radar pulse volume can cause errors of several decibels. The accuracy of measurements also depends upon sampling in time and space, as shown in Figure 5.7 (see also Harrison et al., 2009).

The variability of the reflectivity in the vertical may lead to large differences between the apparent precipitation intensity measured by a radar and the precipitation reaching the ground, particularly as the range increases. Figure 5.8 shows vertical profiles as seen by a radar in convective and widespread rain at various ranges. There may be large deviations from these profiles on particular occasions, but they indicate the percentage of precipitation observed by radar allowing for the Earth's curvature. Figures 5.8b and 5.8c indicate the difficulties of making measurements in the presence of the bright band (where snow melts to form rain) or shallow rain. Orographic growth of rain within around 0.5 km of the ground may also be difficult to measure, as shown in Figure 5.8d.

Equation 5.4 relates measurements of radar reflectivity to measurements of rainfall using an empirical relationship. Battan (1973) lists 69 published relationships which have been found in different rainfall types. Fortunately the relationships are similar for the same type of rainfall, and Table 5.2 gives typical relationships.

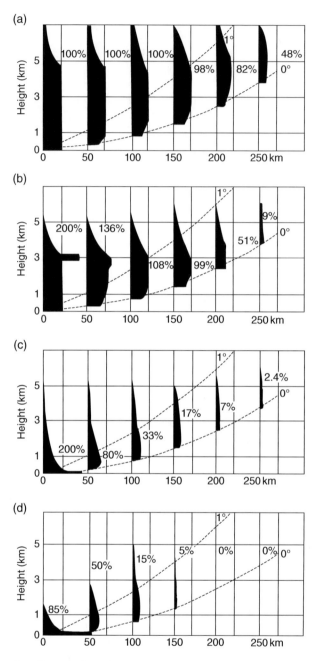

Figure 5.8   Vertical profiles seen by radar at various ranges in (a) convective rain, (b) widespread rain, (c) low level rain or snow, and (d) orographic (very low level) rain. The numbers in each figure give the percentage (referred to the true melted water value, which would be measured at ground level using a raingauge) of rain rate deduced from the maximum reflectivity of the profile. A radar with a 1° beam width is assumed in a flat country. If the radar is lower than nearby obstacles, then much less precipitation can be observed at far ranges. The top and bottom of the main part of a 1° beam elevated at 0.5° are shown as dashed lines labelled 0° and 1° (partly after Joss and Waldvogel, 1990 and Browning and Collier, 1989)

Table 5.2   Typical relationships between Z and R

Equation	Precipitation type	Reference
$140R^{1.5}$	Drizzle	Joss et al. (1970)
$250R^{1.5}$	Widespread rain	Joss et al. (1970)
$200R^{1.6}$	Stratiform rain	Marshall and Palmer (1948)
$31R^{1.71}$	Orographic rain	Blanchard (1953)
$500R^{1.5}$	Thunderstorm rain	Joss et al. (1970)
$486R^{1.37}$	Thunderstorm rain	Jones (1956)
$2000R^{2.0}$	Aggregate snowflakes	Gunn and Marshall (1958)
$1780R^{2.21}$	Snowflakes	Sekhon and Srivastava (1970)

Following work by Zawadzki (1984), Collier (1986), Joss and Waldvogel (1990) and Rosenfeld et al. (1993), it has become clear that uncertainties in the drop size distribution may not be the largest source of errors in radar measurements of rainfall. Doneaud et al. (1984) found an excellent correlation between integrated convective rainfall volume and the area–time integral (ATI) of the area exceeding a given optimum threshold reflectivity value.

5.4.3   Adjustment using raingauges

The most commonly used method of adjusting radar estimates of reflectivity remains the use of the independent measurements from raingauges. Many different techniques have been implemented, from adjustment factors derived from simple applications of the ratios of raingauge to radar values (see for example Collier et al., 1983) to the more complicated procedures building on the work of Doneaud et al. (1984).

A technique which has been applied successfully builds upon the matching of the unconditional probabilities of rainfall intensity obtained from raingauges and radar reflectivity (Rosenfeld et al., 1993; 1994). This method is known as the probability matching method (PMM). Rosenfeld et al. (1994) concluded that 600 mm of accumulated rainfall was required to give an adequate empirical representation of the radar and raingauge rainfall probability distributions. This translates to some 600 samples depending upon the time interval used. However, Seed et al. (1996) found that the PMM makes systematic underestimates of high rainfall rates when based upon only 500 samples. This bias only decreases slowly as the number of samples increases.

The PMM involves converting both radar and raingauge data to dBZ using the generic formula

$$dBZ = 10\log_{10}\left(fV^{p}\right) \tag{5.5}$$

where V is the value of the rain intensity (mm h^{-1}) measured by radar or raingauge, f is a factor whose typical value is 200, and p is a power whose value is typically 1.6.

Recent applications of this technique have used averaging intervals of 5 minutes to 1 hour, for example 9 minutes in Rosenfeld et al. (1994) and 30 minutes in Atencia et al. (2010). The scatter in matching radar data (which are areal data) to raingauge data (which are point data) introduces sampling errors (Kitchen and Blackall, 1992; Villarini et al., 2008).

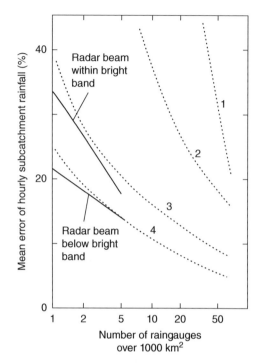

Figure 5.9 Mean error of the hourly rainfall totals in river catchments of average area 60 km² as determined from radar measurements in various kinds of rainfall conditions, plotted as a function of the number density of adjusting raingauge sites (solid lines). Also shown for comparison is the mean error of the hourly subcatchment totals as determined from a network of raingauges in the absence of radar, again plotted as a function of the number density of raingauge sites (dotted lines). The set of four dotted curves represents the measurement errors for the raingauge network in the presence of (1) extremely isolated showers, (2) typical showers, (3) typical widespread rain and (4) extremely uniform rain. For all curves the mean error is defined as the mean value of the difference between the estimated rainfall and the 'optimum estimate' without regard to sign (after Collier, 1977; from Browning, 1978)

The accuracy of hourly rainfall totals over river subcatchments – full curves that can be achieved with an adjusted radar located within 50 km of an area of interest – is shown in Figure 5.9. The dotted curves represent the accuracy achieved with networks of raingauges in the absence of radar. For both sets of curves the accuracy is plotted as a function of raingauge density. The accuracy of the raingauge networks depends critically on the nature of the rain, as shown by the large differences between the four dotted curves. The radar measurements, on the contrary, are nearly independent of rainfall type.

5.4.4 Summary of problem areas associated with radar measurements of precipitation

Figure 5.10 summarizes the errors in radar measurement of surface precipitation which arise from the characteristics of the precipitation being observed. On any one occasion, all these sources of error may not be present or may be of varying significance.

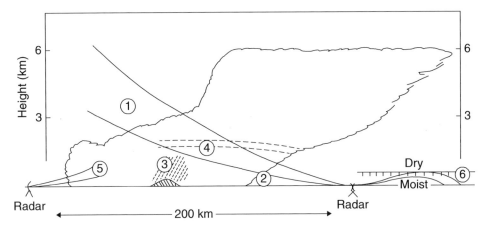

Figure 5.10 Schematic representation of the problems associated with the measurement of precipitation by radar: (1) radar beam overshooting the shallow precipitation at long ranges; (2) low level evaporation beneath the radar beam; (3) orographic enhancement above hills which goes undetected; (4) the bright band; (5) underestimation of the intensity of drizzle because of the absence of large droplets; (6) radar beam bent in the presence of a strong hydrolapse, causing it to intercept land or sea (from Browning and Collier, 1982)

5.4.5   The use of multi-parameter radar

Raindrops are oblate as they fall, with their larger dimensions being on average horizontally oriented. Therefore polarimetric radars were developed to measure the radar reflectivities in two directions (horizontal and vertical), and hence the differential reflectivity Z_{DR} and the differential phase shift between the two polarizations.

Polarimetric radars provide information on the shape and orientation of the radar targets. Development of multi-parameter radar hardware has been slow since the initial production of high speed switches enabling the rapid alternate transmission of horizontally and vertically polarized microwave radiation, described by Seliga and Bringi (1976). However in recent years the recognition of the potential of multi-parameter radar for measuring precipitation has led to commercial developments in this field at all frequencies, and the rapid introduction of polarization radars in operational networks. In what follows we discuss the use of linear polarization in which the radar can transmit pulses which are alternately polarized in the horizontal (H) and vertical (V). There are two copolar returns Z_H and Z_V, and if the radar is dopplerized then the phases of the horizontally and vertically polarized returns ϕ_h and ϕ_v can be estimated. There are four basic parameters: the differential reflectivity Z_{DR}, the copolar correlation ρ_{hv}, the specific differential phase shift K_{DP} and the linear depolarization ratio (LDR). A further parameter, the degree of polarization (DOP), is not commonly used. Pulses may be polarized at 45° with reception at H and V, which has the advantage that no rapid polarization switching between transmitted pulses is needed.

5.4.6   Drop size distributions

Variations in drop size spectra affect the representativeness of the mean drop size. Natural raindrop size spectra are well represented by a normalized gamma function (Ulbrich, 1983) of the form

$$N(D) = N_W f(\mu) \left(\frac{D}{D_0} \right)^{\mu} \exp\left[-(3.67 + \mu) \frac{D}{D_0} \right] \tag{5.6}$$

where

$$f(\mu) = \left[\frac{6}{(3.67)^4} \right] / \left[\frac{(3.67 + \mu)^{\mu+4}}{\Gamma(\mu+4)} \right] \tag{5.7}$$

When $\mu = 0$ with $N_W \equiv N_0$, this reduces to

$$N(D) = N_0 \exp\left[-3.67 \frac{D}{D_0} \right] \tag{5.8}$$

where D_0 is the median volume drop diameter and can be estimated from Z_{DR}; μ is the shape of the drop spectrum; and N_W is a normalized drop concentration. These equations may be expressed in terms of D_M, the mass weighted mean diameter, where D_M is equal to D_0 times the factor $(4 + \mu) / (3.67 + \mu)$. Values of μ range from 2 to near 10.

5.4.7   Rainfall estimation using parametric variables

Testud et al. (2000) summarize the rainfall estimation procedures which have been developed and are relatively robust, adding a new method developed for use with the spaceborne rain radar of the Tropical Rainfall Measurement Mission (TRMM), but proposed for ground-based polarimetric radars. To these we may add a further technique using Z_{DR} and K_{DP}. The techniques used are as follows:

1. The reflectivity-only method as discussed above, using a defined drop size distribution (Marshall and Palmer, 1948) or a gamma distribution (Ulbrich, 1983):

$$R(Z_H) = 2.92 \times 10^{-2} Z_H^{1/1.5} \quad \text{(S/C-band)}$$
$$R(Z_H) = 5.72 \times 10^{-2} Z_H^{0.621} \quad \text{(X-band) for } Z_H \tag{5.9}$$
$$\leq 35\,\text{dBZ (Kim et al., 2010)}$$

2. A two-parameter method as proposed by Chandrasekar et al. (1990):

$$R(Z_H, Z_{DR}) = 1.98 \times 10^{-3} Z_H^{0.97} Z_{DR}^{-1.05} \quad \text{(S-band)}$$
$$R(Z_H, Z_{DR}) = 0.0039 Z_H^{1.07} Z_{DR}^{-5.97} \quad \text{(Bringi and Chandrasekar, 2001)}$$
$$R(Z_H, Z_{DR}) = 17[Z_H(0.00012$$
$$- 0.000041 Z_{DR})]^{0.73} \quad \text{(X-band)(Matrosov, 2010)} \tag{5.10}$$

Figure 5.11 shows images of Z_H and Z_{DR}.

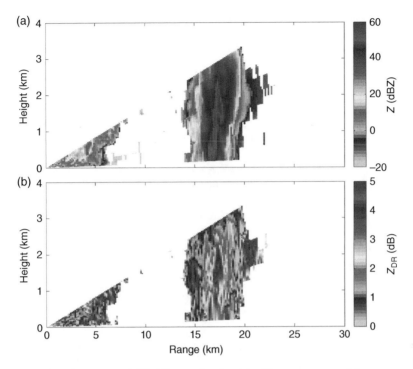

Figure 5.11   (a) Reflectivity and (b) differential reflectivity Z_{DR} reconstructed from measurements made at constant elevation angles by the NCAS mobile X-band dual-polarization Doppler radar on 3 August 2013 at 132740 UTC (from Blyth et al., 2015)

3. A K_{DP} estimator derived by Sachidananda and Zrnic (1986):

$$
\begin{aligned}
R(K_{DP}) &= 37.1K_{DP}^{0.85} \quad \text{(S-band)} \\
R(K_{DP}) &= 16.03K_{DP}^{0.95} \quad \text{(C-band)} \\
R(K_{DP}) &= 17.0K_{DP}^{-0.73} \quad \text{(X-band)(Matrosov, 2010)}
\end{aligned}
\tag{5.11}
$$

For the X-band equation, see also Kim et al. (2010) for the water content W, where

$$
\begin{aligned}
W(Z_H, K_{DP}) &= 0.00383Z_H^{0.55} \quad &\text{for } K_{DP} \le 0.3°\,\text{km}^{-1} \\
&= 0.996K_{DP}^{0.713} \quad &\text{for } K_{DP} > 0.3°\,\text{km}^{-1}
\end{aligned}
$$

4. A K_{DP} and Z_{DR} estimator proposed by Ryzhkov and Zrnic (1995):

$$
R(K_{DP}, Z_{DR}) = 52K_{DP}^{0.96}Z_{DR}^{-0.447}
\tag{5.12}
$$

5. A rain profiling algorithm, known as the ZPHI technique (Testud et al., 2000), based upon a set of three power law relationships between attenuation A and Z,

K_{DP} and A, and R and A, each of which is 'normalized' by N_0^* of the drop size distribution:

$$A = a\left(N_0^*\right)^{1-b} Z^b$$
$$K_{DP} = \alpha\left(N_0^*\right)^{1-b} A^\beta \tag{5.13}$$
$$R = c\left(N_0^*\right)^{1-d} A^d$$

Here a, α, b, β, c and d are constants dependent upon radar frequency (as given in Testud et al., 2000); and N_0^* is the intercept parameter of an exponential or gamma drop size distribution $(0.8 \times 10^7 \, \text{m}^{-4}; \mu = 2)$.

To estimate the performance of polarization techniques in improving rainfall estimates it is necessary to know how variations in the drop size spectra affect the representativeness of the mean drop size inferred from polarization parameters, as noted by Illingworth (2004) and others. As mentioned in section 5.4.6, natural raindrop size spectra may be represented by a normalized gamma function. The liquid water content remains constant if μ changes (Illingworth and Blackman, 2002). Generally the range of μ is from –1 to 5 (Battan, 1973), although others have suggested values of μ from 0 to 15. High values of μ are not usually regarded as reliable.

Each of these algorithms has advantages and disadvantages, but none of them is entirely satisfactory. The ZPHI technique does solve two problems: it is insensitive to attenuation and calibration error, and the physical rain variation is represented through N_0^* variations. Hence, the systematic retrieval of N_0^* leads to some automatic adaptation of the rainfall estimate from one kind of rain to another.

5.4.8   Measurement of snow

When the ratio $\pi D / \lambda$ becomes larger than 0.1, for example for snowflakes or hail, Mie theory of scattering should be used in place of Rayleigh theory. To take account of this effect, an 'equivalent' radar reflectivity factor Z_e is used. For light rain Z_e is the same as Z, but for snowflakes this is not the case as the particles are far from being spherical. The radar cross-section of an irregularly shaped particle composed of a weak dielectric like ice is the same as that of a sphere of the same mass. Hence the exact shape of the particle is immaterial. The radar cross-section of an ice particle is

$$\Delta = \pi^5 \left|K\right|_i^2 D^6 / \lambda^4 \tag{5.14}$$

where $\left|K\right|_i^2$ is the dielectric factor; the subscript i indicates that a value appropriate for ice should be used. The dielectric factor is equal to $\left|\left(\varepsilon_r - 1\right)/\left(\varepsilon_r + 2\right)\right|^2$, where ε_r is the relative permittivity; this is related to the index of refraction n by $\varepsilon_r = n^2$. The diameter implied in equation 5.14 is that of a sphere having the same mass as the particle in question. The equivalent radar reflectivity factor Z_e is defined as

$$Z_e = \frac{\lambda^4}{\pi^5 \left|K\right|_w^2} \tag{5.15}$$

Table 5.3   Example values of equivalent radar reflectivity factor Z_e (dBZ) for rain and snow for two precipitation rates (from Smith, 1984)

	Precipitation rate R(mm h^{-1})	
	1	10
Rain	23	39
Snow	26	48

where the subscript w indicates that the value appropriate for water (approximately 0.93 for the usual meteorological radar wavelengths) is used by convention. Smith (1984) discusses the appropriate value of $|K|_i^2$, pointing out that there are two possible 'correct' values depending upon the particle sizes determined. If the particle sizes used are melted drop diameters, then $|K|_i^2$ is 0.208 and

$$Z_e = 0.224\,Z \tag{5.16}$$

However, if the particle sizes are expressed as equivalent ice sphere diameters, then $|K|_i^2$ is 0.176 and

$$Z_e = 0.189\,Z \tag{5.17}$$

Table 5.3 compares equivalent radar reflectivity factors calculated for precipitation rates of 1 and 10 mm h^{-1}, for rain using the Marshall and Palmer relationship and for snow using the Sekhon and Srivastava (1970) relationship.

Jatila (1973) used a calibration taking the water equivalent of snow collected in a single raingauge to assess the accuracy of radar measurements of snowfall. He found that 60% of the snowfall amounts within about 50 km of the radar site fell in the interval −24% to +32% of the daily amounts of snowfall measured by gauges. Pollock and Wilson (1972), using three radars and an extensive gauge network including 13 weighing precipitation gauges, found similar results within 30 km of the radar sites. Collier and Larke (1978) showed that comparable accuracy can be achieved in hilly terrain provided the height at which melting occurs is known, as shown in Figure 5.12. Browning (1983) pointed out that snowfall development below the radar beam due to orographic effects can lead to significant underestimates of the surface snowfall.

5.4.9 Measurement of hail

On observing hail the radar backscattered power is no longer proportional to the sixth power of the particle size, and Mie scattering theory is applicable as the particle diameters are larger than one-tenth of the radar wavelength. For hail alone a simple relationship between reflectivity and hail diameter D_H can be

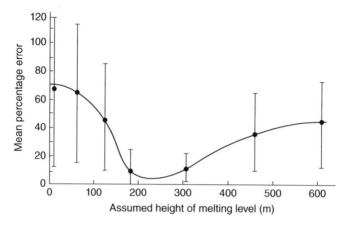

Figure 5.12   The dependence of the mean percentage error, regardless of sign, of estimates of snow depth over areas of 100 km² derived from radar data, on the assumed height of the melting level (from Collier and Larke, 1978)

obtained (Hardaker and Auer, 1994). Assuming Rayleigh scatter in the hail at C-band,

$$Z_e = 10\log_{10}\left(561 D_H^{2.6}\right)$$ (5.18)

The measurement of maximum value of Z_e provides a method of hail detection (Geotis, 1963). Waldvogel et al. (1979) suggested the following simple criterion for hail detection:

$$H_{45} > H_0 + 1.4\,\text{km}$$ (5.19)

where H_{45} is the height of the 45 dBZ contour, which must exceed the height H_0 of the 0 °C level by more than 1.4 km.

Hail cannot be identified unambiguously from the observed values of Z and Z_{DR}, but its presence will raise Z and depress Z_{DR}. Smyth et al. (1999) suggested that the best technique to identify hail is to continually monitor Z, Z_{DR} and the forward propagation differential phase ϕ_{DP} along each ray (where K_{DP} is the range derivative of the forward propagation phase) and to use the failure of consistency as an indication of the presence of hail. The bright band can be identified by the values of LDR being in the range –14 to –18 dB. Melting graupel is associated with lower values of LDR, of about –22 to –26 dB.

5.4.10   Precipitation type

Straka et al. (2000) and Meischner et al. (2004) provide reviews of the range of values of Z, Z_{DR}, K_{DP}, LDR and ρ_{HV} expected for hail, graupel, rain, rain and wet hail mixtures, and snow crystals and aggregates. Figure 5.13 shows an example of the classification. Liu and Chandrasekar (2000) have extended this classification using fuzzy logic. However, Illingworth (2004) points out that difficulties remain in identifying different forms of ice crystals.

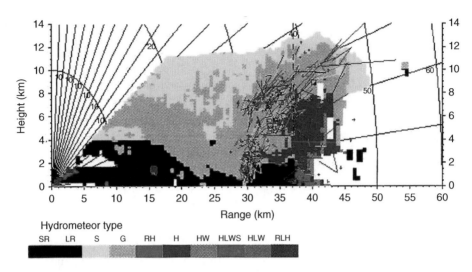

Figure 5.13 Vertical section of the supercell storm of 21 July 1998 at 1741 UTC along 232° azimuth from the POLDIRAD radar, DLR, Germany. ITF flashes are plotted for a 30 s interval around the nominal scanning time. Hydrometeor or particle colour codes: SR, small rain; LR, large rain; S, snow; G, graupel; RH, rain–hail mixture; H, hail; HW, hail wet; HLWS, hail large wet spongy; HLW, hail large wet; RLH, rain/large-hail mixture) (from Meischner et al., 2004) (*see plate section for colour representation of this figure*)

5.5   Soil moisture

5.5.1   *Approaches*

There are two approaches to measuring soil moisture:

1. Soil water content, which is expressed as a ratio of the mass of water contained in a sample of soil either to the dry mass of the soil sample, or to the original volume of the sample. These two expressions are linearly related by a coefficient known as the dry bulk density of the soil being sampled.
2. Soil water potential, which expresses the potential energy of the water contained in the soil, and is composed of gravitational, pressure and osmotic components. In most cases, the soil–water mixture can be considered locally homogeneous, and the osmotic potential becomes negligible.

The soil moisture must be measured at many test points in an area under examination because of its high spatial variability. A summary of methods is given in WMO (1994) and Herschy (1999).

5.5.2   *Gravimetric method*

The gravimetric method is the only direct method of measuring soil moisture. It involves collecting a soil sample, weighing the sample before and after drying it, and calculating its moisture content. The soil sample is considered to be dry when its weight remains constant at a temperature of 105 °C. Many different types of sampling equipment may be used. For example, the simplest equipment for soil moisture sampling is the hand auger – essentially an aluminium pipe used to sample to depths as great as 17 metres. Alternatively a soil sampling tube, core barrel, or drive sampler can be used, which offer an advantage in soil sampling because volumetric samples can be obtained for calculating moisture content by volume.

The open-drive sampler consists of a core barrel 50 millimetres in inside diameter and 100 millimetres long, with extension tubes 25 millimetres in diameter and 1.5 metres long for sampling at depth. Brass cylinder liners, 50 millimetres in length, are used to retain the undisturbed core samples. The samples are removed from the core barrel by pushing a plunger.

5.5.3   Electrical resistance method

The electrical resistance of a block of porous material in moisture equilibrium with the soil can provide a measure of soil moisture. Two electrodes are fixed in the block, which is made of such materials as plaster of Paris, nylon or fibreglass. The contact resistance remains constant, and once the block has been placed in the soil and has reached moisture equilibrium, it adjusts to further changes in soil moisture after a brief lag time. Soil moisture blocks are generally considered most dependable in the low moisture content range. Their suitability for measuring moisture content is restricted by hysteretic effects, and by the fact that calibration depends on the density and temperature of the soil.

5.5.4   Neutron method

The neutron method indicates the amount of water per unit volume of soil. This method is based on the principle of measuring the slowing of neutrons emitted into the soil from a fast neutron source (Greacen, 1981). The energy loss is much greater in neutron collisions with atoms of low atomic weight and is proportional to the number of such atoms present in the soil. The effect of such collisions is to change a fast neutron to a slow neutron. Hydrogen, which is the principal element of low atomic weight found in the soil, is largely contained in the molecules of the water in the soil. The number of slow neutrons detected by a counter tube after emission of fast neutrons from a radioactive source tube is electronically indicated on a scale. The soil volume measured by this method is bulb shaped, and has a radius of 1 to 4 metres according to the moisture content and the activity of the source.

5.5.5   Gamma ray attenuation method

The intensity of a gamma ray that passes through a section of soil undergoes an exponential decrease that principally depends on the apparent density of the soil, on the water contained in the soil, and on the coefficients of attenuation of the soil water, which are constants. The method consists of concurrently lowering a gamma ray source, usually caesium-137, and a gamma ray detector (scintillator photomultiplier) down a pair of parallel access tubes that have been installed in the soil. At each measurement level the signal can be translated into the apparent wet density of the soil; or, if the apparent dry bulk density of the soil is known, the signal can be converted into a measure of the volumetric soil moisture content. This is not specific to water alone.

5.5.6 COSMOS-UK

A cosmic-ray method for measuring area-average soil moisture at the hectometre horizontal scale has been described by Zreda et al. (2008; 2012). The stationary cosmic-ray soil moisture probe measures neutrons that are generated by cosmic rays within air and soil and other materials, moderated by mainly hydrogen atoms located primarily in soil and water.

A new UK investment by the Natural Environment Research Council (NERC) is enabling the Centre for Ecology and Hydrology (CEH) to set up the COsmic-ray Soil Moisture Observing System UK (COSMOS-UK) (Figure 5.14), a new network that will deliver real-time weather monitoring and field scale measurements of soil moisture across the United Kingdom. The measurement principle is very similar to the neutron probe developed at CEH Wallingford, with the advantage that the COSMOS system relies on naturally occurring neutrons, rather than an artificial source.

Each COSMOS-UK station is equipped with a range of research-quality meteorological and soil monitoring sensors. Large area soil moisture (at a spatial scale of approximately 700 m) is provided by the COSMOS probes. Point measurements of soil moisture are also available from two near-surface time domain transmissometry (TDT) probes and at three depths from a profile probe. Soil temperature is measured by the near-surface TDT probes and at five depths by a soil temperature profile. Two soil heat flux plates provide the soil heat flux, an important term in the surface energy balance. Net radiation is calculated from the four components recorded individually (incoming and outgoing short wave and long wave radiation). An automatic weather station provides the following meteorological variables: wind speed and wind direction; atmospheric pressure; air temperature and relative humidity; and precipitation, measured using a weighing rain gauge. A phenological camera gives a visual record of the land surface (e.g. state of vegetation, snow cover) and atmospheric visibility. Each station is powered from a battery and solar panel. Data are logged and telemetered to CEH in real time, where they are processed and made available via the COSMOS-UK web interface.

5.5.7 Dielectric methods

The apparent dielectric constant (permittivity) of a volume of soil varies with the amount of moisture it contains. The soil is a complex mixture of particles of differing chemical compositions in a matrix of air and water. The dielectric constants of soil particles range from about 2 to 7, and those of air and water are 1 and 80 respectively. When the dry bulk density of a soil remains constant, its apparent dielectric constant allows the conversion of the dielectric measurements to a measure of volumetric soil moisture content. There are two methods, as follows.

Time domain reflectometry (TDR) measures the speed of a microwave pulse between a pair of waveguides that have been placed in the soil; this is a function of the

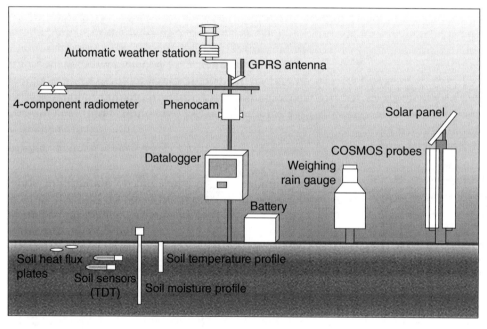

Figure 5.14   COSMOS station (courtesy CEH)

apparent dielectric permittivity of the soil–water–air mixture between the wave guides (Topp et al., 1980). As the pulse speed increases, the permittivity decreases, which indicates a decrease in the moisture content of the soil.

The *capacitance method* uses a sensor which consists of electrodes embedded in the soil, known as an electronic oscillator, together with a frequency meter and connecting cables. The electrodes and their adjacent soil form a capacitor with a capacitance that is a function of the permittivity of the soil, giving the soil moisture content. The varying capacitance can be measured by the frequency changes of the signal across the capacitor using an appropriate calibration. Figure 3.3 shows an example of the soil moisture measured using such a probe.

5.5.8 Tensiometric method

A tensiometer consists of a porous point, cup or similar connected through a tube to a pressure measuring device. The system is filled with water, and the water comes into equilibrium with the moisture in the surrounding soil. Water flows out of the point as the soil dries, and creates greater tension, or flows back into the point as the soil becomes wetter and has less tension. These changes in pressure or tension are indicated on a measuring device. Multiple tensiometers may be located at several depths to allow the estimation of a soil moisture profile. A calibration curve is needed.

5.5.9 Satellite remote sensing

Satellite remote sensing, discussed in Chapter 6, is the only way to obtain measurements that provide very wide area integrated information. Satellite-borne radar can also provide measurements over intermediate areas.

5.6 Evaporation and evapotranspiration

Evaporation pans and lysimeters (see Chapter 3) are used in networks. Many different types of evaporation pan are in use, some with square sections and some with circular. Daily evaporation is computed as the difference in water level between two measurements corrected for any precipitation during the period.

Evapotranspiration is estimated using soil evaporimeters and lysimeters by the water- or heat-budget methods, by the turbulent-diffusing method or by various empirical formulae (see for example WMO, 1976). Transpiration of vegetation is estimated as the difference between measured evapotranspiration and contemporaneously measured evaporation from the soil.

It is not possible to measure evapotranspiration directly from a large area under natural conditions. Over a long period a water-budget approach may be used from a basin in which all items of inflow and outflow, except evapotranspiration, can be measured (see WMO, 1985). Alternatively energy-budget and aerodynamic and pan evaporation techniques may be used (see WMO, 1966).

5.7 Flow measurement: basic hydrometry

The discharge equation for flow in open channels and pipes is

$$Q = VA \tag{5.20}$$

where Q is the discharge in m³s⁻¹, V is the velocity in m s⁻¹ and A is the wetted area in m². The technique is based upon estimating velocity and area either directly or indirectly. Most stream flow stations use the current meter to measure velocity, area being directly measured by a sounding rod or line or other techniques. In rivers with a stable bed, area can usually be related to a function of stage, although velocity and area can also be measured directly at ultrasonic stations. Electromagnetic stations may measure velocity and area indirectly, relying upon the relation

$$V = \sqrt{(2gH)} \tag{5.21}$$

where H is the total head over the weir in metres. Methods include the Pitot tube, open channel orifices, sluice gates and flow in pipes. Doppler systems measure flows directly.

The general discharge equation for weirs and flumes is also based on the velocity–area principle. Figure 5.15 shows that $V = \sqrt{(2gH/3)}$ and $A = 2H/3$, and from equation 5.20

$$Q = \left(\frac{2}{3}\right)^{3/2} C_d b \sqrt{(g)} H^{3/2} \tag{5.22}$$

where Q is the discharge in m³s⁻¹, b is the length of crest in metres, g is the acceleration due to gravity in m s⁻², and C_d is the coefficient of discharge.

Current meters can be classified generally into those having vertical axis rotors, known as cup type, and those having horizontal axis rotors, known as propeller type. The cup-type current meter consists of a rotor revolving about a vertical shaft. The rotor has six usually conical cups fixed at an equal angle on a ring mounted on the vertical shaft. The propeller-type current meter consists of a propeller revolving about a horizontal shaft with or without a vane and a means of attaching the instrument.

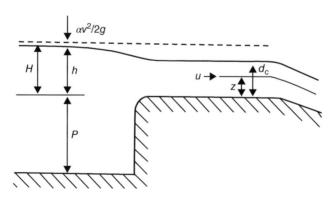

Figure 5.15 Diagram to show the principle and theory of the weir equation of discharge: the parameters relate to equations 5.21 and 5.22 (see Herschy, 1999)

5.8   Measuring stream discharge

The discharge of a river is measured by dividing its cross-section with spaced verticals as shown in Figure 5.16. The verticals are spaced to ensure an adequate sample of both velocity distribution and bed profile. The discharge is derived from the sum of the products of velocity, depth and distance between the verticals. The spacing and number of verticals are critical for an accurate measurement. For a channel width of 0.0–0.5 m Herschy (1995) recommends three to four verticals, and for a channel width greater than 10 m he recommends greater than 20.

5.8.1   The stage-discharge curve

A simple stage-discharge curve can be expressed as

$$Q = C(h + a)^n \tag{5.23}$$

where Q is the discharge, h is the stage or gauge height, C and n are constants, and a is the value of stage at zero flow. This equation may be rewritten as

$$\log Q = \log C + n \log(h + a) \tag{5.24}$$

This represents a straight line on double logarithmic paper, where the slope of the line is n. The geometry or shape of the river channel section is reflected in the slope: for example, for a rectangular channel section $n = 3/2$, for a concave section of parabolic shape $n = 2$, and for a triangular or semicircular section $n = 5/2$.

Care must be taken in extrapolating stage-discharge curves. Herschy (1995; 1999) suggests that the most satisfactory procedure, referred to as the stage–area–velocity method, makes use of curves of discharge, area and mean velocity plotted

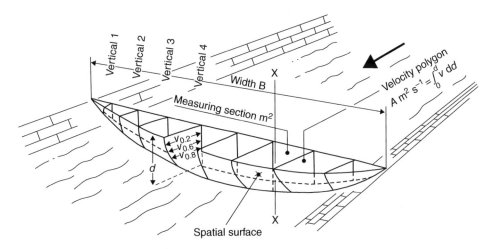

Figure 5.16   Schematic view of measuring cross-section. The cross-section is divided into segments by spacing verticals at a sufficient number of locations across the channel. The discharge is derived from the sum of the products of velocity, depth and distance between the verticals (from Herschy, 1999)

against stage on the same graph. The method depends upon the characteristic of the rate of increase of mean velocity at the higher stages decreasing rapidly, so that the stage/mean-velocity curve can be extended with acceptable error. Extrapolated discharges are found from the multiplication of mean velocity and area as shown in Figure 5.17.

Other methods include the Manning equation,

$$Q = n^{-1}AR^{2/3}S^{1/2} \tag{5.25}$$

and the Chezy equation,

$$Q = ACR^{1/2}S^{1/2} \tag{5.26}$$

where n is Manning's roughness coefficient, C is the Chezy roughness coefficient and A is the cross-section wetted area. The hydraulic radius R is given by

$$R = A/P \tag{5.27}$$

where P is the wetted perimeter, S is the slope of the water surface and Q is the discharge in $m^3 s^{-1}$. The average velocity is therefore

$$v = n^{-1}R^{2/3}S^{1/2} \quad m s^{-1} \quad (\text{Manning}) \tag{5.28}$$

$$v = CR^{1/2}S^{1/2} \quad m s^{-1} \quad (\text{Chezy}) \tag{5.29}$$

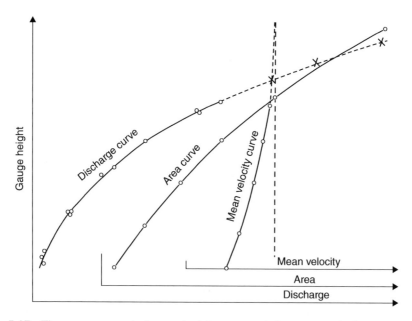

Figure 5.17 The stage–area–velocity method for extrapolating a stage-discharge curve (from Herschy, 1999)

5.8.2   *Automated moving boat methods*

A propeller-type current meter is suspended from a boat at about 1 m below the surface, and the boat crosses the channel normal to the stream flow (Smoot and Novak, 1969). As the boat moves, an echo sounder records the geometry of the cross-section and the current meter resolves the stream and boat velocities. The velocity measurement taken at each of the sampling verticals in the cross-section is a vector quantity which is the relative velocity past the meter (see Herschy, 1999).

The broadband acoustic Doppler current profiler determines the velocity by measuring the Doppler shift between acoustic signals received off sediment particles in the flow. The magnitude and direction of the velocity in a water column and the depth of the flow are measured. The system is usually installed on a boat which measures its own relative position as it moves across the channel.

A further method is based upon the motion of water cutting across the vertical component of the Earth's magnetic field and an electromotive force (EMF) being so induced in the water. The EMF is measured by electrodes at each side of the river, and is directly proportional to the average velocity of the cross-sectional flow. Other methods involve the use of floats and tracers injected into the river, and are summarized in Herschy (1999).

5.9   Brief overview of modern telemetry

5.9.1   *Ground-based telemetry links*

Special data networks are now available for the transmission of digital data. Modems and computer links are provided by many companies, and the links may be dedicated to a user or available on a dialup basis. A disadvantage of telephone lines may be the costs involved, particularly if links are required for continuous access to data, as may be the case for flood or severe weather warnings. Nowadays cable TV and telephone lines using fibre optic technology provide ready access to the World Wide Web.

5.9.2   *VHF and UHF radio links*

Where telephone lines are not available, VHF (30–300 MHz) or UHF (300 MHz to 3 GHz, 10 cm wavelength) radio links offer alternative communication methods. However, permission to use these frequencies must be received from the appropriate regulatory body to avoid interference with other users. Generally UHF frequencies are the most widely used because they have a greater bandwidth and therefore a greater number of channels. HF radio (short wavelength) was used widely when telephone lines were not available. It propagates by successive reflections of the radio waves between the ionosphere and the ground, and can therefore cover large distances. However, the technique may suffer from fading due to weather interference and sunspot activity. Therefore satellite communication links have largely replaced this method.

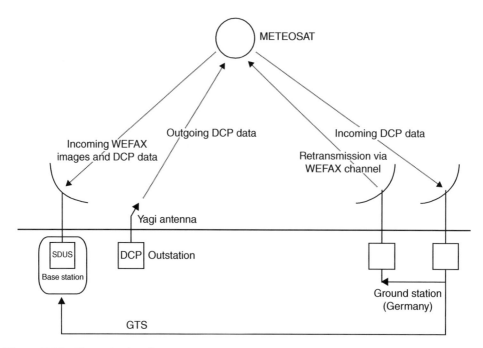

Figure 5.18   The path taken from an outstation, via METEOSAT, through its ground station and back to the user through the retransmission WEFAX route and over the WMO GTS network (from Strangeways, 1999)

5.9.3   Satellite links

Many geostationary satellites such as the METEOSAT, GEOS, GOMS and INSAT now have the facility to collect remotely sensed images and telemeter the data to ground stations. Likewise, the commercially operated network satellites spaced around the Earth, notably Inmarsat, also provide data transmission services. Figure 5.18 shows the data collection platform (DCP) system provided by METEOSAT. A METEOSAT outstation transmitter operates at a frequency of between 402.01 and 402.20 MHz and is directed to the satellite by a Yagi antenna.

Appendix 5.1   Combining dissimilar estimates by the method of least squares

The true value of a variable can be estimated in several different ways using the information obtained from the different measuring instruments. The information can be used to obtain a single estimate by applying the method of least squares (see for example Stephenson, 1965). This estimate will be better than the individual estimates because it will be more precise.

Assume observations $x_1, x_2, \ldots, x_n$ are made of an unknown quantity μ; then

$$x_i = \mu + e_i \qquad (i = 1, \ldots, n) \tag{5.30}$$

where $e_1, e_2, \ldots, e_n$ are the measurement errors. The sum of the squared errors is

$$S = \sum_{i=1}^{n}(x_i - \mu)^2 \qquad (5.31)$$

and hence

$$\frac{dS}{d\mu} = -2\sum_{i=1}^{n}(x_i - \mu) \qquad (5.32)$$

The least squares estimate of μ is obtained by minimizing S with respect to μ. This is achieved by putting $dS/d\mu = 0$. Then

$$\sum(x_i - \hat{\mu}) = 0 \qquad (5.33)$$

$$\hat{\mu} = \sum x_i/n = \bar{x} \qquad (5.34)$$

Therefore the sample mean is also the least squares estimate of μ. The basic assumption necessary to apply the least squares method in this way is that the precision of the observations should be the same. When the instruments are not equally precise, the method must be modified to give more weight to the observations with the smaller precision.

Assume that readings are taken from n instruments and that σ_i is the precision of the ith instrument. Then a set of constants $P_1, P_2, \ldots, P_n$ is chosen so that

$$\sigma_1^2 P_1 = \sigma_2^2 P_2 = \cdots = \sigma_n^2 P_n \qquad (5.35)$$

For example a set of constants may be as follows:

$$P_i = \frac{\sigma_k^2}{\sigma_i^2} \qquad (i = 1, \ldots, n) \qquad (5.36)$$

where σ_k^2 is the square of the smallest precision. Therefore P_k is the largest constant and is equal to 1. In order to obtain the least squares estimates, the difference between each measurement and its expected value is squared and then multiplied by the appropriate P constant. The adjusted sum of squares is generated by adding up these quantities, and the least squares estimates of the unknown parameters are found by minimizing the adjusted sum of squares. Chatfield (1983) gives examples.

Summary of key points in this chapter

1. The precision of measurements refers to their reproducibility in space, time and quantity, whereas a measurement may be quantitatively accurate but imprecise, and vice versa.
2. Non-recording gauges consist of open receptacles with vertical sides, i.e. right cylinders having various sizes of orifices and height.

3. Recording gauges are based upon weighing, float and tipping buckets, and a disdrometer measures the drop size distribution and velocity of falling hydrometeors.

4. An aerodynamic rain collector, although being advantageous over the standard 5 inch gauge, does suffer from outsplash in heavy rain.

5. The depth of lying fresh snow is measured on open ground with a graduated ruler or scale.

6. The water equivalent of a snowfall is the amount of liquid precipitation contained in that snowfall using weighing or melting, raingauges, radioisotope gauges, and natural gamma radiation surveys.

7. Errors in raingauge measurements include losses due to wind, wetting of the internal walls of the collector, evaporation, outsplash and insplash, and blowing and drifting snow.

   For a radar wavelength λ and a spherical raindrop with diameter D, a backscattering cross-section $\sigma_b(D)$ and a total attenuating cross-section $\sigma_a(D)$ are proportional to D^6/λ^4, assuming Rayleigh scattering (small drops). The radar reflectivity factor is given by $Z = \int_0^{D_{max}} N(D)D^6 dD$, where $N(D)$ is the drop size distribution within the resolution cell.

8. The reflectivity Z is related to the rate of rainfall R by the empirical relationship $Z = AR^B$, where A and B depend on the rainfall. Values often used are $A = 200$ and $B = 1.6$, although many other values appear in the literature.

9. Errors in radar measurements may arise from a number of problems including ground echoes, screening (occultation), anomalous propagation, attenuation of the radar beam, and interaction of the radar beam with the bright band (when snow melts to form rain).

10. Radar estimates of reflectivity may be adjusted using raingauge values of rainfall.

11. Polarimetric radars provide information on the shape and orientation of drops, and variations in drop size spectra affect the representativeness of the mean drop size. Polarimetric parameters provide methods of measuring rainfall depending upon the radar wavelength.

12. The Mie theory of scattering is applied when $\pi D/\lambda$ becomes larger than 0, for example for snowflakes or hail. Comparable accuracy can be achieved for measurements of snowfall as for rainfall, provided the height at which melting occurs is known.

13. For hail the radar backscattered power is no longer proportional to the sixth power of the target particle size. However, simple relationships between reflectivity and hail diameter have been obtained.

14. Polarimetric parameters have been used to estimate precipitation type.

15. Soil moisture may be estimated using measurements of soil water content or soil water potential, employing gravimetric, electrical resistance, neutron, gamma ray attenuation, dielectric, tensiometric, cosmic-ray and satellite-based methods.

16. Evaporation pans and lysimeters are used to measure evaporation and evapotranspiration.
17. The discharge equation for flow in open channels is $Q = VA$, where Q is the discharge ($m^3 s^{-1}$), V is the velocity ($m \ s^{-1}$) and A is the wetted area (m^2). Stage-discharge curves may be derived.
18. Modern telemetry uses VHF, UHF and satellite links.

Problems

1. Define precision and accuracy and illustrate the difference.
2. What are the main components of systematic errors in measurements of rainfall using gauges?
3. Specify the main errors in radar measurements of precipitation.
4. Define the reflectivity factor Z and give examples of Z–R relationships for stratiform and convective rainfall.
5. What does parametric radar provide? List the polarimetric parameters that may be measured.
6. Give a formula for the natural raindrop size spectrum.
7. Give a formula for the relationship between rainfall and specific differential phase for an X-band radar.
8. What is the difference between Rayleigh and Mie scattering?
9. Give typical values for rainfall and reflectivity for rain and snow.
10. Give typical values of the linear depolarization ratio for the bright band and melting graupel.
11. Describe the neutron and cosmic-ray methods of measuring soil moisture.
12. Give the discharge equation for flow in an open channel, and derive the velocity assuming the depth of the open channel is 10 m.
13. Give the equation for a simple stage-discharge curve.
14. Give the velocity derived from the Manning equation, defining the parameters in the equation.
15. Outline modern telemetry.

References

Atencia, A., Llasat, M.C., Garrote, L. and Mediero, L. (2010) Effect of radar rainfall time resolution on the predictive capability of a distributed hydrological model. *Hydrol. Earth Syst. Sci. Discuss.* 1(7), 7995–8043.

Battan, L.J. (1973) *Radar Observation of the Atmosphere*. University of Chicago Press, Chicago.

Blanchard, D.C. (1953) Raindrop size distribution in Hawaiian rains. *J. Meteorol.*, 10, 457–473.

Blyth, A.M., Bennett, L.J. and Collier, C.G. (2015) High-resolution observations of precipitation from cumulonimbus clouds. *Meteorol. Appl.*, 22, 75–89.

Bringi, V.N. and Chandrasekar, V. (2001) *Polarimetric Doppler Weather Radar*. Cambridge University Press, Cambridge.

Browning, K.A. (1978) Meteorological applications of radar. *Rep. Prog. Phys.*, 41, 761–806.

Browning, K.A. (1983) Air motion and precipitation growth in a major snowstorm. *Q. J. R. Meteorol. Soc.*, 109, 225–242.

Browning, K.A. and Collier, C.G. (1982) An integrated radar–satellite nowcasting system in the UK. Chapter 1.5 in *Nowcasting*, ed. K.A. Browning, pp. 47–61. Academic Press, London.

Browning, K.A. and Collier, C.G. (1989) Nowcasting of precipitation systems. *Rev. Geophys.*, 27, 345–370.

Chandrasekar, V., Bringi, V.N., Balakrishnan, N. and Zrnic, D.S. (1990) Error structure of multiparameter radar and surface measurement of rainfall. III: Specific differential phase. *J. Atmos. Oceanic Technol.*, 7, 621–629.

Chatfield, C. (1983) *Statistics for Technology*, 3rd edn. Chapman and Hall, London.

Collier, C.G. (1977) Radar measurements of precipitation. Proceedings of the WMO Technical Conference on Instrumentation and Methods of Observation, Hamburg, 27–30 July, WMO Technical Document 480, pp. 202–207. World Meteorological Organization, Geneva.

Collier, C.G. (1986) Accuracy of rainfall estimates by radar. I: Calibration by telemetering raingauges. *J. Hydrol.*, 83, 207–223.

Collier, C.G. (1996) *Applications of Weather Radar Systems: A Guide to Uses of Radar Data in Meteorology and Hydrology*, 2nd edn. Wiley-Praxis, Chichester.

Collier, C.G. (2000) Precipitation estimation and forecasting. WMO Operational Hydrology Report 467, WMO 887. World Meteorological Organization, Geneva.

Collier, C.G. and Larke, P.R. (1978) A case study of the measurement of snowfall by radar. *Q. J. R. Meteorol. Soc.*, 104, 615–621.

Collier, C.G., Larke, P.R. and May, B.R. (1983) A weather radar correction procedure for real-time estimation of surface rainfall. *Q. J. R. Meteorol. Soc.*, 109, 589–608.

Doneaud, A.A., Nisov, S.I., Priegrutz, D.L. and Smith, P.L. (1984) The area–time integral as an indicator for convective rain volumes. *J. Climate Appl. Met.*, 23, 555–561.

Fabry, F., Bellon, A., Duncan, M.R. and Austin, G.L. (1994) High resolution rainfall measurements by radar for very small basins: the sampling problem re-examined. *J. Hydrol.*, 161, 415–428.

Folland, C.K. (1988) Numerical models of the raingauge exposure problem, field experiments and an improved collector design. *Q. J. R. Meteorol. Soc.*, 114, 1485–1516.

Geotis, S.G. (1963) Some radar measurements of hailstorms. *J. Appl. Meteorol.*, 2, 270–275.

Greacen, E.L. (ed.) (1981) Soil water assessment by neutron method. CSIRO Special Publication. CSIRO, Melbourne.

Gunn, K. and Marshall, J. (1958) The distribution with size of aggregate snowflakes. *J. Meteorol.*, 15, 452–461.

Hardaker, P.J. and Auer, A.H. (1994) The separation of rain and hail using single polarisation radar echoes and IR cloud-top temperatures. *Meteorol. Appl.*, 1, 201–204.

Harrison, D.L., Scovell, A.W. and Kitchen, M. (2009) High-resolution precipitation estimates for hydrological users. *Proc. Inst. Civil Eng. Water Manag.*, 162, 125–135.

Herschy, R.W. (1995) *Streamflow Measurement*, 2nd edn. Chapman and Hall, London.

Herschy, R.W. (ed.) (1999) *Hydrometry: Principles and Practices*, 2nd edn. Wiley, Chichester.

Illingworth, A. (2004) Improved precipitation rates and data quality by using polarimetric measurements. Chapter 5 in *Weather Radar Principles and Advanced Applications*, ed. P. Meischner, pp. 130–166. Springer, Berlin.

Illingworth, A. and Blackman, T.M. (2002) The need to represent raindrop size spectra as normalized gamma distributions for the interpretation of polarization radar observations. *J. Appl. Meteorol.*, 41, 1578–1583.

Jatila, E. (1973) Experimental study of the measurement of snowfall by radar. *Geophysica*, 12, 1–10.

Jones, D.M.A. (1956) Rainfall drop-size distribution and radar reflectivity. Research Report 6, Urbana Meteorology Laboratory, Illinois State Water Survey.

Joss, J. and Waldvogel, A. (1967) Ein Spektograph fur Niederschlagstrophen mit automatischer Auswertung. *Pure Appl. Geophys.*, 68, 240–246.

Joss, J. and Waldvogel, A. (1990) Precipitation measurement and hydrology: a review. Battan Memorial and Radar Conference, *Radar in Meteorology*, ed. D. Atlas, Chapter 29a, pp. 577–606. American Meteorological Society, Boston.

Joss, J., Schran, K., Thams, J.C. and Waldvogel, A. (1970) On the quantitative determination of precipitation by radar. Wissanschaftliche Mitteilung 63. Eidenossische Kommission zum Studium der Hagelbildung und dert Hagelabwehr, Zurich.

Kim, D.-S., Maki, M. and Lee, D.-I. (2010) Retrieval of three-dimensional raindrop size distribution using X-band polarimetric radar data. *J. Atmos. Oceanic Technol.*, 27, 1265–1285.

Kitchen, M. and Blackall, R.M. (1992) Representativeness errors in comparisons between radar and gauge measurements of rainfall. *J. Hydrol.*, 134, 13–33.

Liu, H.P. and Chandrasekar, V. (2000) Classification of hydrometeors based on polarimetric radar measurements: development of fuzzy logic and neuro-fuzzy systems, and *in situ* verification. *J. Atmos. Oceanic Technol*, 17, 140–164.

Marshall, J.S. and Palmer, W. (1948) The distribution of raindrops with size. *J. Meteorol*, 5, 165–166.

Matrosov, S.Y. (2010) Evaluating polarimetric X-band radar rainfall estimators during HMT. *J. Atmos. Oceanic Technol.*, 27, 122–134.

Meischner, P., Dotzek, N., Hagen, M. and Holler, H. (2004) Understanding severe weather systems using Doppler and polarisation radar. Chapter 6 in *Weather Radar: Principles and Advanced Applications*, ed. P. Meischner, pp. 167–198. Springer, Berlin.

Norbury, J.R. (1974) A rapid-response raingauge for microwave attenuation studies. *J. Technol. Atmos.*, 8, 245–251.

Pollock, D.M. and Wilson, J.W. (1972) Basin precipitation: land and lake. IFYGL Technical Plan, vol. 1, pp. 107–112.

Probert-Jones, J.R. (1962) The radar equation in meteorology. *Q. J. R. Meteorol. Soc.*, 88, 485–495.

Rosenfeld, D., Wolff, D.B. and Atlas, D. (1993) General probability-matched relation between radar reflectivity and rain rate. *J. Appl. Meteorol.*, 32, 50–72.

Rosenfeld, D., Wolff, D.B. and Amitai, E. (1994) The window probability matching method for rainfall measurements with radar. *J. Appl. Meteorol.*, 33, 682–693.

Ryzhkov, A.V. and Zrnic, D.S. (1995) Comparison of dual-polarisation radar estimators of rain. *J. Atmos. Oceanic Technol.*, 12, 249–256.

Sachidananda, M. and Zrnic, D.S. (1986) Differential propagation phase shift and rainfall estimation. *Radio Sci.*, 21, 235–247.

Schonhuber, M., Lammer, G. and Randeu, W.L. (2007) One decade of imaging precipitation measurement by 2D-video-disdrometer. *Adv. Geosci.*, 10, 85–90.

Seed, A.W., Nicol, J., Austin, G.L., Stow, C.D. and Bradley, S.G. (1996) The impact of radar and raingauge sampling errors when calibrating a weather radar. *Meteorol. Appl.*, 3, 43–52.

Sekhon, R.S. and Srivastava, R.V. (1970) Snow size spectra and radar reflectivity. *J. Atmos. Sci.*, 27, 229–267.

Seliga, T. and Bringi, V.N. (1976) Potential use of radar differential reflectivity measurements at orthogonal polarisations for measuring precipitation. *J. Appl. Meteorol.*, 15, 69–75.

Sevruk, B. (1992) Snow cover measurements and areal assessment of precipitation and soil moisture. Operational Hydrology Report 35, WMO 749. World Meteorological Organization, Geneva.

Shaw, E.M. (1994) *Hydrology in Practice*, 3rd edn. Chapman and Hall, London.

Smith, P.L. (1984) Equivalent radar reflectivity factors for snow and ice particles. *J. Clim. Appl. Meteorol.*, 23, 1258–1260.

Smoot, G.F. and Novak, C.E. (1969) Measurement of discharge by the moving boat method. Water Resources Investigations, Book 3, Chapter A.11. US Geological Survey, Washington DC.

Smyth, T.J., Blackman, T.M. and Illingworth, A.J. (1999) Observations of oblate 'hail' using dual polarisation radar and implications for hail-detection schemes. *Q. J. R. Meteorol. Soc.*, 125, 993–1016.

Stephenson, G. (1965) *Mathematical Methods for Science Students*. Longmans, Green, Glasgow.

Straka, J.M., Zrnic, D.S. and Ryzhkov, A.V. (2000) Bulk hydrometeor classification and quantification using polarimetric radar data: synthesis of relations. *J. Appl. Meteorol.*, 39, 1341–1372.

Strangeways, I.C. (1999) Transmission of hydrometric data by satellite. Chapter 9 in *Hydrometry: Principles and Practices*, 2nd edn, ed. R.W. Herschy, pp. 245–264. Wiley, Chichester.

Strangeways, I.C. (2004) Improving precipitation measurement. *Int. J. Climatol.*, 24, 1443–1460.

Testud, J., Bouar, E.L., Obligis, E. and Ali-Mehenni, M. (2000) The rain profiling algorithm applied to polarimetric weather radar. *J. Atmos. Oceanic Technol.*, 17, 332–356.

Topp, G.C., Davis, J.L. and Annan, A.P. (1980) Electromagnetic determination of soil water content: measurement in coaxial transmission lines. *Water Resour. Res.*, 16, 574–582.

Ulaby, F.T., Moore, R.K. and Fung, A.K. (1981) *Microwave Remote Sensing*, vols 1–3. Addison-Wesley, Reading, MA.

Ulbrich, C.W. (1983) Natural variations in the analytical form of the raindrop size distribution. *J. Clim. Appl. Meteorol.*, 22, 1764–1775.

Villarini, G., Mandapaka, P.V., Krajewski, W.F. and Moore, R.J. (2008) Rainfall and sampling uncertainties: a raingauge perspective. *J. Geophys. Res.*, 113. doi 10.1029/2007JD009214.

Waldvogel, A., Federer, B. and Grimm, P. (1979) Criteria for the detection of hail cells. *J. Appl. Meteorol.*, 18, 1521–1525.

WMO (1966) Measurement and estimation of evaporation and evapotranspiration. Technical Note 83, WMO 201, pp. 95–115. World Meteorological Organization, Geneva.

WMO (1976) The CIMO internation evaporimeter comparisons. WMO 449. World Meteorological Organization, Geneva.

WMO (1982) Methods of correction for systematic error in point precipitation measurement for operations. B. Sevruk. Operational Hydrology Report 21, WMO 589. World Meteorological Organization, Geneva.

WMO (1985) Casebook on operational assessment of areal evaporation. Operational Hydrology Report 22, WMO 635. World Meteorological Organization, Geneva.

WMO (1994) Guide to hydrological practices, 5th edn. WMO 168. World Meteorological Organization, Geneva.

Zawadzki, I.I. (1984) Factors affecting the precision of radar measurements of rain. 22nd Conference on Radar Meteorology, Zurich, Switzerland, pp. 254–256. American Meteorological Society, Boston.

Zreda, M., Desilets, D., Ferre, T.P.A. and Scott, R.L. (2008) Measuring soil moisture content non-invasively at intermediate spatial scale using cosmic-ray neutrons. *Geophys. Res. Lett.*, 35, L21402. doi 10.10219/2008GL 035655.

Zreda, M., Shuttleworth, W.J., Zeng, X., Zweck, C., Desilets, D., Franz, T. and Rosolem, R. (2012) COSMOS: the COsmic-ray Soil Moisture Observing System. *Hydrol. Earth Syst. Sci.*, 16, 4079–4099.

6

Satellite-Based Remote Sensing

6.1 Overview of satellite remote sensing

In the previous chapter methods of measuring parts of the hydrological cycle using ground-based instrumentation were discussed. However, some of the parameters which must be known in order to model and predict watershed processes may be derived from satellite measurements of the electromagnetic radiation which the ground and the water below, upon and above the ground reflect or emit.

In spite of the considerable effort which has been put into the development of these measurement systems, many of them, although by no means all, are not yet viable for operational use compared with ground-based systems. Current plans to implement new satellites will undoubtedly change this situation dramatically, and in this chapter the past, present and future capabilities of satellite systems will be outlined. The use of satellite data for weather forecasting is a significant and increasing subject area, and will not be discussed here.

Over the last 50 years the development of satellite systems in low polar orbits, starting with the Television Infrared Observation Satellites (TIROS) and the later NIMBUS satellites, has provided opportunities to develop measurement systems with coverage over large areas where conventional observations are sparse or absent. These techniques, however, suffer from limitations imposed by poor temporal resolution due to the orbit geometry (600–800 km altitude), measurements over specific areas only being available every 6 or 12 hours. The advent of geostationary satellite orbits (altitude 36,000 km) in the late 1960s and early 1970s, GOES and METEOSAT, rectified this temporal resolution problem, but initially provided less spatial resolution. In recent years high spatial resolution has become possible, down to 1 metre for visible and infrared measurements and down to 1 kilometre for active microwave systems.

Table 6.1 lists the main satellite systems which have provided, continue to provide and will provide platforms on which instruments are mounted to give data of hydrometeorological significance. Early satellites carried instruments which were only capable of making measurements at visible and infrared wavelengths, but by the early 1970s radiometers were able to provide data in several microwave bands, giving a capability to measure rainfall, soil moisture, evaporation and transpiration,

Hydrometeorology, First Edition. Christopher G. Collier.
© 2016 John Wiley & Sons, Ltd. Published 2016 by John Wiley & Sons, Ltd.
Companion website: www.wiley.com/go/collierhydrometerology

Table 6.1 Satellite systems used for hydrometeorological studies

Satellite	Type*	Series status†	Operator	Years of operation	Main instruments (not necessarily for all in a series)‡
GOES series 1–13	G	✓	NOAA, USA	1998–present	VIS, IR, NIR, WV
METEOSAT 1–10	G	✓	ESA/EUMETSAT	1978–present	VIS, IR, NIR, WV, radiation budget
NIMBUS 1–7	O	✗	NOAA, USA	1969–1978	Spectrometer (SIRS); interferometer (IRIS), SMMR (NIMBUS-7)
NOAA 1–19	O	✓	NOAA, USA	1970–present	Radiometer (HIRS, SCAMS, SMMR)
METOP A, B, C	O	✓	ESA/EUMETSAT	METOP-A 2006–present METOP-B 2012–present METOP-C 2017	VIS, IR, WV, sounder, thematic mapper (TM1)
Landsat 1–5	O	✓	NOAA, USA	1972–present	MSS, radiometer
SPOT 1–7	O	✓	French Earth Resources	1985–present	High resolution VIS, IR
METEOR 1–3 M	O	✓	Russian Space Agency	1969–present	VIS, IR, WV
GMS 1–5; ATSAT	G	✓	Japan Met. Agency	1978–present	VIS, IR, WV
TIROS 1–15	O	✓	NOAA, USA	1960–present	VIS, IR, WV
TIROS-N	O	✓	NOAA, USA	1978–present	VIS, IR, AMSU
INSAT 1–11; GSAT	G	✓	Indian Met. Agency	1982–present	VIS, IR, WV
Fengyen 1–4	G	✓	China Aerospace Science & Tech. Corp	2002–present	VIS, IR
MAO DFH series	O	✗	China Aerospace Science & Tech. Corp.	1970–present (?)	VIS, IR
ERS 1, 2	O	✗	ESA	1989–2011	SAR, SCATT, ATSR, radar altimeter
ENVISAT	O	✓	ESA	2002–present	SAR, SCATT, ATSR, radar altimeter
RADARSAT 1, 2	O	✓	Canadian Dept. Energy, Mines & Resources	1995–present	SCATT, SAR, AVHRR, MOMS
SEASAT	O	✗	NOAA, USA	1978–1978	SAR, radar altimeter

Mission	Type[*]	Series status[†]	Operator	Period	Instruments[‡]
Heat Capacity Mapping Mission (HCMM)	O	✗	NASA	1978–1980	VIS, IR, NIR
TRMM	O	✓	NASA	1998–present	VIS, IR, microwave precipitation radar
DMSP/SSMI and SSMIS	O	✓	US Dept. of Defense	F08–18 1987–present	VIS, IR, microwave imager, sounder
TERRA	O	✓	NASA	2000–present	ASTER, MODIS, CERES
AQUA	O	✓	NASA	2002–present	MODIS, AMSR-E
Himawari-8	G	✓	Japan Met. Agency	2014–present	16-channel imager
Soil Moisture and Ocean Salinity (SMOS)	O	✓	ESA	2009–present	MIR, L-band (21 cm) radiometer

[*] Type: G, geostationary; O, polar orbit.

[†] Series status: ✓ operating; ✗ not operating.

[‡] Instruments: AMSR-E, advanced microwave scanning radiometer, Earth observing system; AMSU, advanced microwave sounding unit; ASTER, advanced spaceborne thermal emission and reflection radiometer; ATSR, along-track scanning radiometer; AVHRR, advanced very high resolution radiometer; CERES, clouds and the Earth's radiant energy system; EMSR, electronically scanning microwave radiometer; HIRS, high resolution infrared radiation sounder; IR, infrared; IRIS, infrared interferometer spectrometer; MIR, microwave imaging radiometer; MODIS, moderate resolution imaging spectroradiometer; MOMS, modular opto-electronic multispectral scanner; MSS, multispectral scanner; NIR, near infrared; SAR, synthetic aperture radar; SCAMS, scanning microwave systems; SCATT, scatterometer; SIRS, satellite infrared spectrometer; SMMR, scanning multichannel microwave radiometer; SSMI, special sensor microwave imager; VIS, visible; WV, wave mode.

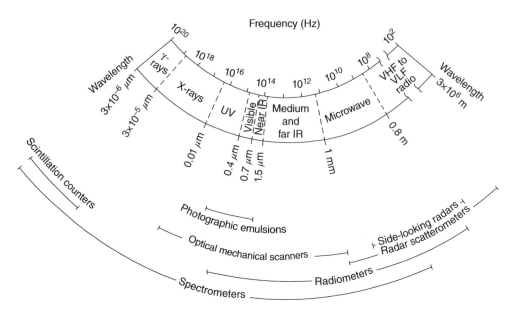

Figure 6.1   Electromagnetic spectrum and satellite instrumentation associated with specific wavelength intervals (after Hall and Martinec, 1985)

and to provide atmospheric profiles of temperature and humidity. Figure 6.1 shows the range of frequencies which have been employed.

Much of the impetus for these satellite developments has been the need to satisfy the requirements of meteorology, although the Landsat, the MODIS instrument and ERS satellites have enabled techniques appropriate to hydrology to be explored. The Landsat programme is the world's longest continuously acquired collection of space-based, moderate resolution, land remote sensing data. The website can be accessed at earthengine.google.org. The availability of data with high (compared with some conventional measurements) spatial and temporal resolution has led to new approaches to watershed modelling (Van de Griend and Engman, 1985). Parameters may now be measured in detail over wide areas, although the complementary nature of ground-based and space-based measurements, particularly when assessing rainfall amount, has been recognized (Collier, 1985 and others).

Most of the current satellite systems employ several different types of instrument. In general, however, they may be divided into systems using visible and infrared wavelengths and those using microwave wavelengths, either passively (receiving microwaves generated at the Earth's surface or within the atmosphere) or actively (receiving microwave output from the satellite system which is reflected or retransmitted from the surface or the atmosphere, including clouds).

Presently important missions for hydrological studies are the ASTER and MODIS sensors aboard the TERRA platform, the AMSR-E and MODIS aboard the AQUA platform, the SMOS (Soil Moisture and Ocean Salinity) mission of ESA, and the future Landsat series. In 1999 NASA launched TERRA which became operational in 2000. This was designed to improve understanding of the carbon and energy fluxes. Instruments on this satellite included ASTER for land surface studies, MODIS for monitoring large scale changes in the biosphere, and CERES for cloud and radiant energy monitoring.

In 2002 NASA launched the AQUA mission aimed at collecting data on the Earth's water cycle. This mission carries six instruments in an orbit of 705 km, and includes AMSR-E, which is a passive microwave radiometer at frequencies from 6.9 to 80 GHz. The spatial resolution ranges from 5.4 to 56 km. Further details of the ASTER and AMSR-E instruments have been described by Jong et al. (2008).

6.2 Surface scattering of electromagnetic radiation

Thermal radiation is emitted by all objects at temperatures above absolute zero, and is the radiation which is detected by most passive remote sensing systems. Generally a hot object will distribute its emission over a range of wavelengths and directions. Real materials do not behave as black bodies, and therefore *emissivity* ε is introduced, which is the actual radiance of a body at a temperature T to the black body value. The emissivity is usually dependent on wavelength. Hence the spectral radiance of a real object is

$$L_\lambda = \varepsilon(\lambda) L_{\lambda p} \qquad (6.1)$$

where $L_{\lambda p}$ is the black body radiance and subscript p refers to Planck. A body which is a good emitter (high ε) must also be a good absorber of radiation. *Reflectivity* $r = 1 - \varepsilon$, and therefore emissivity lies between 0 and 1.

The thermal emissivity of a surface is related to its reflectance properties. Figure 6.2 shows a well collimated beam of radiation of flux density $F(\text{W m}^{-2})$ measured perpendicularly to the direction of propagation incident on a surface at angle θ_0. A component of this radiation is scattered into the solid angle $d\Omega_1$, in a direction specified by the angle θ_1. θ_0 is often called the incidence angle, and its complement $(\pi/2) - \theta_0$ the depression angle.

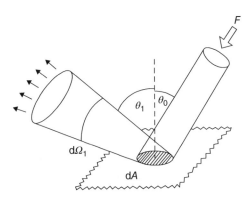

Figure 6.2 Radiation, initially of flux density F, is incident at angle θ_0 on an area dA and is then scattered into solid angle $d\Omega_1$ in a direction θ_1. The azimuthal angles φ_0 and φ_1 are omitted for clarity (from Rees, 1993)

Table 6.2   Typical values of albedo (from Schanda, 1986; Rees, 1993)

Material	Albedo (%)
Water (naturally occurring)	1–10
Water (pure)	2
Forest	5–10
Crops (green)	5–15
Urban areas	5–20
Grass	5–30
Soil	5–30
Cloud (low)	5–65
Lava	15–20
Sand	20–40
Ice	25–40
Granite	30–35
Cloud (high)	30–85
Limestone	35–40
Snow (old)	45–70
Snow (fresh)	75–90
Global average	≈35

Taking the outgoing radiance of the surface E as a result of its illumination L_1 (W m^{-2}sr^{-1}) in the direction (θ_1, φ_1), then the *bidirectional reflectance distribution function* (BRDF) R is defined as

$$R = L_1/E \qquad (6.2)$$

The subscript 1 denotes the scattering direction. The reflectivity r is referred to as the *albedo* and is related to the emissivity as follows:

$$r = 1 - \varepsilon \qquad (6.3)$$

Table 6.2 gives typical values of albedo.

There are two basic approaches to calculating BRDF (see Rees, 1993):

1. The *small perturbation method*, in which the interaction of the incident field with the surface is used to calculate the outgoing field in the vicinity of the surface.
2. The *facet model*, in which the surface is imagined to be composed of planar facets each of which is tangential to the true surface. The facets must be more than one wavelength across.

The above considers the reflectance from rough surfaces as depending upon the incidence and viewing directions and not the wavelength. However as shown in Figure 6.3 this is not appropriate. In the figure the simplified spectral albedos of various materials in the visible and near infrared wavebands are shown. It is clear that, for example, operating at about 1.2 µm (near infrared) only results in difficulty in discriminating between a number of materials. This problem is reduced if three or four channels are used between 0.5 and 1.2 µm.

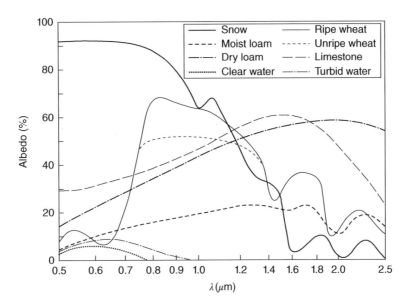

Figure 6.3 Typical spectral albedos of various materials in the visible and near infrared (from Rees, 1993)

There is an abrupt change in the reflectance of vegetation at about $0.7\,\mu$m, which is used by a number of techniques for assessing the amount of vegetation present in an image (Harris, 1987). The high and uniform reflectance of snow in the visible and near infrared is caused by the wavelength independent scattering events at the interfaces between ice and air. Hence under favourable conditions snow can be detected when illuminated only by moonlight (Foster, 1983). The reflectance properties of rock are less distinctive than those of vegetation, and the reflectance properties of soil are dependent upon the type of rock from which the soil is derived and the content of vegetable matter and water.

The emissivities of various materials in the thermal infrared region are shown in Figure 6.4. In this figure it is clear that multi-frequency measurements in passive microwave radiometry from 1 to 36 GHz are very useful in identifying various materials.

6.3   Interaction of electromagnetic radiation with the atmosphere

Radiation makes one journey through the atmosphere, or two journeys for systems that detect reflected radiation. Each time radiation passes through the atmosphere it experiences both absorption and scattering. The sum of these two energy losses is referred to as *attenuation*. Also the refractive index of the atmosphere differs from unity. All these effects are dependent upon wavelength.

The extent of attenuation is defined logarithmically in decibels. The *optical depth* τ (sometimes called the *opacity*) gives the extent to which a given layer of material attenuates the intensity of radiation passing through it, and is defined as

$$\tau = \ln\left(I_1/I_2\right) \tag{6.4}$$

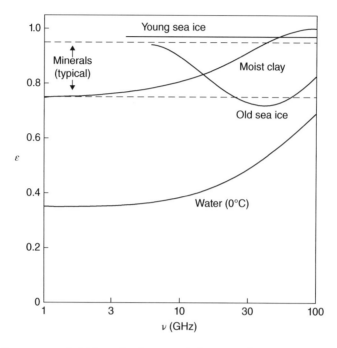

Figure 6.4 Microwave emissivities ε in the normal direction of various materials at a range of frequencies ν (from Rees, 1993)

where I_1 is the incident (unattenuated) intensity and I_2 is the output (attenuated) intensity. The total optical depth of the atmosphere at a wavelength of 0.55 μm is about 0.3 under normal conditions. Here 0.1 is contributed by atmospheric gases and 0.2 by particulate matter. Figure 6.5 shows the total atmospheric attenuation for a vertical ray. A detailed explanation of the shape of this figure is contained in Rees (1993). It is the basis of remote sensing techniques used for sounding the atmosphere. By tuning the observation frequency the height of a specific layer may be varied and parameters such as temperature and pressure may be measured as a function of height (see for example Harries, 1994). The radiation balance of the Earth and planets is discussed in Appendix 6.1.

6.4   Visible and infrared data

Satellite data at visible frequencies allow estimates to be made of a variety of parameters important to hydrological modelling, such as the fraction of impervious area in a river catchment, or the fraction of water surfaces or of polluted area (see for example McGinnis et al., 1979). Flood inundation areas may also be analysed, provided the data resolution is adequate (Ramamoorthi and Rao, 1985). Indeed it is also possible in some areas to infer likely underground water resources from such data (Salman, 1983).

Nevertheless, visible and infrared data have been most commonly used to make estimates of rainfall, snowfall and soil moisture. These estimates are made by relating the reflectivity of clouds or the Earth's surface in the visible wavelengths, the thermal

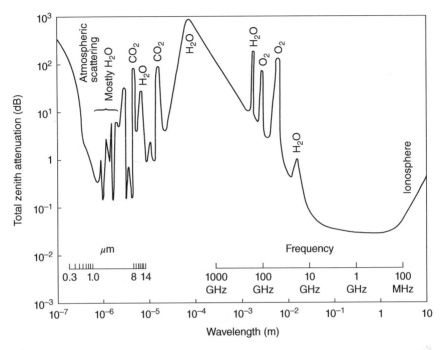

Figure 6.5 Total zenith attenuation by the atmosphere under normal conditions (simplified) (from Rees, 1993)

infrared radiation emitted by a surface, or microwave radiation emitted by a surface, to independent measurements of the parameter. The relationship between the wavelength of maximum emitted radiance λ_{max} and the absolute temperature T of the radiating surface is given by *Wien's displacement law*, which states that the product of absolute temperature and wavelength is constant for a *black body* (perfect emitter at all wavelengths):

$$\lambda_{max} T = w \tag{6.5}$$

where w is Wien's displacement constant.

Infrared radiant emission is given by the *Stefan–Boltzmann law*, which relates the total radiant emittance from a black body to the fourth power of the absolute temperature:

$$W = \varepsilon \sigma T^4 \tag{6.6}$$

where W is the radiant power per unit area, ε is the emissivity of the surface, σ is the Stefan–Boltzmann constant, and T is the absolute temperature (K). Of all common natural surfaces, water is most like a black body, and emissivities may vary from 0 to less than 1.0 and depend upon wavelength. The *thermal inertia P* is the resistance of a material to temperature change, given by

$$P = (kc\rho)^{1/2} \tag{6.7}$$

where k is the thermal conductivity, c is specific heat and ρ is the density.

The thermal infrared temperature that is measured from a satellite is modified by the atmosphere. The radiation measured when looking down vertically is partly made up of reflected radiation from the sky in the clear sky situation. If clouds are present then rapid changes in the radiant emission from a surface can occur.

6.4.1   Precipitation

Two types of technique for estimating rainfall have been developed, these being known as *cloud indexing* and *life history* methods. (Snowfall as opposed to snow cover has not been examined in any depth.)

Cloud indexing was the first technique developed for rainfall estimation. A rainfall coefficient or cloud index was evaluated from features of the satellite cloud field defined by visible or infrared images, for example brightness or texture. These indices were then related, via regression equations, to raingauge observations of rainfall (Barrett, 1970; Follansbee, 1973; 1976; Follansbee and Oliver, 1975). These techniques were most successful for rainfall over periods of days or months, partly because satellite data from polar orbiter platforms were only available every 6 or 12 hours. More recently, similar techniques have been developed for use with geostationary data (Delbeato and Barrell, 1985), but in general they only function well for convective cloud, as opposed to frontal cloud. Figure 6.6 shows the derived relationship over south-east Australia between 90 minute rainfall totals and cloud top temperature. Pattern recognition techniques using textural or radiance features are also being assessed (see for example Wu et al., 1985).

Stout et al. (1979) were able to estimate the rainfall produced by convective clouds from the sum of the area of the clouds and the rate of change of that area. That is, the volumetric rain rate for a particular cloud is given by

$$R_v = a_0 A_c + a_1 \frac{dA_c}{dt} \tag{6.8}$$

where A_c is the area of the cloud, dA_c/dt is the rate of change of cloud area, and a_0 and a_1 are empirical coefficients. This type of technique, known as a life history procedure, requires satellite images at frequent time intervals which can only be provided by geostationary satellites.

Variants on the Stout et al. technique have been produced by Griffith et al. (1976; 1978) (also described in Woodley et al., 1980; Scofield and Oliver, 1977a; 1977b; see also Scofield, 1984). In the latter technique precipitation is favoured by high cloud brightness (low cloud top temperature), and heavy rainfall is favoured by cold cloud top temperatures and the growth and merging of clouds. Improvements to deal more adequately with storms having warm tops have been proposed by Scofield (1981). The addition of spectral and textural information from geostationary satellite data to the evolutionary data has been investigated by Martin and Howland (1986) and found to produce useful improvements in the tropics.

A typical example of a rainfall field derived from both geostationary and polar orbiting satellites using infrared satellite data and these types of techniques is shown in Figure 6.7.

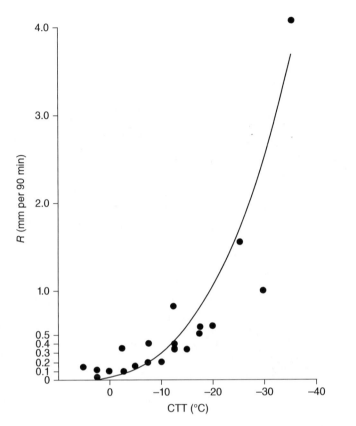

Figure 6.6 Observed surface rainfall 90 minute totals R and cloud top temperatures CTT over south-east Australia for cumuliform cloud or cloud in the convective sector of extratropical cyclones; the points are derived from several studies (adapted from Delbeato and Barrell, 1985)

Negri and Adler (1981) found that thunderstorm top ascent rates are correlated with maximum storm radar reflectivities, and the minimum black body temperature observed during the lifetime of a storm is correlated with the maximum volumetric storm rainfall. In general this technique only produces acceptable estimates of convective rainfall. However, Negri et al. (1984) have concluded that the Griffith–Woodley technique is unnecessarily complicated for daily rainfall estimation. Doneaud et al. (1985) have proposed a simple procedure in which the area–time integral (ATI) of cloud areas over the lifetime of a storm is related to the total rain volume. Investigations using this technique with radar data have been discussed in Chapter 5. Again this technique seems to work well only for convective rainfall.

6.4.2 Snow depth

Seasonal changes in snow cover can be monitored regularly from satellite visible and thermal infrared images (Rango et al., 1977; Rango and Martinec, 1979; Dey et al., 1979; 1983), although accurate classification of snow types and distribution requires the use of multispectral data as discussed in section 6.5. The use of these techniques

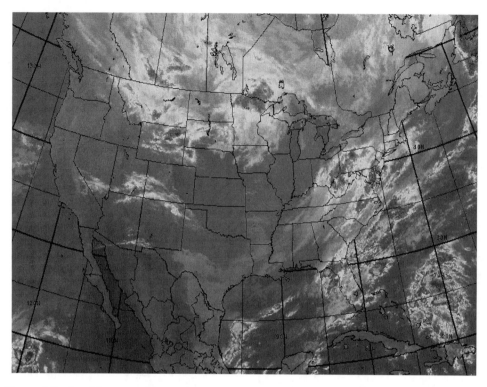

Figure 6.7   Rainfall estimates from infrared satellite data, 1630 GMT 8 October 2012: the US satellite images displayed are infrared (IR) images. Warmest (lowest) clouds are shown in white; coldest (highest) clouds are displayed in shades of yellow, red and purple. Imagery is obtained from the GOES and METEOSAT geostationary satellites, and the two US Polar Orbiter (POES) satellites. POES satellites orbit the Earth 14 times each day at an altitude of approximately 520 miles (870 km) (courtesy WSI Corporation) (*see plate section for colour representation of this figure*)

in the USSR has been reviewed by Deleur (1980), and elsewhere by Rango and Itten (1976) and Rango and Peterson (1980).

Measurement techniques based upon the use of visible and/or infrared images are limited to sensing the surface conditions of the snow (Martinec and Rango, 1981). As the snow melts it is very important for hydrologists to assess the potential runoff. This was discussed in Chapter 4. Odegaard et al. (1980) demonstrated the use of high resolution satellite data with *in situ* surface observations in hydrological studies of snowmelt for hydro-electric generation. One problem in measuring the extent of snow cover from a satellite is the masking produced by clouds. However, it is possible to distinguish between clouds and snow cover by using a range of spectral bands as discussed in section 6.5.

6.4.3   Soil moisture and evapotranspiration

The relationship of the soil spectral reflectance to the water content depends upon several variables in addition to the water content, such as the soil texture and organic content. Hence, the use of reflected energy, either visible or infrared, is not practical.

Nevertheless, the diurnal range of surface temperature or the measurement of the difference between the air temperature and the temperature within a crop may be related to soil moisture (Idso et al., 1975a; 1975b; Wetzel et al., 1984). An experimental technique for evaluation of the diurnal temperature range has been developed using measurements from the Heat Capacity Mapping Mission from 1978 to 1980 (Carlson et al., 1981). A similar technique has been evaluated by Klaassen and Van Den Berg (1985) for the calculation of evapotranspiration derived from surface temperatures (see also Gurney and Camillo, 1984).

6.5   Multispectral data

The use of single images, or sequences of images, of either visible or infrared data has been discussed in the previous section. However, with the possible exception of rainfall measurement techniques, procedures have not proved reliable enough for operational implementation. Improvements can be obtained by combined use of data from several spectral bands.

6.5.1   Precipitation

Infrared sensors provide information on temperature, and thus indirectly on the heights of the tops of clouds. On the other hand visible sensors provide information on the depth of clouds and their geometry and composition. Combination of this information enables recognition of high cloud tops associated with deep clouds which are likely to produce significant rainfall (Barrett and Martin, 1981). Early work on this type of technique was carried out by Lethbridge (1967), Dittberner and Vonder Haar (1973) and Reynolds et al. (1978). However, problems arising from registration errors between visible and infrared images, instrument calibration, and time difference between images and illumination geometry caused the results of the early work to be less encouraging than researchers had expected. Lovejoy and Austin (1979a) demonstrated that these problems could be overcome.

Several investigations using models of convective processes aimed to tackle the problems from a physical, rather than a statistical, point of view (Gruber, 1973; Wylie, 1979). These techniques are able to define rain areas. However, the estimation of rainfall amounts is more problematic, and ground-truth data such as that provided by radar may be needed for calibration of the satellite data when making estimates of rain amounts, or reliable estimates of rain area. Tsonis and Isaac (1985) have proposed a scheme which uses radar data to define areas of rain and no rain for calibration and subsequently uses the satellite visible and infrared data only to delineate areas of rain; no quantitative estimates are produced.

6.5.2   Cloud recognition

The latest generation of satellites provide data with high spatial and cloud top temperature resolution. Unfortunately these data are not adequate on their own to allow the automated analysis of different cloud types. The combination of data from different spectral bands can alleviate the problems.

Liljas (1982) evolved a technique using data from three AVHRR (Table 6.1) spectral channels as follows:

Visible (VIS)	0.55–0.9 μm	(TIROS-N)
	0.58–0.68 μm	(NOAA-6, -7)
Near infrared (NIR)	0.725–1.1 μm	
Infrared (IR)	10.5–11.5 μm	

The technique is similar to that proposed by Barrett and Grant (1978), and is based upon the assumption that different cloud types, land surfaces and water surfaces have different radiational properties in different parts of the electromagnetic spectrum. Signatures were derived of different cloud types within boxes in three-dimensional intensity space as shown in Figure 6.8a. The technique has been assessed over southern Scandinavia with encouraging results (Figures 6.8 b and 6.8c).

6.5.3 *Snow*

The problem of separating cloud and snow cover on visible and infrared images was mentioned in section 6.4.2. The thematic mapper on Landsat 4 and 5 (Table 6.1) provided data in seven wavelength bands as follows:

Band 1	0.45–0.52 μm
Band 2	0.52–0.60 μm
Band 3	0.63–0.69 μm
Band 4	0.76–0.90 μm
Band 5	1.55–1.75 μm
Band 6	10.40–12.50 μm
Band 7	2.08–2.3 μm

Bands 1 to 5 and band 7 are in the visible, near infrared and middle infrared wavelength regions, and have a spatial resolution of 30 m. Band 6, a thermal infrared band, has a resolution of 120 m. Figure 6.9 shows the typical spectral reflectance curve for snow, showing saturation levels for the visible and near infrared bands of the thematic mapper. Band 5 may be used to distinguish between clouds and snow as the reflectance of snow decreases markedly at a wavelength of about 1.5 μm, whereas the reflectance of clouds remains high.

6.6   Passive microwave techniques

The microwave radiation (roughly 10 to 0.1 cm wavelength) thermally emitted by an object is called its brightness temperature T_B. The microwave radiation emitted from a black body is proportional to its physical temperature T, but, as for infrared

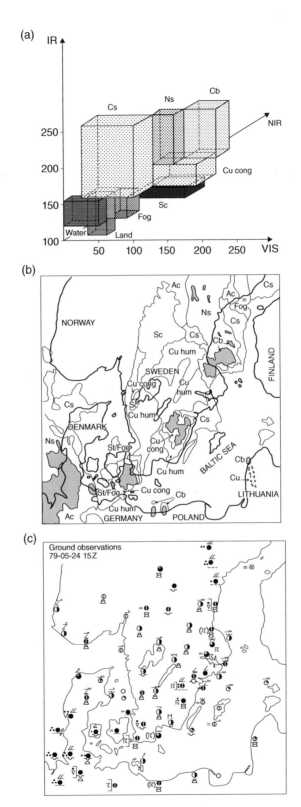

Figure 6.8 Illustrating (a) a classification in a three-dimensional intensity space of cloud, fog, land and water; (b) cloud analysed using this classification at 1500 UTC 24 May 1975 over southern Scandinavia; and (c) the weather observed about 1.5 hours after the satellite passage (after Liljas, 1982)

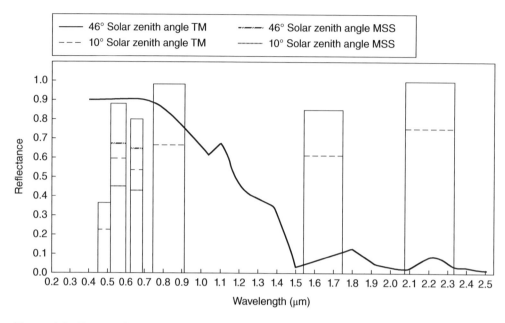

Figure 6.9 Typical reflectance curve for snow (O'Brien and Munis, 1975) showing saturation levels for visible and near infrared bands of the thematic mapper (TM) and multispectral scanner (MSS) (adapted from Salomonson and Hall, 1979 by Hall and Martinec, 1985)

emission (section 6.2), most real surfaces or objects emit only a fraction of the radiation a black body would emit at the same physical temperature. This fraction is the emissivity ε of the object, given, for microwave wavelengths, by

$$T_B = \varepsilon T \tag{6.9}$$

The emissivity may be determined by measuring the brightness temperature using a radiometer mounted on a satellite, and measuring the physical temperature by some independent means. However, the emissivity usually varies with wavelength, and therefore by measuring the brightness temperature at more than one microwave wavelength it is possible to either determine the physical temperature from the radiometric measurements or estimate more than one physical property. Techniques for deriving these properties from a set of brightness temperature measurements at multiple wavelengths or path angles are known as *inversion techniques*.

Inversion techniques have been developed for the retrieval of atmospheric and surface parameters, such as temperature and humidity profiles and rainfall. A comprehensive review of these procedures is contained in Ulaby et al. (1986) from which the following is adapted. Assuming that the brightness temperature $T_B(\lambda_i)$ is linearly related to the required parameter represented by the function $g(z)$, where z is the vertical distance from the radiometer, then

$$T_B(\lambda_i) = T_{BGD}(\lambda_i) + \int w(\lambda_i, z) g(z) dz \qquad i = 1, 2, \dots, n \tag{6.10}$$

where T_{BGD} is the background radiometric temperature corresponding to the Earth's surface; $w(\lambda, z)$ is the weighting function of $g(z)$. If observations are made at an angle other than zenith or nadir (vertically under the satellite) depending on the platform, then z is replaced with $z \sec \theta$. If the relationship is not linear, then it is still possible to use a similar formulation to that given by equation 6.6, although the weighting function is more complex. A linear relationship is usual for microwave wavelengths, with non-linear relationships being more usual for wavelengths closer to the infrared region. T_{BGD} is usually very variable because the emission by the Earth's surface varies spatially and temporally. Indeed, in estimating soil moisture content or snow depth, for example, the use of the isothermal equation 6.5 is not applicable. Account must be taken of a radiative function of the type given in equation 6.6, which describes the transfer of radiation from a given point within the material making up the Earth's surface layer to the ground–atmosphere interface. A weighted average of the physical temperature over the depth of the surface layer T is then used to calculate the emissivity:

$$T_B = \varepsilon T \qquad (6.11)$$

The depth known as the *skin depth*, from which 63% of the radiation emanates, depends upon the material in the surface layer. At 1.55 cm wavelength this depth is several millimetres for water, soils and first-year sea ice, several centimetres for sea ice which is several years old, and several metres for dry snow (Zwally and Gloersen, 1977).

6.6.1 Precipitation

The use of the microwave region of the electromagnetic spectrum to make measurements of precipitation has the advantage over visible and infrared techniques of detecting the precipitation itself, with clouds being wholly or partially transparent.

One of the first instruments to be used meteorologically to measure the natural radiation in the microwave region was ESMR-5 (electrically scanning microwave radiometer) launched on NIMBUS-5 in 1972. Simple comparisons of the NIMBUS-7 SMMR imagery operating at 19.35 GHz (1.55 cm wavelength) with imagery from visible and infrared wavelength radiometers, ground-based radars and conventional meteorological observations established over water an association of areas of relatively warm brightness temperature with areas of rainfall (Theon, 1973; Wilheit et al., 1973; 1977). Curves of 19.35 GHz brightness temperatures and rainfall rate were published by Wilheit et al. (1977) using a model based upon the concept of many optically thin layers bound on top by a variable freezing level.

In this work several problems were identified, notably that rainfall was rarely uniform over the whole field of view of the instrument and that the relationship between rainfall rate and brightness temperature was non-linear (Spencer et al., 1983). Improvements to the technique, including the use of both vertical and horizontal polarization for measurements over land, have been proposed by Wilheit (1975). Nevertheless, measurements over the land remain much more difficult due to variability of the emissivity, and hence the background radiation measured by the satellite instrument.

In an attempt to overcome the problems of non-linearity mentioned above, and ambiguities introduced between emissions due to rain and those due to non-precipitating cloud, Spencer (1986) proposed a technique based upon analysis of radiometric data gathered from the NIMBUS-7 SMMR (Table 6.1) at a frequency of 37 GHz (0.81 cm wavelength). This radiometer system provides dual polarized data at five different wavelengths (0.81, 1.43, 1.66, 2.00 and 4.54 cm), and can be used to detect a wide range of snow and ice features as well as rainfall. Figure 6.10 shows how the difference between brightness temperatures from the horizontal and vertical polarized radiation at 37 GHz is related to the mean brightness temperature and rainfall rate for convective rainfall over the United States.

Spencer (1986) suggests that 37 GHz can provide the best compromise for measuring rainfall over the oceans between an easily detectable precipitation signal (such as at 19.35 GHz) and cloud penetration ability. However, it was noted that the scattering signal at 37 GHz is not as strong as the emission signal of light precipitation at 19 GHz. Hence the 37 GHz technique may be better for observing heavy convective precipitation, and observations at 19 GHz may relate better to measurements of light precipitation. Plans were prepared to include a scanning microwave radiometer on the US

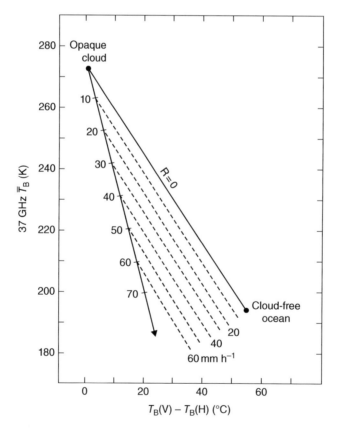

Figure 6.10 Bipolarization (horizontal/vertical) plot of rainfall rates with the observed relationship between radar rainfall rates and SMMR brightness temperature T_B during the summer over the United States (after Spencer, 1986)

Polar Platform Space Station (McElroy and Schneider, 1985), a project known in Europe as the Columbus Project. However, Columbus was cancelled in 1991. The US space station FREEDOM was announced in 1984, but was never constructed; it evolved into the International Space Station, the first module of which was launched in 1998. Meteorological experiments have been carried out over the last 10 years, although routine meteorological measurements have been made from other spacecraft, as noted in Table 6.1. Berg et al. (2012) describe the 25 years of microwave brightness temperature data provided by the SSM/I and SSMIS instruments flying on the DMSP satellites.

6.6.2 Global Precipitation Climatology Project (GPCP)

The GPCP is an international project drawing upon input data sets and techniques from a number of investigators. A global complete monthly analysis of surface precipitation at a 2.5° latitude × 2.5° longitude resolution is available from January 1979 to the present. It is a merged analysis incorporating precipitation estimates from low orbit satellite microwave data (available from 1987), geosynchronous infrared data and surface raingauge data. Version 1 of the data set was described by Huffman et al. (1997), and version 2 describing new data sets has been described by Adler et al. (2003).

6.6.3 Global Precipitation Measurement mission (GPM)

The Global Precipitation Measurement mission (GPM) is an international network of satellites that provide the next-generation global observations of rain and snow. Building upon the success of the Tropical Rainfall Measuring Mission (TRMM), the GPM concept centres on the deployment of a 'core' satellite carrying an advanced radar/radiometer system to measure precipitation from space and serve as a reference standard to unify precipitation measurements from a constellation of research and operational satellites (Figure 6.11). GPM, initiated by NASA and the Japan Aerospace Exploration Agency (JAXA) as a global successor to TRMM, comprises a consortium of international space agencies, including the Centre National d'Études Spatiales (CNES), the Indian Space Research Organization (ISRO), the US National Oceanic and Atmospheric Administration (NOAA), the European Organization for the Exploitation of Meteorological Satellites (EUMETSAT), and others. The GPM Core Observatory was launched on 27 February 2014 at 1337 EST from Tanegashima Space Center, Japan.

6.6.4 Snow depth

The electrical properties which produce a particular dielectric constant (a measure of the response to an applied electric field) of a substance govern the microwave emission from the substance. Within a substance, changes in the dielectric constant,

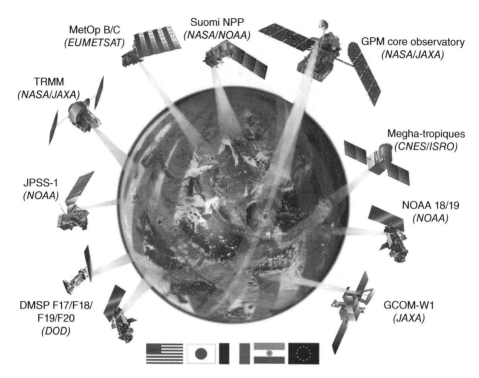

Figure 6.11   Global Precipitation Mission (GPM) Constellation. Suomi NPP, National Polar-orbiting Partnership (2011–); JPSS-1, NOAA Joint Polar Satellite System (2017–); GCOM-W1, Global Change Observation Mission (2012–); Megha-Tropiques, French Water Cycle and Energy Budget in the Tropics (2011–) (see also Table 6.1) (from pmm.nasa.gov/GPM) (*see plate section for colour representation of this figure*)

which is a function of temperature and frequency, cause scattering of the natural microwave emission or reflection of an incident microwave transmission. Hence in considering the electrical properties of snow, account must be taken of the presence of ice, air and water within the snowpack. This has led to the development of models based upon randomly spaced scattering spheres (see for example Chang et al., 1976). Analysis of the response of this type of model to microwaves using SMMR data led Kunzi et al. (1982) to conclude that microwave sensors could offer improved snow cover information over optical sensors. This work is supported by the results of several studies noted by Ulaby et al. (1986). Figure 6.12 shows a comparison by Rango et al. (1979) of brightness temperature versus snow depth at two different polarizations and two different wavelengths. The brightness of temperature shows a stronger sensitivity to snow depth at a wavelength of 0.81 cm than at 1.55 cm.

Foster et al. (1980) found that the relationship between snow depth and brightness temperature varied from one area to another, and therefore it is prudent to derive relationships specific to an area of interest. However, the use of dual polarization at two different frequencies may provide more widely applicable relationships, and Kunzi et al. (1982) derived formulae for snow cover, snow depth and water equivalent, and snowmelt.

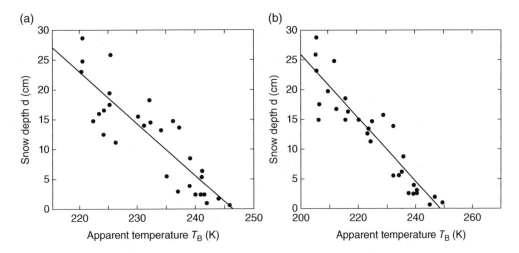

Figure 6.12 Brightness temperature versus snow depth measured at (a) $\lambda = 1.55$ cm, horizontal polarization, by the NIMBUS-5 ESMR, and (b) $\lambda = 0.81$ cm, vertical polarization, by the NIMBUS-6 ESMR (after Rango et al., 1979 from Ulaby et al., 1986)

6.6.5 Sea ice and sea surface temperature

The first satellite observations of the sea surface temperature and sea ice were made using ESMR data from NIMBUS-5 (Table 6.1) (see for example Gloersen and Salomonson, 1975). Zwally and Gloersen (1977) provided impressive microwave images of the Antarctic and Arctic in both summer and winter. Recent examples of sea ice climatology are shown in Figure 6.13.

The apparent temperature of the sea surface T_S when viewed from a satellite at angle θ is given by

$$T_S = T_B e^{-\tau \sec \theta} + \Gamma T_D e^{-\tau \sec \theta} + T_U \tag{6.12}$$

where τ is the total atmospheric opacity, T_D and T_U are the downward and upward atmospheric brightness temperatures, Γ is the surface reflectivity, and T_B is the average brightness temperature of the sea surface. In general T_B may consist of contributions from sea water, first-year ice and multi-year ice (of quite a different composition, due to melting and refreezing, to first-year ice). The accuracy of estimates of ice concentration will depend upon the extent to which the satellite field of view contains open water, and the extent to which it is possible to calibrate the measurements using thermal infrared estimates of the ice temperature in cloud-free regions. The NOAA polar orbiting satellites (POES) have been collecting sea surface temperature data for over 22 years, and with sea ice data this information is now an integral part of ocean forecasting systems (Martin et al., 2007).

6.6.6 Soil moisture and evapotranspiration

For measuring soil moisture, longer wavelengths are used to minimize atmospheric effects, and microwaves have the advantage of being able to penetrate vegetation. Jackson et al. (1981) describe how thermal microwave emission from soils is generated

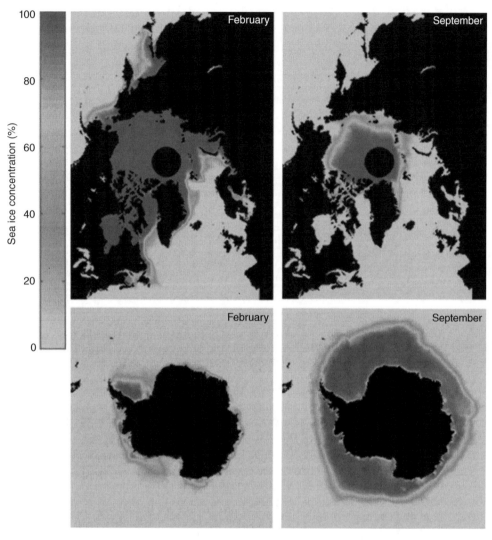

Figure 6.13 Arctic (top panels) and Antarctic (bottom panels) sea ice concentration climatologies from 1978 to 2002: approximate seasonal winter maximum (February in the Arctic and September in the Antarctic) and minimum (September in the Arctic and February in the Antarctic) levels (image provided by National Snow and Ice Data Center, University of Colorado, Boulder, CO) (*see plate section for colour representation of this figure*)

within the soil volume, the amount of energy at a point depending upon the dielectric properties (primarily moisture in the case of soil) and the temperature of the soil. As the energy propagates upwards through the soil it is affected by the dielectric gradients (soil moisture) along the path of propagation. As the energy crosses the surface boundary, it is reduced by the effective transmission coefficient (the emissivity), which is determined by the dielectric characteristics of the soil near the surface.

The soil layer which determines this transmission coefficient is the skin depth mentioned earlier, and is a few tenths of a wavelength thick, for example 2–5 cm for a wavelength of 21 cm. This depth is the effective soil moisture sampling depth of this technique. Wang (1983) concluded that a significant problem in estimating soil moisture from passive microwave measurements is the uncertainties introduced by variations of surface roughness. Ulaby et al. (1986) suggest that active microwave (radar) sensors are best able to meet the required spatial resolution.

6.7   Active (radar) microwave techniques

6.7.1   Synthetic aperture radar

Measurements of the intensity of brightness of microwave energy generated in a radar mounted on a satellite, and reflected or scattered from the atmosphere or the Earth's surface, may be made on receipt back at the satellite antenna. The amount of backscattered energy is related to the surface roughness, orientation, slope and dielectric constant of the material of which the surface is composed. It also depends upon any rainfall intercepted by the radar beam, as short wavelengths are attenuated by heavy rainfall.

Figure 6.14 shows the geometry of the SEASAT synthetic aperture side-looking radar (SAR) (Table 6.1). This sensor used the motion of the satellite to synthesize a large aperture antenna, and operated at 1.275 GHz (23.5 cm wavelength). Hall and Martinec (1985) point out that the length of the electromagnetic wave with respect to the size of the terrain feature determines whether a surface appears rough or smooth at a particular wavelength. Smooth water surfaces are excellent specular reflectors, and because they are not normally viewed at right angles to the SAR, they specularly reflect all the microwave energy into space. However, horizontal and vertical surfaces may appear to the SAR as corner reflectors and return a large part of the energy back

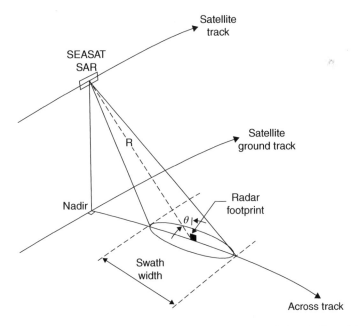

Figure 6.14   Geometry of SEASAT SAR (adapted from Fu and Holt, 1982 by Hall and Martinec, 1985)

to the antenna. In Figure 6.14 the image resolution in the across-track direction d_r is given by (Fu and Holt, 1982)

$$d_r = \frac{c\tau}{2\sin\theta} \qquad (6.13)$$

where τ is the effective pulse width and c is the speed of light. The maximum resolution achieved by the SEASAT SAR was 25 m × 25 m, and the image area was 100 km × 4000 km. SEASAT also carried a radar altimeter, operating at 13.56 GHz (2.2 cm wavelength), for measuring the roughness of the oceans.

Unfortunately SEASAT only operated for 100 days, but this was long enough to indicate the potential of the data for measuring soil moisture (Ulaby et al., 1983; Brisco et al., 1983; see also Ulaby et al., 1986).

The launch of ERS-1 and later ERS-2 operating at 5.7 cm wavelength, giving a maximum resolution of 125 m, provided much SAR data with which to take these techniques further. Davie and Kelly (1997) demonstrated that processing could optimize noise reduction, reducing the resolution to 100 m, which was considered to give a good representation of field size in a catchment area. However the main restriction on the quality of the data is that for single satellite operation the return period is 35 days. Nevertheless it is a straightforward procedure to define a retrieval algorithm if one has access to a data set that includes both backscatter coefficients and *in situ* soil moisture content measurements.

Fox et al. (2000) describe an algorithm for retrieving soil moisture content (SMC) over the west Pennine moors of north-west England. Comparisons using this technique throughout the summer of 1996 are shown in Figure 6.15. Although the satellite measurements show a high degree of accuracy, the limitation imposed

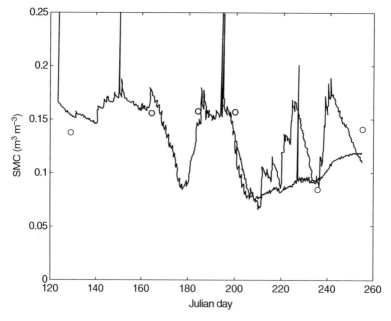

Figure 6.15 Comparison of soil moisture content as measured by a capacitance probe (fainter line), lysimeter (darker line) and ERS-1 SAR (circles); spikes show instrumental noise

by the satellite return period is evident. A simple water balance equation allows one to determine the evapotranspiration (ET) as follows:

$$ET = R - P - \left(SMC_t - SMC_{t-1}\right) \tag{6.14}$$

where R is the rainfall in the period $t - 1$ to t and P is the percolation out of the soil column.

More recently the Canadian Space Agency (CSA) RADARSAT 1/2 SAR images have been used to study tropical cyclone morphology and dynamics (Li et al., 2013). ESA and CSA will launch the Sentinel-1 and RADARSAT Constellation Mission SAR operational missions in 2013–16.

6.7.2   Radar systems

Radar systems similar to ground-based radars offer considerable potential. For example Atlas et al. (1982) state that there is no doubt that 'satellite active microwave systems could provide precision global distributions of rain rate given no constraints on the number of spacecraft, antenna size, and power'. Problems of radar beam filling and the unwanted effects of radar echoes from the ground might be overcome by using sufficiently large antennas. However, a practical antenna size requires the use of short wavelengths which are attenuated by heavy rainfall.

In addition, because of antenna cost and size constraints, the use of non-attenuating wavelengths (greater than 5 cm) would provide a field of view too broad to resolve significant features of precipitation. Unfortunately the return from the ground, beam filling and reflectivity gradients produce narrow beam swaths, particularly at off-nadir angles. The use of a shorter attenuating wavelength does increase the resolution, and at higher rainfall rates improves the sensitivity. However at moderate and high rainfall rates, the rapid increase in attenuation limits the achievable penetration into a storm. Meneghini and Kozu (1990) discuss these problems further.

One possible solution, proposed by Atlas et al. (see also Inomata et al., 1981), is to use two wavelengths, one attenuating and the other not. A further problem is that the signal averaging time needed to obtain a reliable estimate of the radar echo power implies either a very high radar pulse repetition frequency or a very slow scan. The following possibilities were put forward by Atlas et al. (1982):

1. Use of a modified radar altimeter operating at 13.5 GHz (2.2 cm wavelength) in conjunction with infrared images or a passive microwave radiometer. Unfortunately this would be restricted to nadir measurements only.
2. Use of a short wavelength scanning radar operating at two wavelengths (0.86 cm and 3 cm say). This would enable a narrow beam width to be obtained with a practical size antenna. Unfortunately severe attenuation would occur in rainfall of only moderate intensity, limiting the accuracy which could be achieved.
3. Use of a surface target attenuation radar, taking the scattering properties of the surface as a calibration for attenuation measurements. Unfortunately the scattering properties are influenced by soil moisture, and this problem would have to be overcome.

Table 6.3   TRMM precipitation-related instruments (after Liu et al., 2012)

Instrument name	Band frequencies and wavelengths	Spatial resolution (km)		Swath width (km)	
		Pre-boost	Post-boost	Pre-boost	Post-boost
Visible and infrared scanner (VIRS)	5 channels: 0.63, 1.6, 3.75, 10.8, 12 μm	2.2	2.4	720	833
TRMM microwave imager (TMI)	5 frequencies: 10.7, 19.4, 21.3, 37, 85.5 GHz	4.4 (at 85.5 GHz)	5.1 (at 85.5 GHz)	760	878
Precipitation radar (PR)	13.8 GHz	4.3 (vertical 250 m)	5 (vertical 250 m)	215	247
Lightning imaging sensor (LIS)	0.7774 μm	3.7	4.3	580	668

Of these options the first has been adopted for the NASA Tropical Rainfall Measuring Mission (TRMM) launched in November 1997, which included the first satellite precipitation radar.

6.7.3   Tropical Rainfall Measuring Mission (TRMM)

TRMM is a joint USA–Japan satellite mission providing the first detailed and comprehensive data set of four-dimensional distributions of rainfall and latent heating over the tropical and subtropical oceans and continents (40°S–40°N). There are five instruments on board TRMM as given in Table 6.3. The precipitation radar (PR) is an electronically scanning K_u-band (13.8 GHz, 2.2 cm wavelength) radar that measures the backscattered echo from precipitation and cloud particles. The rain retrievals are based on application of reflectivity/rain-rate (Z_e–R) relationships after correcting the reflectivity measurements for attenuation (Adler et al., 2001). The reflectivities are derived from a hybrid method (Iguchi et al., 2000) that is based on the Hitschfeld–Bordan iterative attenuation correction scheme which performs well at low attenuations, and the surface reference method for which the relative error decreases with increasing path attenuation (Iguchi and Meneghini, 1994). The sampling rate over the United States is slightly less than 11 hours (Simpson et al., 1996). Kozu et al. (2001) have shown the calibration stability to be within 0.8 dB. The precipitation measurements from TRMM have been evaluated using ground-based measurements (Wang and Wolff, 2012), but these measurements have also been used for determining ground radar calibration biases (Anagnostou et al., 2001). An example of the TRMM monthly rainfall images is shown in Figure 6.16. Liu et al. (2012) outline the availability of precipitation data and services from TRMM.

6.8   The surface energy balance system (SEBS)

Jong et al. (2008) outline the surface energy balance system (SEBS), in which a number of variables, parameters and remote sensing methods are used. Optical and thermal sensors provide information on aspects of the surface energy budget

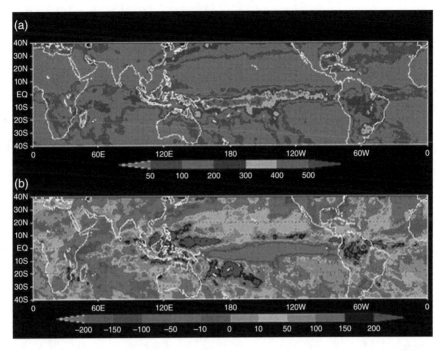

Figure 6.16   (a) TRMM monthly mean rainfall (mm) January and February 1998 and (b) the difference in two-month mean rainfall between 1998 and 1999 (GSFC, NASA) (*see plate section for colour representation of this figure*)

(Su, 2002). Modelling is used to assess soil moisture content and its spatial distribution. By coupling the soil moisture balance to the surface energy balance it is possible to incorporate satellite observations to estimate topsoil moisture content. Satellite observations provide estimates of albedo, emissivity, surface temperature, fractional vegetation cover, leaf area index, height of vegetation and roughness height (see earlier in this chapter). These observations are combined with meteorological measurements and radiation data within a numerical model framework.

6.9   Summary of satellite measurement issues

Over the last 40 years or so rapid advances in measuring hydrometeorological parameters from space have been made. Indeed, most of the input requirements to watershed models are likely to be satisfied, at least partially, by satellite-borne instrumentation, as predicted by Collier (1985). However, those parameters which vary from day to day, for example snow cover, soil moisture, evapotranspiration and precipitation, require fully backed-up operational measurement systems.

Since several systems will provide measurements of the same parameters across a wide part of the electromagnetic spectrum, it is important to assess the relative merits and cost-effectiveness of each technique. Figure 6.17 summarizes the interactions of passive microwaves with precipitation clouds and the surface using two wavebands,

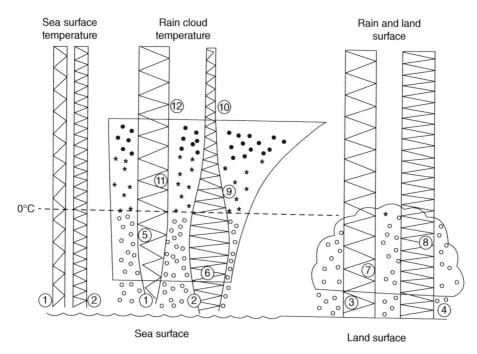

Figure 6.17   The interactions of high (e.g. 85 GHz) and low (e.g. 19 GHz) frequency passive micro-wave with precipitation clouds and the surface. The width of the vertical columns represents the intensity or temperature of the upwelling radiation. In this figure the illustrated features and their demarcations are: (a) the small emissivity of sea surface for both low (1) and high (2) frequencies; (b) the large emissivity of land surface for both low (3) and high (4) frequencies; (c) the emission from cloud and raindrops, which increases with vertically integrated liquid water for the low frequency (5), but saturates quickly for the high frequency (6); (d) the signal of the water emissivity at the low frequency is masked by the land surface emissivity (7); (e) the saturated high frequency emission from the rain (8) is not distinctly different from the land surface background (4); (f) ice precipitation parti-cles aloft backscatter down the high frequency emission (9), causing cold brightness temperatures (10), regardless of surface emission properties; (g) the ice lets the low frequency emission upwell unimpeded (11), allowing its detection above cloud top as warm brightness temperature (12) (from Rosenfeld and Collier, 1999)

shorter (85 GHz) and longer (19 GHz). The physical principles have been used to formulate a large number of rain estimation methods as noted previously.

By far the largest amount of effort has been put into the measurement of precipita-tion, in particular rainfall, and evaluations of the measurement techniques have been carried out (e.g. Johnstone et al., 1985; Yumaz et al., 2005). Table 6.4 presents a summary of some of the accuracies reported. Two points emerge from this table. First, most of the assessments of the accuracy of rainfall totals have been made as integrations over areas greater than 10^3 km^2. However, some techniques are capable of measurements over areas as small as 4 km^2. Second, most of the studies have been concerned in the main with the measurement of convective rainfall rather than frontal rainfall.

The first point is reinforced by the work of Augustine et al. (1981), who found that the best correlations between satellite estimates of rainfall made using the

Table 6.4 Summary of the performance of satellite rainfall estimation techniques

Technique	Area over which estimates are assessed (km²)	Period of integration (h)	Approximate percentage accuracy (%)	Sample references describing techniques (rainfall types)
Cloud indexing	10^5	24	122	Follansbee and Oliver (1975) (convective/frontal)
Cloud texture, neural networks	16	0.5	50	Hsu et al. (1997); Sorooshian et al. (2000)
Life history	10^4	1	85	Griffith et al. (1978) (convective)
	10^5	24	55	
	10^4	0.5	50	Wylie (1979) (convective)
	6×10^3	0.5	65	Stout et al. (1979)
Bi-spectral	10^5	0.5–2	49	Lovejoy and Austin (1979a; 1979b) (convective/frontal)
Passive microwave	10^3	24	70	Lovejoy and Austin (1980) (convective/frontal) Wilheit et al. (1973); Spencer et al. (1983)
Active microwave	10^3	monthly	20	Wang and Wolff (2012) (convective)

Griffith–Woodley technique and raingauge/radar data were achieved for 6 hour periods over 9350 km² (the largest area considered); the worst correlations were found for half-hour rainfall totals over areas of 55 km². The most accurate rainfall estimates in all rainfall types would appear to be obtained by the Lovejoy and Austin bi-spectral technique with an error of around 49% for rainfall totals of 0.5 to 1 hour over areas of 10^5 km². Other techniques (Griffith and Woodley, 1981), tuned to making measurements in convective rainfall, may attain this accuracy over areas of around 10^4 km². The accuracy of passive microwave techniques is in the region of 60% over 400 km². Active microwave techniques give monthly rainfall totals of 20% over 10^3 km².

Rainfall amount is not measured with uniform accuracy. Some techniques, for example that of Scofield (1982), are fine tuned to measure heavy convective rainfall likely to produce flash floods. However, no one technique cannot be expected to measure all rainfall amounts with the same accuracy. For example Wylie and Laitsch (1983) found that up to 30% of daily rainfall amounts at a point were correctly detected for rainfall amounts greater than 10 mm, yet only up to 15% of daily rainfall amounts were correctly detected for rainfall amounts greater than 25 mm. The procedure used was simple, and more complex procedures are likely to improve on this performance, but nevertheless the difference between rainfall amounts is likely to remain.

Clearly the different satellite measurement techniques all have their advantages and disadvantages. It is unlikely that one satellite system for the measurement of rainfall will be universally accepted, although active microwave techniques may offer the most promise. The same is true for the determination of snow cover, soil moisture and evapotranspiration, although in these cases there is more convergence of techniques. Table 6.1 shows clearly that plans exist for several satellites with SAR instrumentation. At the same time there will be consolidation of less advanced systems to provide operational measurements, and it is this development which will have the most immediate impact on hydrometeorology.

Appendix 6.1   Radiation balance

Assume that the Earth has no atmosphere; then the mean surface equilibrium temperature is the balance between the incoming and the outgoing radiation. Take F as the flux of solar radiation at a distance R of the Earth from the Sun, a as the fractional planetary albedo, r as the Earth's radius, and T_e as the equilibrium temperature of the planet. Then from the Stefan–Boltzmann law, the Earth's emitted radiation is given by

$$E_e = 4\pi r^2 \sigma T_e^4 \tag{6.15}$$

where σ is the Stefan–Boltzmann constant. The absorbed radiation from the Sun is

$$E_a = \frac{\pi r^2 (1-a) F}{R^2} \tag{6.16}$$

Then

$$4\pi r^2 \sigma T_e^4 = \frac{\pi r^2 (1-a) F}{R^2}$$

$$T_e^4 = \frac{(1-a) F}{4\sigma R^2} \tag{6.17}$$

Here the term $F/R^2 = S_0$ for $R = 1$ is the solar constant at the Earth, and has a value of about 1370 W m^{-2}. For a mean albedo of the Earth of 0.3, $T_e = 256\,\text{K}$. The values of T_e for several planets (Harries, 1994) are as in Table 6.5. Also shown are the mean surface temperatures T_s. Note the importance of the greenhouse effect on Venus, and its minor importance on Mars.

Table 6.5   Examples of values of equilibrium temperature and mean surface temperature for selected planets (Harries, 1994)

Planet	Relative distance from the Sun R	Mean albedo a	Equilibrium temperature T_e (K)	Surface temperature T_s (K)
Venus	0.72	0.77	227	750
Earth	1.00	0.30	256	280
Mars	1.52	0.15	216	240
Jupiter	5.20	0.58	98	134

Summary of key points in this chapter

1. Some of the parameters which must be known to model and predict watershed processes may be derived from satellite measurements.

2. Over the last 50 years the development of satellite systems in low polar orbits has provided opportunities to develop measurement systems with coverage over large areas.

3. By the early 1970s radiometers were able to provide data in several microwave bands in addition to measurements at visible and infrared wavelengths.

4. Most of the current satellite systems employ several different types of instrument.

5. Thermal radiation is emitted by all objects at temperatures above absolute zero.

6. Generally a hot object will distribute its emission over a range of wavelengths.

7. The spectral radiance of a real object is $L_\lambda = \varepsilon(\lambda)L_{\lambda p}$, where $L_{\lambda p}$ is the black body radiance and subscript p refers to Planck; $\varepsilon(\lambda)$ is the emissivity at wavelength λ.

8. The thermal emissivity of a surface is related to its reflectance properties.

9. The bidirectional reflectance distribution function (BRDF) is L_1/E, where the subscript 1 denotes the scattering direction and L_1 is the outgoing radiance of the surface as a result of illumination.

10. The reflectivity is referred to as the albedo.

11. There are two approaches to calculating BRDF, namely the small perturbation method and the facet model.

12. Operating at about 1.2 μm (near infrared) only results in difficulty in discriminating between a number of materials. The use of three or four channels reduces this problem.

13. The optical depth τ (sometimes called the opacity) gives the extent to which a given layer of material attenuates the intensity of radiation passing through it.

14. Wien's displacement law is $\lambda_{max}T = w$, where w is Wien's displacement constant, λ_{max} is the wavelength of maximum emitted radiance, and T is the absolute temperature of the radiating surface.

15. The Stefan–Boltzmann law relates the total radiant emittance from a black body to the fourth power of the absolute temperature.

16. The thermal inertia is the resistance of a material to temperature change.

17. Clouds may cause rapid changes in the radiant emission from a surface.

18. Techniques for estimating rainfall are cloud indexing and life history methods.

19. The area–time integral (ATI) of cloud areas over the lifetime of a storm is related to the total rain volume.

20. Seasonal changes in snow cover can be monitored from satellite visible and thermal infrared images.

21. The relationship of the soil spectral reflectance to the water content depends upon several variables in addition to the water content, such as soil texture and organic content.

22. Infrared sensors provide information on temperature, and thus indirectly on the height of the tops of clouds, whereas visible sensors provide information on the depths of clouds. Therefore estimates of precipitation may be made.

23. The microwave radiation thermally emitted by an object is called its brightness temperature, which may be used to determine emissivity. However, the emissivity varies with wavelength and therefore more than one microwave wavelength is used.

24. The skin depth is the depth from which 63% of the radiation emanates from the material in the surface layer.

25. The use of the microwave region of the electromagnetic spectrum to make measurements of precipitation has the advantage of detecting the precipitation when clouds are present.

26. Curves of 19.35 GHz brightness temperatures and rainfall rate may be used to derive precipitation. Also dual polarized 37 GHz radiometer data at different wavelengths may be used to detect a wide range of snow and ice features as well as rainfall.

27. The Global Precipitation Climatology Project is an international project drawing upon a range of input data sets and techniques.

28. The Global Precipitation Measurement mission is a constellation of satellite systems, and uses a range of techniques including active radar to measure precipitation.

29. The apparent temperature of the sea surface can be measured by satellite instruments.

30. Measurements of the intensity of brightness of microwave energy can be made using a radar mounted on a satellite (synthetic aperture radar, SAR). Such measurements are used to estimate soil moisture content.

31. The Tropical Rainfall Measuring Mission (TRMM) involved several instruments mounted on a co-latitude (40°S–40°N) orbit satellite, including a precipitation radar operating at K_u-band (13.8 GHz, 2.2 cm wavelength).

32. The surface energy balance system (SEBS) uses a number of variables, parameters and remote sensing methods to measure the coupling of soil moisture balance to the surface energy balance.

33. Different satellite measurement techniques all have their advantages and disadvantages for measuring rainfall.

Problems

1. Name four current satellite systems operating in geostationary and polar orbits.
2. Define emissivity and give the formulae for reflectivity, bidirectional reflectance distribution function (BRDF) and albedo.
3. What are typical values of albedo for water, forest and green crops?
4. Outline two basic approaches to calculating BRDF.
5. Draw a graph of albedo versus wavelength for snow.
6. Define optical depth and thermal inertia.
7. Give Wien's displacement law and the Stefan–Boltzmann law.
8. Outline the cloud indexing and life history methods of estimating rainfall, giving a formula for the volumetric rain rate for a particular cloud.
9. Describe inversion techniques for deriving properties of clouds.
10. What is skin depth?
11. Describe what passive microwave frequencies are used to estimate precipitation, and outline how such frequencies might be used.
12. Outline the GPCP and the GPM missions.
13. Give the apparent temperature of the sea surface when viewed from a satellite at an angle of θ degrees.
14. Describe both passive and active satellite radar techniques, giving specific examples.
15. Outline the Tropical Rainfall Measuring Mission.
16. Outline the Surface Energy Balance System.
17. Summarize the interactions of passive microwaves with precipitation clouds and the surface using two wavebands.
18. Summarize the accuracy of satellite rainfall estimation techniques.

References

Adler, R.F., Huffman, G.J., Bolvin, D.T., Curtis, S. and Nelkin, E.J. (2001) Tropical rainfall distributions determined using TRMM combined with other satellite and raingauge information. *J. Appl. Meteorol.*, 39, 2007–2023.

Adler, R.F., Huffman, G.J., Cheng, A., Ferraro, R., Xu, P.-P., Janowtak, J., Rudolf, B., Schneider, U., Curtis, S., Bolvin, D., Gruber, A., Susskind, J., Arkin, P. and Nelkin, E. (2003) The Version-2 Global Precipitation Climatology Project (GPCP) monthly precipitation analysis (1979–present). *J. Hydrometeorol.*, 4, 1147–1167.

Anagnostou, E.N., Morales, C.A. and Dinku, T. (2001) The use of TRMM precipitation radar observations in determining ground radar calibration biases. *J. Atmos. Oceanic Technol.*, 18, 616–628.

Atlas, D., Echweman, J., Meneghini, R. and Moore, R.K. (1982) The outlook for precipitation measurements from space. *Atmos. Ocean*, 20, 50–61.

Augustine, J.A., Meitin, J.G., Griffith, C.G. and Woodley, W.L. (1981) An objective evaluation of the Griffith/Woodley rain estimation technique. 4th Conference on Hydrometeorology, 7–9 October, Reno, NV, pp. 134–140. American Meteorological Society, Boston.

Barrett, E.C. (1970) The estimation of monthly rainfall from satellite data. *Mon. Weather Rev.*, 98, 322–327.

Barrett, E.C. and Grant, C.K. (1978) An appraisal of Landsat 2 cloud imagery and its implications for the design of a future meteorological observing system. *J. Brit. Interplanet. Soc.*, 31, 3–10.

Barrett, E.C. and Martin, D.W. (1981) *The Use of Satellite Data in Rainfall Monitoring*. Academic Press, London.

Berg, W., Kummerow, C., Sapiano, M., Rodriguez-Alvarez, N. and Weng, F. (2012) A fundamental climate data record of microwave brightness temperature data from 25 years of SSM/I and SSMIS observations. *Gewex WCRP News*, August, 4–5.

Brisco, B., Ulaby, F.T. and Dobson, M.C. (1983) Spaceborne SAR data for land-cover classification and change detection. Digest IEEE International Geoscience Remote Sensing Symposium (IGARSS'83), San Francisco, 31 August to 2 September.

Carlson, T.B., Dodd, J.K., Benjamin, S.G. and Cooper, J.N. (1981) Satellite estimation of the surface energy balance, moisture availability and thermal inertia. *J. Appl. Meteorol.*, 20, 67–87.

Chang, T.C., Gloersen, P., Schmugge, T., Wilheit, T.T. and Zwally, H.J. (1976) Microwave emission from snow and glacier ice. *J. Glaciol.*, 17, 23–39.

Collier, C.G. (1985) Remote sensing for hydrological forecasting. In Facets of Hydrology, vol. II, ed. J.C. Rodda, pp. 1–24. Wiley, Chichester.

Davie, T. and Kelly, R. (1997) Spatial resolution and variation of soil moisture as detected by ERS-1 SAR imagery. Proceedings of the British Hydrological Society 6th National Hydrometeorology Symposium, Salford, 15–18 September 1997, pp. 5.11–5.17. Institute of Hydrology, Wallingford.

Delbeato, R. and Barrell, S.L. (1985) Rain estimation in extratropical cyclones using GMS imagery. *Mon. Weather Rev.*, 113, 747–755.

Deleur, M.S. (1980) *Satellite Methods of Studying Snow Cover* [in Russian]. Gridrometeoizdat, Leningrad.

Dey, B., Gregory, A.F. and Moore, H. (1979) Application of satellite images for monitoring snowline in the Yukon and North West territories. *Polar Record*, 19, 145–154.

Dey, B., Goswami, D.C. and Rango, A. (1983) Utilization of satellite snow-cover observations for seasonal stream flow estimates in the Western Himalayas. *Nord. Hydrol.*, 3, 257–266.

Dittberner, G.J. and Vonder Haar, T.H. (1973) Large scale precipitation estimates using satellite data: application to the Indian Monsoon. *Arch. Met. Geophys. Biokl. B*, 21, 317–334.

Doneaud, A.A., Miller, J.R., Johnson, L.R., Vonder Haar, T.H. and Laybe, P. (1985) The area-time-integral technique to estimate convective rain volume over areas applied to satellite data: a preliminary investigation. Preprints 6th Conference on Hydrometeorology, 29 October to 1 November, pp. 238–245. American Meteorological Society, Boston.

Follansbee, W.A. (1973) Estimation of average daily rainfall from satellite cloud photographs. NOAA Technical Memorandum NESS 44, Washington, DC.

Follansbee, W.A. (1976) Estimation of daily precipitation over China and the USSR using satellite imagery. NOAA Technical Memorandum NESS 81, Washington, DC.

Follansbee, W.A. and Oliver, V.J. (1975) A comparison of infrared imagery and video pictures in the estimation of daily rainfall from satellite data. NOAA Technical Memorandum NESS 62, Washington, DC.

Foster, J.L. (1983) Night-time observations of snow using visible imagery. *Int. J. Remote Sens.*, 4, 785–791.

Foster, J.L., Rango, A., Hall, D.K., Chang, A.T.C., Allison, L.J. and Diesen, B.C. III (1980) Snowpack monitoring in North America and Eurasia using passive microwave satellite data. *Remote Sens. Environ.*, 10, 205–298.

Fox, N.I., Saich, P. and Collier, C.G. (2000) Estimating the surface water and radiation balance in an upland area from space. *Int. J. Remote Sens.*, 21, 2985–3002.

Fu, L. and Holt, B. (1982) Seasat views oceans and sea ice with synthetic aperture radar. JPL 81–120. NASA Jet Propulsion Laboratory, Pasadena, CA.

Gloersen, P. and Salomonson, V.V. (1975) Satellites: new global observing techniques for ice and snow. *J. Glaciol.*, 15, 373–389.

Griffith, C.G. and Woodley, W.L. (1981) The estimation of convective precipitation from GOES imagery with the Griffith/Woodley technique. Precipitation Measurements from Space Workshop Report, ed. D. Atlas and O.W. Thiele, NASA Goddard Space Flight Center, pp. D154–D158.

Griffith, C.G., Woodley, W.L., Browner, S., Teijeiro, J., Maier, M., Martin, D.W., Stout, J. and Sikdar, D.N. (1976) Rainfall estimation from geosynchronous satellite imagery during the daylight hours. NOAA Technical Report ERL 356-WPM07, Boulder, CO.

Griffith, C.G., Woodley, W.L., Grube, P.G., Martin, D.W., Stout, J. and Sikdar, D.N. (1978) Rain estimation from geosynchronous satellite imagery: visible and infrared studies. *Mon. Weather Rev.*, 106, 1153–1171.

Gruber, A. (1973) Estimating rainfall in regions of active convection. *J. Appl. Meteorol.*, 12, 110–118.

Gurney, R.J. and Camillo, P.J. (1984) Modelling daily evapotranspiration using remotely sensed data. *J. Hydrol.*, 69, 305–324.

Hall, D.K. and Martinec, J. (1985) Remote Sensing of Ice and Snow. Chapman and Hall, London.

Harries, J.E. (1994) *Earthwatch: The Climate from Space.* Wiley, Chichester.

Harris, R. (1987) *Satellite Remote Sensing.* Routledge & Kegan Paul, London.

Hsu, K., Gao, X., Sorooshian, S. and Gupta, H.V. (1997) Precipitation estimation from remotely sensed information using artificial neural networks. *J. Appl. Meteorol.*, 36, 1176–1190.

Huffman, G.J. and coauthors (1997) The Global Precipitation Climatology Project (GPCP) combined precipitation datasets. *Bull. Am. Meteorol. Soc.*, 78, 36–50.

Idso, S.B., Jackson, R.D. and Reginato, R.J. (1975a) Detection of soil moisture by remote surveillance. *Am. Sci.*, 63, 549–557.

Idso, S.B., Schmugge, T.J., Jackson, R.D. and Reginato, R.J. (1975b) The utility of surface temperature measurements for the remote sensing of soil water status. *J. Geophys. Res.*, 80, 3044–3049.

Iguchi, T. and Meneghini, R. (1994) Intercomparison of single frequency methods for retrieving a vertical rain profile from airborne or spaceborne radar data. *J. Atmos. Oceanic Technol.*, 11, 1507–1516.

Iguchi, T., Meneghini, R., Awaka, J., Kozu, T. and Okamoto, K. (2000) Rain profiling algorithm for TRMM precipitation radar data. *Remote Sens. Appl. Earth Atmos. Ocean*, 25, 973–976.

Inomata, H., Okamoto, K., Ajima, T., Masuko, H., Yohidada, S. and Gugono, M. (1981) Remote sensing of rainfall rates using airborne microwave rain-scatterometer/radiometer. Proceedings 15th International Symposium on Remote Sensing of Environment, Ann Arbor, MI, USA.

Jackson, R.D., Schmugge, T.J., Nicks, A.D., Coleman, G.A. and Engman, E.T. (1981) Soil moisture updating and microwave remote sensing for hydrological simulation. *Hydrol. Sci. Bull.*, 26, 39/1981, 305–319.

Johnstone, K.J., Hogg, W.D., Hanssen, A.J., Niitsoo, A. and Polavarapu, V.L. (1985) An evaluation of some satellite rainfall estimation techniques. Preprints 6th Conference on Hydrometeorology, 29 October to 1 November, pp. 233–237. American Meteorological Society, Boston.

Jong, S. De, Kwast, H. Van Der, Addink, E. and Su, B. (2008) Remote sensing for hydrological studies. Chapter 15 in *Climate and the Hydrological Cycle*, IAHS Special Publication 8, pp. 297–320. IAHS Press, CEH Wallingford.

Klaassen, W. and Van Den Berg, W. (1985) Evapotranspiration derived from satellite observed surface temperatures. *J. Clim. Appl. Meteorol.*, 24, 412–424.

Kozu, T. and coauthors (2001) Development of precipitation radar onboard the Tropical Rainfall Measuring Mission (TRMM). *IEEE Trans. Geosci. Remote Sens.*, 39, 102–116.

Kunzi, K.F., Patil, S. and Rott, H. (1982) Snow cover parameters retrieved from Nimbus-7 Scanning Multichannel Microwave Radiometer (SMMR) data. *IEEE Trans. Geosci. Remote Sens.*, GE-20, 452–467.

Lethbridge, M. (1967) Precipitation probability and satellite radiation data. *Mon. Weather Rev.*, 95, 487–490.

Li, X., Zhang, J.A., Yang, X., Pichel, W.G., DeMaria, M., Long, D. and Li, Z. (2013) Tropical cyclone morphology from spaceborne synthetic aperture radar. *Bull. Am. Meteorol. Soc.*, 94, 215–230.

Liljas, E. (1982) Automated techniques for the analysis of satellite cloud imagery. In Nowcasting, ed. K.A. Browning, pp. 167–176. Academic Press, London.

Liu, Z., Ostrenga, D., Teng, W. and Kempler, S. (2012) Tropical Rainfall Measuring Mission (TRMM) precipitation and services for research and applications. *Bull. Am. Meteorol. Soc.*, 93, 1317–1325.

Lovejoy, S. and Austin, G.L. (1979a) The delineation of rain areas from visible and IR satellite data for GATE and mid-latitudes. *Atmos. Ocean*, 17, 77–92.

Lovejoy, S. and Austin, G.L. (1979b) The sources of error in rain amount estimating schemes for GOES visible and IR satellite data. *Mon. Weather Rev.*, 107, 1048–1054.

Lovejoy, S. and Austin, G.L. (1980) The estimation of rain from satellite-borne radiometers. *Q. J. R. Meteorol. Soc.*, 106, 255–276.

Martin, D.W. and Howland, M.R. (1986) Grid history: a geostationary satellite technique for estimating daily rainfall in the tropics. *J. Clim. Appl. Meteorol.*, 25, 184–195.

Martin, M.J., Hines, A. and Bell, M.J. (2007) Data assimilation in the FOAM operational short-range ocean forecasting system: a description of the scheme and its impact. *Q. J. R. Meteorol. Soc.*, 133, 981–995.

Martinec, J. and Rango, A. (1981) Areal distribution of snow water equivalent evaluated by snow cover monitoring. *Water Resour. Res.*, 17, 1480–1488.

McElroy, J.H. and Schneider, S.R. (1985) The Space Station Polar Platform: integrating research and operational missions. NOAA Technical Report NESDIS 19, US Dept. of Commerce, Washington, DC.

McGinnis, D.F. Jr, Scofield, R.A., Schneider, S.R. and Berg, C.P. (1979) Satellites as an aid to water resources managers. American Society of Civil Engineers Convention, Boston, MA, 2–6 April, preprint 3486.

Meneghini, R. and Kozu, T. (1990) *Spaceborne Weather Radar*. Artech, Boston.

Negri, A.J. and Adler, R.F. (1981) Relation of satellite-based thunderstorm intensity to radar-estimated rainfall. *J. Appl. Meteorol.*, 20, 288–300.

Negri, A.J., Adler, R.F. and Wetzel, P.J. (1984) Rain estimation from satellites: an examination of the Griffith Woodley Technique. *J. Clim. Appl. Meteorol.*, 23, 102–116.

O'Brien, H.W. and Munis, R.H. (1975) Red and near-infrared spectral reflectance of snow. In *Operational Applications of Satellite Snowcover Observations*, ed. A. Rango, Proceedings Workshop, South Lake, Tahoe, CA, 18–20 August, NASA SP-391, pp. 345–360. NASA, Washington, DC.

Odegaard, H.A., Anderson, T. and Ostrem, G. (1980) Applications of satellite data for snow mapping in Norway. NASA CP 2116, pp. 92–106.

Ramamoorthi, A.S. and Rao, P.S. (1985) Inundation mapping of the Sahibi river flood of 1977. *Int. J. Remote Sens.*, 6(3, 4), 443–445.

Rango, A. and Itten, K.I. (1976) Satellite potentials in snowcover monitoring and runoff prediction. *Nord. Hydrol.*, 7, 209–230.

Rango, A. and Martinec, J. (1979) Application of a snowmelt-runoff model using Landsat data. *Nord. Hydrol.*, 10, 225–238.

Rango, A. and Peterson, R. (eds) (1980) *Operational Application of Satellite Snowcover Observations.* NASA Conference Publication 2116.

Rango, A., Salomonson, V.V. and Foster, J.L. (1977) Seasonal streamflow estimation in the Himalayan region employing meteorological satellite snow cover observations. *Water Resour. Res.*, 13, 109–112.

Rango, A., Chang, A.T.C. and Foster, J.L. (1979) The utilization of spaceborne microwave radiometers for monitoring snowpack properties. *Nord. Hydrol.*, 10, 25–40.

Rees, W.G. (1993) *Physical Principles of Remote Sensing.* Cambridge University Press, Cambridge.

Reynolds, D.W., Vonder Haar, T.H. and Grant, L.O. (1978) Meteorological satellites in support of weather modification. *Bull. Am. Meteorol. Soc.*, 59, 269–281.

Rosenfeld, D. and Collier, C.G. (1999) Estimating surface precipitation. Chapter 4.1 in *Global Energy and Water Cycles*, ed. K.A. Browning and R.J. Gurney, pp. 124–133. Cambridge University Press, Cambridge.

Salman, A.A.B. (1983) Using Landsat imagery interpretation for underground water prospection around Qena Province, *Egypt. Int. J. Remote Sens.*, 4, 179–189.

Salomonson, V.V. and Hall, D.K. (1979) A review of Landsat-D and other advanced systems relative to improving the utility of space data in water resources management. In *Operational Applications of Satellite Snowcover Observations*, ed. A. Rango and R. Peterson, Proceedings Workshop, Sparks, NV, 16–17 April, NASA CP-2116, pp. 281–296. NASA, Washington, DC.

Schanda, E. (1986) *Physical Fundamentals of Remote Sensing.* Springer, Berlin.

Scofield, R.A. (1981) Analysis of rainfall from flash flood producing thunderstorms using GOES data. Proceedings IAMAP Symposium, Hamburg, 25–29 August, ESA SP-175, pp. 51–58.

Scofield, R.A. (1982) A satellite technique for estimating rainfall from flash flood producing thunderstorms. Proceedings International Symposium on Hydrometeorology, ed. A.I. Johnson and R.A. Clark, 13–17 June, Denver, CO, pp. 121–120. American Water Resources Association, Bethesda, MD.

Scofield, R.A. (1984) Satellite-based estimates of heavy precipitation. Recent Advances in Civil Space Remote Sensing, *SPIE* 481, pp. 84–91.

Scofield, R.A. and Oliver, V.J. (1977a) A scheme for estimating convective rainfall from satellite imagery. NOAA Technical Memorandum NESS 86, Washington, DC.

Scofield, R.A. and Oliver, V.J. (1977b) Using satellite imagery to estimate rainfall from two types of convective systems. Preprints 11th Technical Conference on Hurricanes and Tropical Meteorology, 13–16 December, Miami Beach, FL, pp. 204–211. American Meteorological Society, Boston.

Simpson, J.C., Kummerow, C., Tao, W.-K. and Adler, R.F. (1996) On the Tropical Rainfall Measuring Mission (TRMM). *Meteorol. Atmos. Phys.*, 60, 19–36.

Sorooshian, S., Hsu, K., Gao, X., Gupta, H.V., Iman, B. and Braithwaite, D. (2000) Evolution of PERSIANN system satellite-based estimates of tropical rainfall. *Bull. Am. Meteorol. Soc.*, 81, 2035–2046.

Spencer, R.W. (1986) A satellite passive 37-GHz scattering-based method for measuring oceanic rain rates. *J. Clim. Appl. Meteorol.*, 25, 754–766.

Spencer, R.W., Hinton, B.B. and Olson, W.S. (1983) Nimbus-7 37 GHz radiances correlated with radar rain-rates over the Gulf of Mexico. *J. Clim. Appl. Meteorol.*, 22, 2095–2099.

Stout, J.E., Martin, D.W. and Sikdar, D.N. (1979) Estimating GATE rainfall with geosynchronous satellite images. *Mon. Weather Rev.*, 107, 505–598.

Su, Z. (2002) The Surface Energy Balance (SEBS) for estimation of turbulent heat fluxes. *Hydrol. Earth Syst. Sci.*, 6, 85–99.

Theon, J. (1973) A multi-spectral view of the Gulf of Mexico from Nimbus-5. *Bull. Am. Meteorol. Soc.*, 54, 934–937.

Tsonis, A.A. and Isaac, G.A. (1985) On a new approach for instantaneous rain area delineation in the mid-latitudes using GOES data. *J. Clim. Appl. Meteorol.*, 24, 1208–1218.

Ulaby, F.T., Dobson, M.C. and Brunfeldt, D.R. (1983) Improvements of moisture estimation accuracy of vegetation-covered soil by combined active/passive microwave remote sensing. *IEEE Trans. Geosci. Remote Sens.*, GE-21, 300–307.

Ulaby, F.T., Moore, R.K. and Fung, A.K. (1986) *Microwave Remote Sensing: Active and Passive. Vol. III: From Theory to Applications.* Artech, Boston.

Van de Griend, A.A. and Engman, E.T. (1985) Partial area hydrology and remote sensing. *J. Hydrol.*, 81, 211–251.

Wang, J.R. (1983) Passive microwave sensing of soil moisture content: the effects of soil bulk density and surface roughness. *Remote Sens. Environ.*, 13, 329–344.

Wang, J. and Wolff, D.B. (2012) Evaluation of TRMM rain estimates using ground measurements over Central Florida. *J. Appl. Meteorol. Climatol.*, 51, 926–940.

Wetzel, P.J., Atlas, D. and Woodward, R.H. (1984) Determining soil moisture from geosynchronous satellite infrared data: a feasibility study. *J. Clim. Appl. Meteorol.*, 23, 375–391.

Wilheit, T.H. (1975) The electrically scanning microwave radiometer (ESMR) experiment. In Nimbus-6 User Guide, pp. 87–108. NASA Goddard Space Flight Center.

Wilheit, T.H., Theon, J., Shenk, W. and Allison, L. (1973) Meteorological interpolation of the images from Nimbus-5 Electrically Scanned Microwave Radiometer. NASA X-651-73-189, Goddard Space Flight Center.

Wilheit, T.H., Chang, A.T.C., Rao, M.S.V., Rodgers, E.B. and Theon, J.S. (1977) A satellite technique for quantitatively mapping rainfall rates over the oceans. *J. Appl. Meteorol.*, 16, 551–560.

Woodley, W.L., Griffith, C.G., Griffin, J.S. and Stromatt, S.C. (1980) The inference of GATE convective rainfall from SMS-1 imagery. *J. Appl. Meteorol.*, 19, 388–408.

Wu, R., Weinman, J.A. and Chin, R.T. (1985) Determination of rainfall rates from GOES satellite images by pattern recognition technique. *J. Atmos. Oceanic Technol.*, 2, 314–330.

Wylie, D.P. (1979) An application of a geostationary satellite rain estimation technique to an extra-tropical area. *J. Appl. Meteorol.*, 18, 1640–1648.

Wylie, D.P. and Laitsch, D. (1983) The impacts of different satellite data on rain estimation schemes. *J. Appl. Meteorol.*, 22, 1270–1281.

Yumaz, K.K., Hogue, T.S., Hsu, K.-L., Sorooshian, S., Gupta, H.V. and Wagener, T. (2005) Intercomparison of raingauge, radar and satellite-based precipitation estimates. *J. Hydrometeorol.*, 6, 497–517.

Zwally, H.J. and Gloersen, P. (1977) Passive microwave images of the polar regions and research applications. *Polar Record*, 18, 431–450.

7

Analysis of Precipitation Fields and Flood Frequency

7.1 Introduction

An analysis of precipitation produces a gridded field from available observations and *a priori* information. This is a difficult problem because of the wide range of spatial and temporal scales present in a rainfall field, in addition to its discontinuous nature. Therefore, the quality of a precipitation analysis can vary significantly according to data availability and to the spatial and temporal resolutions considered. In particular, for small scale precipitating systems and over sparse observational networks, the analysis relies heavily on the basis of the chosen spatial interpolation technique. However, the spatial distribution of surface precipitation on short timescales (less than one day) is important for the evaluation of high resolution numerical weather prediction (NWP) models, for hydrological forecasts (including flood predictions), and for soil moisture analysis (including agricultural water management and drought monitoring).

7.2 Areal mean precipitation

High resolution measurements of the spatial and temporal structure of precipitation use data from distrometers, raingauges, radar and satellite procedures. Point measurements are often a poor representation of area-average precipitation, and it can be dangerous to assume that they are representative, particularly for short duration samples and for convective storms. Nevertheless a number of techniques have been developed for the estimation of areal precipitation.

The *isohyetal* method involves constructing isohyets across a catchment based upon measured values, measuring or computing the area between isohyets to provide weights to be used when calculating the average value, and then creating a weighted average. The areas between isohyets and the catchment boundary are measured with a planimeter. The areal rainfall is calculated from the product of the areas a_i in Figure 7.1. In this figure the areal rainfall is given by $\sum_{i=1}^{5} a_i R_i / A$, where A is

Hydrometeorology, First Edition. Christopher G. Collier.
© 2016 John Wiley & Sons, Ltd. Published 2016 by John Wiley & Sons, Ltd.
Companion website: www.wiley.com/go/collierhydrometerology

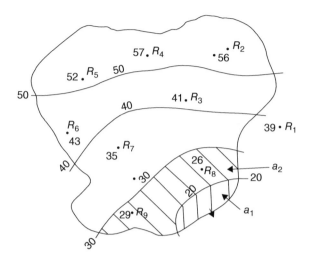

Figure 7.1   Isohyetal method of estimating areal precipitation (from Shaw, 1994)

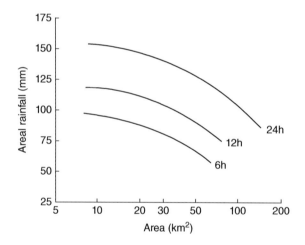

Figure 7.2   Curves showing the typical relationship between maximum storm depth, area covered and duration in hours (from Shaw, 1994)

the total catchment area and R_i is the mean rainfall between the isohyets. The areal rainfall should be determined by analysing a number of storms to derive depth–area relationships for different durations, as shown in Figure 7.2.

The *triangle* method involves constructing a network of triangles as near equilateral as possible with one of the gauges in the network at each apex of each triangle. The area of each triangle, or portion of each triangle, is summed to calculate the average value. The *Thiessen* method comprises constructing a network of polygons by drawing the perpendicular bisector of the line joining each pair of gauges. The area of each polygon, or portion of each polygon, is weighted in order to estimate the areal average precipitation, as indicated in Figure 7.3. This is the most popular method of estimating areal precipitation.

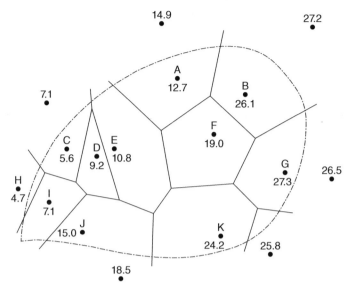

Figure 7.3   Example of the calculation of area-average precipitation using the Thiessen method (from Sumner, 1988 and Shuttleworth, 2012)

7.3   Spatial and temporal storm analysis

7.3.1   Spatial statistical analyses

Unfortunately real data sets are usually limited in some way or other. For example the length of record restricts direct statistical analysis of likely extremes, or the full range of precipitation types is not included. Nevertheless, statistical methodologies are useful for the spatial analyses of precipitation as a means for studying the dominant precipitation producing mechanisms and prevailing moisture flows in an area. Shuttleworth (2012) provides a comprehensive summary of techniques. For example, if an analysis was made of the correlation between precipitation measured at a point and that measured at progressively increasing distances, it might be possible to deduce information about average storm size and typical storm separation.

A *statistical interpolation* (SI) technique is often used. Various other techniques for spatial interpolation of precipitation have been used in previous studies, such as *kriging* (Finkelstein, 1984; Creutin et al., 1988; Crossley, 2001), the *successive correction method* (Bussières and Hogg, 1989; Weymouth et al., 1999), and *optimum interpolation* (Bhargava and Danard, 1994; Garand and Grassotti, 1995). Usually, kriging and successive correction do not need a background field to perform the analysis. However, SI performs the analysis on the difference between an observation and the corresponding background value. The SI technique requires the specification of error statistics associated with each unit of information used to perform the analysis, namely the observations and the background field. As a consequence, the SI technique provides an estimate of analysis error at every grid point.

Given a set of N precipitation observations y_a^i and a background field x_b^j available on a grid of M points, the statistical interpolation technique provides an analysis x_a^i at the model grid point j as:

$$x_a^i = x_b^i + \sum w^{ij}\left(y_a^i - y_b^i\right) \tag{7.1}$$

where y_b^i is the background field spatially interpolated to the observation point i and the summation is over $i = 1$ to N. Writing (7.1) as

$$\delta x^j = \sum w^{ij}\delta y^i \tag{7.2}$$

states that the correction at a given grid point (analysis increment δx^j) is a linear combination of the differences between the model and the observation values at the observation point (background departure δy^i). The weights w^{ij} are obtained by minimizing the variance of the analysis error. The $N \times M$ matrix of weights $\mathbf{W}$ is obtained by solving the following linear system (Daley, 1991):

$$\left(\mathbf{B} + \mathbf{O}\right)\mathbf{W} = \mathbf{b} \tag{7.3}$$

where $\mathbf{B}$ is the covariance matrix of background errors between points at observation locations, $\mathbf{O}$ is the covariance matrix of observation errors, and $\mathbf{b}$ is the error covariance vector between an observation i and a background value j.

A practical solution of equation 7.3 is obtained by reducing the number of observations that can influence the analysis at a given grid point to nearby stations (16 in practice). The horizontal correlations of errors are assumed to be homogeneous (they are only a function of the distance between two points, i.e. $\mathrm{cor}\left(r_i - r_j\right) = \mathrm{cor}\left(r_i - r_j\right)$) and isotropic (they do not depend upon direction, i.e. $\mathrm{cor}\left(i,j\right) = \mathrm{cor}\left(|r_i - r_j|\right)$. Following studies by Mitchell et al. (1990) for humidity analyses, Garand and Grassotti (1995) for satellite precipitation analysis, and Brasnett (1999) for snow depth, the covariances of background error are modelled by a second-order autoregressive function,

$$\mathbf{B}^{ij} = \sigma_b^2\left(1 + \frac{r_{ij}}{L}\right)\exp\left(-\frac{r_{ij}}{L}\right) \tag{7.4}$$

where r_{ij} is the distance between two points i and j, L is a characteristic horizontal scale, and σ_b is the standard deviation of background error (Daley, 1991). The matrix of observation errors is assumed to be diagonal (uncorrelated observation errors),

$$\mathbf{O}^{ij} = \sigma_o^2\delta_{ij} \tag{7.5}$$

where σ_o is the standard deviation of the observation error. The analysis is outlined by Mahfouf et al. (2007).

High resolution measurements of the spatial and temporal structure of precipitation using distrometers and raingauges (e.g. Hosking and Stow, 1987) and radar (e.g. Smith et al., 1994; Bradley and Smith, 1994) have provided detailed descriptions of the structure of rainfall patterns.

Real data sets are usually limited in some way or other. Often the length of record restricts direct statistical analysis of likely extremes, or the full range of precipitation types is not included. Raingauge records may suffer from sampling errors which, particularly for short duration rainfall, may introduce significant biases. Hence stochastic models of precipitation, some of which allow more than one cell type to be represented in the same storm, have been studied, examples being described by Krajewski and Georgakakos (1985), Cox and Isham (1988) and Cowpertwait (1994).

Zawadzki (1973) used ground-based radar data to analyse mesoscale convective rainfall fields statistically. He found that such fields were time homogeneous for periods shorter than 40 minutes, and that spatial fluctuations at a fixed time can be obtained from the statistics of fluctuations in time at a fixed location by converting time into space through the velocity of the storm system. This statistical relationship is known as *Taylor's hypothesis* (Taylor, 1938).

An alternative approach to the statistical description of precipitation may be characterized in terms of the *fractal dimension*, that is, unsmoothed shapes which show scaling behaviour (Lovejoy, 1982; Lovejoy and Mandelbrot, 1985). The property of *self-similarity* implies that, even in periods where the precipitation process seems to behave without major irregularities, if one changes the scale of observation one is able to detect fluctuations which look statistically similar to those of more erratic periods except for a factor of scales.

It is very important to develop realistic multidimensional models of precipitation if the true statistical description of mid-latitude systems is to be reproduced. Elliptical dimensional sampling and functional box counting are used to analyse radar rainfall data. Other statistical models have been developed (Kavvas and Herd, 1985) which reproduce both mesoscale and synoptic scale features of mid-latitude rainfall.

A plot of correlation coefficient can reveal evidence of the prevailing direction of rain-bearing winds and of the location of features in the landscape that are associated with atmospheric activity that generates precipitation. A schematic diagram showing the hypothetical variation of correlation coefficient is shown in Figure 7.4. The statistical three-dimensional description of a rainfall surface is given in Appendix 7.1.

7.3.2   *Temporal analyses*

Analysis of variations in precipitation with time at a specified location is undertaken in many ways depending upon the application. The mean values of annual, seasonal or monthly precipitation are only statistically useful when the probability distribution they follow is normally distributed. Over short periods such as a day, rainfall data are always positively skewed. Trends and oscillations in average precipitation over shorter periods are therefore hard to identify.

One way to smooth out variations in a time series, and to identify trends more obviously, is to use a *running mean*. The mean is taken over an odd number of values centred on the value of interest, i.e. over $(2n+1)$ where $n = 0, 1, 2, \ldots$. Significant and sustained shifts in precipitation can be demonstrated by accumulating the deviations about the period mean. Deviations above (positive) or below (negative) the period mean are accumulated through the data period, either as absolute amounts or as percentages of the mean precipitation.

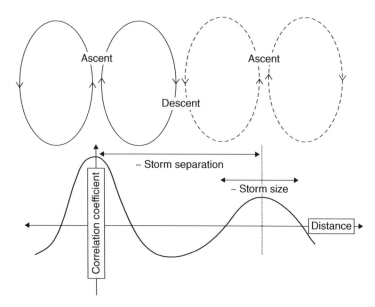

Figure 7.4 Schematic diagram of the hypothetical variation in correlation coefficient versus distance for long term average precipitation relative to one location if produced by randomly occurring convective storms over a moist, flat, featureless plain (from Shuttleworth, 2012)

An alternative approach is the *double mass curve*, which reveals the long term sustained trends. Here the analysis involves accumulating and plotting total precipitation throughout the data period. Any distinct and sustained change is revealed by a change in slope of the resulting graph. The accumulated precipitation total plotted for each value is given for the precipitation time series, as shown in the example in Figure 7.5. In this example the analysis is used to identify a raingauge significantly different from other nearby gauges.

7.3.3   Oscillations in precipitation

Precipitation time series may include periodicity at several frequencies. Identification of the periodicity can be made using serial autocorrelation, or harmonic (Fourier) analysis. This involves fitting a mathematical function F of time t to a time series $A(t)$, such that

$$F(t) = \bar{A} + \sum_{k=1}^{N} A_k \cos\left(kt - \phi_k\right) \tag{7.6}$$

where A_k and ϕ_k are the amplitude and phase assigned to the kth harmonic in the harmonic series so that it represents F. The maximum value of N is determined by the requirement that there is at least one full sinusoidal cycle of the corresponding term in the time period for which data are available. The contribution of each harmonic term to the total variance may be expressed as the ratio of the total variance. Note that not every harmonic may be associated with a physical process.

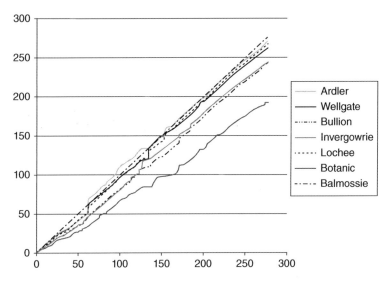

Figure 7.5   Double mass analysis for the accumulated rainfall in millimetres for several raingauge sites, confirming that there is a problem at the Botanic Gardens gauge (courtesy CityRainNet Project)

7.3.4   *Conditional probabilities*

Precipitation data commonly reveal seasonal dependency, and short term persistence due to the persistence in weather conditions. This persistence may be measured at a particular location and can be quantified using simple stochastic techniques. For example, a long time series of precipitation may be analysed to count the number of times a 'wet day' or 'rain day' follows a preceding period comprising a specified number of days with rain or a specified number of dry days. In this way a set of *conditional probabilities* is derived that characterize the level of persistence in the data record.

7.3.5   *Kriging*

Kriging is a statistical method that uses the *variogram* of the rainfall field, i.e. the variance between pairs of points at different distances apart, to optimize the raingauge weightings to minimize the estimation error (see for example Bastin et al., 1984). This uses optimal interpolation based on regression against observed values of surrounding data points weighted according to spatial covariance values. Ordinary kriging estimates the unknown value using weighted linear combinations of the available sample as follows (partly after Chao-yi Lang, Department of Computer Science, Cornell University, people.ece.cornell.edu/land/OldStudentProjects/cs490-94to95/clang/kriging.html):

$$\hat{v} = \sum_{j=1}^{n} w_j v \qquad \sum_{i=1}^{n} w_i = 1 \qquad\qquad (7.7)$$

where v is the unknown value and w_i and w_j are the weighted linear combinations of the available samples. The error of the ith estimate, r_i, is the difference between the estimated value and the true value at that same location:

$$r_i = \hat{v}_i - v_i \qquad (7.8)$$

The average error of a set of k estimates is:

$$m_{\bar{r}} = \frac{1}{k}\sum_{i=1}^{k} r_i = \frac{1}{k}\sum_{i=1}^{k}(\hat{v}_i - v_i) \qquad (7.9)$$

The error variance is:

$$\delta_R^2 = \frac{1}{k}\sum_{i=1}^{k}(r_i - m_R)^2 = \frac{1}{k}\sum_{i=1}^{k}\left[\hat{v}_i - v_i - \frac{1}{k}\sum_{i=1}^{k}(\hat{v}_i - v_i)\right]^2 \qquad (7.10)$$

Unfortunately, this equation cannot be used because the true values $v_1, \ldots, v_k$ are unknown. In order to solve this problem, a stationary random function is applied that consists of several random variables $V(x_i)$, where x_i is the location of the observed data for $i > 0$ and $i \leq n$, and n is the total number of observed data values. The unknown value at the location x_0 which is to be estimated is $V(x_0)$. The estimated value represented by a random function is:

$$\tilde{V}(x_0) = \sum_{i=1}^{n} w_i V(x_i) \qquad (7.11)$$
$$R(x_0) = \tilde{V}(x_0) - V(x_0)$$

The error variance is

$$\tilde{\delta}_R^2 = \tilde{\delta}^2 + \sum_{i=1}^{n}\sum_{j=1}^{n} w_i w_j \tilde{C}_{ij} - 2\sum_{i=1}^{n} w_i \tilde{C}_{i0} + 2\mu\left(\sum_{i=1}^{n} w_i - 1\right) \qquad (7.12)$$

$\tilde{\delta}^2$ is the covariance of the random variable $v(x_0)$ with itself, and it is assumed that all of the random variables have the same variance; μ is the Lagrange parameter (multiplier) (see for example Heath, 2005).

In order to get the minimum variance of error, the partial first derivatives of equation 7.12 are calculated for each w and the result is set to 0. For example, differentiation with respect to w_1 gives

$$\frac{\partial\left(\tilde{\delta}_R^2\right)}{\partial w_1} = 2\sum_{j=1}^{n} w_j \tilde{C}_{1j} - 2\tilde{C}_{10} + 2\mu = 0 \qquad (7.13)$$
$$\sum_{j=1}^{n} w_j \tilde{C}_{1j} + \mu = \tilde{C}_{10}$$

All of the weights w_i can be represented as

$$\sum_{j=1}^{n} w_j \tilde{C}_{ij} + \mu = \tilde{C}_{i0} \quad \text{for each } i, 1 \le i \le n \tag{7.14}$$

We can get each weight w_i through equation 7.14. When this value is obtained, the value located in x_0 can be estimated.

Some advantages of kriging are:

1. It helps to compensate for the effects of data clustering by assigning individual points within a cluster using less weight than for isolated data points (or, by treating clusters more like single points).
2. It can be used to generate a map of the standard error of the estimates that indicates where additional gauges would be most beneficial.
3. It gives estimates of the estimation error (the kriging variance), along with estimates of the linear regression estimator. (However, the error map is basically a scaled version of a map of distance to the nearest data point, so it is not unique.)
4. The availability of the estimation error provides a basis for the stochastic simulation of possible realizations of the estimated values.

7.3.6   Accuracy of the precipitation products

As described by Mahfouf et al. (2007), computing the standard deviations of the distributions of differences between radar, rain gauges and a *global environmental multiscale* (GEM) model, for non-zero values in log-transformed space, it is possible to assess the relative quality of these three independent precipitation estimates (rain gauge σ_o; GEM model σ_b; radar σ_r). Indeed, if $\sigma_b^2 + \sigma_r^2 = \lambda, \sigma_b^2 + \sigma_o^2 = \mu$ and $\sigma_r^2 + \sigma_o^2 = v$, then each individual variance is written as:

$$\sigma_b^2 = 0.5(\lambda + \mu - v)\sigma_r^2 = 0.5(\lambda - \mu + v)\sigma_o^2 = 0.5(-\lambda + \mu + v) \tag{7.15}$$

For example, using $\lambda = 0.995$ $(98,395)$, $\mu = 0.974(5179)$ and $v = 0.789(1715)$, based on the different sample sizes shown in brackets, the following values of errors are produced:

$$\sigma_b = 0.77, \quad \sigma_o = 0.62, \quad \sigma_r = 0.64 \tag{7.16}$$

This estimation of background error is compatible with an independent estimate $(\sigma_b = 0.71)$. The observation error is comparable to that deduced from the comparison between two collocated measurements $(\sigma_o = 0.52)$ for which the representativeness error (point value versus grid average) is not included. The estimates above indicate that radar and raingauges are of comparable accuracy, and both are slightly more accurate than the GEM model.

From the description of the SI method (section 7.3.1), three important parameters are required to specify the statistics of background and observation errors, namely the standard deviations of the background error σ_b, the observation error σ_o, and the

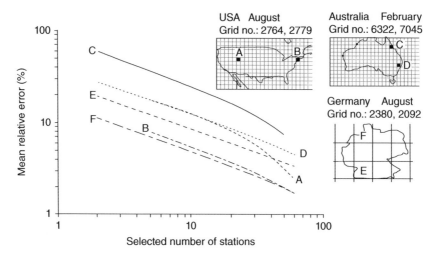

Figure 7.6   Mean relative error (%) as a function of the selected number of stations per grid box in Australia for February and in Germany and the United States for August 1987 (from Rudolf et al., 1984)

characteristic length scale L, of the correlation model defined by equation 7.4. A similar procedure is carried out when implementing a spatial interpolation using other geostatistical techniques such as kriging. Figure 7.6 shows an example of the dependency of the mean relative error calculated for 2.5° grid boxes of monthly rainfall on the number of raingauge stations in Australia, Germany and the United States. A limit of 1 mm is set for the mean absolute monthly error in order to remove very low values. This analysis is from the Global Precipitation Climatology Centre (Rudolf et al., 1994).

7.4   Model storms for design

Design storms are precipitation patterns defined for use in the design of a hydrologic system. This can be defined by a hyetograph (time distribution of rainfall) or an isohyetal map (spatial distribution of rainfall). As a result of varying storm rainfall profiles combined with different initial catchment wetness, all design storms do not produce floods having the same frequency of occurrence. Figure 7.7 shows design profiles for storms with 90 and 75 percentile points of profile peakedness as contained in the UK Natural Environment Research Council *Flood Studies Report* (*FSR*), vol. 1 (NERC, 1975).

The *Flood Estimation Handbook* (*FEH*), vol. 1 (Institute of Hydrology, 1999) describes two approaches to flood frequency estimation, one based upon statistical analysis of flood peak data and the other based upon the *FSR* rainfall–runoff method. In order to combine considerations of areal rainfall depths over a range of areas and varying durations of heavy falls, a more complex analysis of storm patterns is required. The technique of depth–area–duration (DAD) analysis enables the maximum falls of rain for different durations over a range of areas to be determined.

Another approach is *extreme value analysis*. A series of extreme values is constructed by selecting the highest (or lowest) observation from a set of data. An extreme value

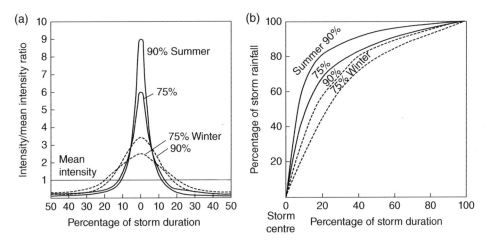

Figure 7.7 Design profiles for storms with 90 and 75 percentile points of profile peakedness (summer and winter) (*Flood Studies Report*, vol. 1, NERC, 1975)

distribution is usually expressed in terms of the cumulative distribution function $F(x)$ plotted against x, giving an S-shaped curve. $F(x)$ is usually transferred to a new variable y known as the *reduced variate* so that the cumulative probability distribution may be represented by a straight line.

A general solution to the extreme value problem was described by Jenkinson (1955) as

$$X = x_0 + \frac{\alpha\left[1 - exp\left(-ky\right)\right]}{k} \tag{7.17}$$

where y is defined from $F(x) = exp\left[-exp\left(-y\right)\right]^3$. The curves represented by equation 7.17 depend upon the value of k. Figure 7.8 show three curves for values of $k = 0, k < 0$ and $k > 0$. These curves are known as *Fisher–Tippett distribution* types I, II and III respectively (Fisher and Tippett, 1928). The type I distribution is often known as a *Gumbel distribution* (Gumbel, 1958) (see Appendix 7.2). Annual extremes of daily and hourly rainfall are usually best fitted by a type II distribution, but there may be significant departures due to sample size, precipitation type, storm organization and movement, and topographic influences.

7.5   Approaches to estimating flood frequency

There are three principal methods of estimating flood rarity:

1. comparison of the recorded peak discharge with a measured flood frequency curve, i.e. the rainfall–runoff method as in the UK *Flood Studies Report* (NERC, 1975)
2. statistical analysis of flood peak data and flood frequency curves, as in the UK *Flood Estimation Handbook* (Institute of Hydrology, 1999)
3. relating the return periods of the flood, the effective rainfall and the soil moisture deficit (Clark, 1995; 2007).

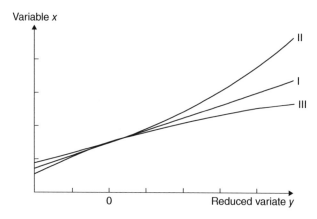

Figure 7.8 Fisher–Tippett distributions types I, II and III

Table 7.1 Flood frequency terminology

Annual probability (%)	Annual chance
10	1 in 10
5	1 in 20
2	1 in 50
1.3	1 in 75
1	1 in 100
0.5	1 in 200
0.1	1 in 1000

The choice of which method to use is important as the estimate produced governs the size and feasibility of a flood alleviation scheme, and may determine the risk of flooding. In the *FEH* it is suggested that a gauged record as long as the target return period is required, which is usually not practical. The statistical approach requires that data are pooled from a group of catchments judged to be similar in characteristics, but this may be difficult to achieve. In the case of the *Flood Studies Report* the results may be uncertain due to the assumptions made in the design storm. Finally a water balance approach depends upon knowledge of runoff volume, soil moisture deficit and percolation, all of which are unlikely to be known accurately. The flood producing rainfall is derived from a back analysis using a non-linear flow model based upon a unit hydrograph approach. In spite of the difficulties that arise from the estimate of the flood producing flow, and its relationship to the storm rainfall, estimates can be made with some confidence by comparing the results derived from historic events.

Floods have the same probability of occurrence each year. A risk-based approach has largely replaced the traditional and potentially misleading term 'return period', as shown in Table 7.1. However, area-average precipitation for storms subdivided on the basis of either return period or duration are still employed (see for example Niemczynowicz, 1982). Point–area relationships may be extended by deriving average *areal reduction factors* (ARFs) as shown in Figure 7.9.

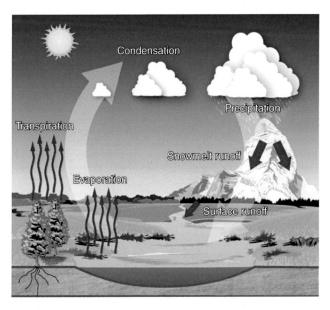

Figure 1.1 Simplified representation of the hydrological cycle (NWS Jetstream NOAA, USA, www.srh.noaa.gov/jetstream/atmos/hydro.htm)

Hydrometeorology, First Edition. Christopher G. Collier.
© 2016 John Wiley & Sons, Ltd. Published 2016 by John Wiley & Sons, Ltd.
Companion website: www.wiley.com/go/collierhydrometerology

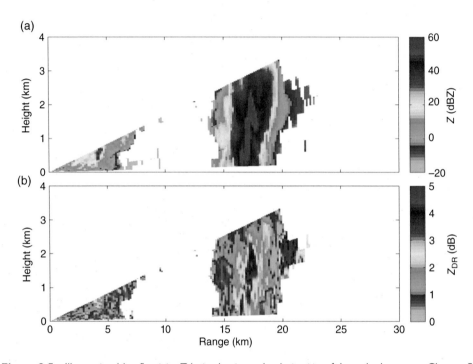

Figure 2.5 Illustrating (a) reflectivity Z (using horizontal polarization of the radar beam, see Chapter 5) and (b) differential reflectivity Z_{DR} (the ratio of horizontal and vertical polarizations of the radar beam, see Chapter 5) reconstructed from measurements made at constant elevation angles by the NCAS mobile X-band dual polarization Doppler radar on 3 August 2013 at 1327 UTC (from Blyth et al., 2015)

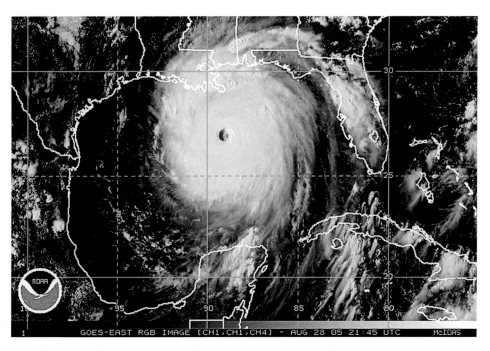

Figure 2.11 Hurricane Katrina over the Gulf of Mexico at its peak, 28 August 2005; category 5, highest winds 175 mph (280 km h⁻¹), lowest pressure 902 mbar (hPa)

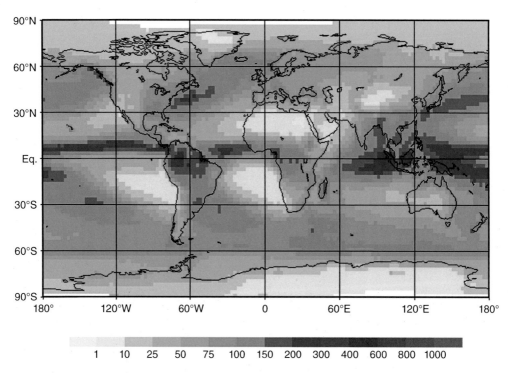

Figure 2.14 Globally averaged annual precipitation 1980–2004 Jan–Dec (mm per month) from the Global Precipitation Climatology Project (GPCP) Version 2 (source: GPCC Visualizer)

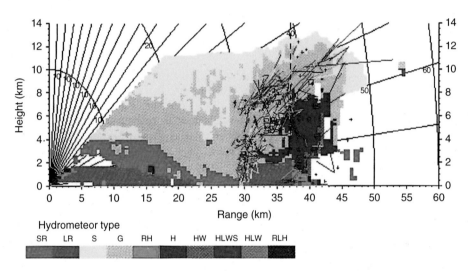

Figure 5.13 Vertical section of the supercell storm of 21 July 1998 at 1741 UTC along 232° azimuth from the POLDIRAD radar, DLR, Germany. ITF flashes are plotted for a 30s interval around the nominal scanning time. Hydrometeor or particle colour codes: SR, small rain; LR, large rain; S, snow; G, graupel; RH, rain–hail mixture; H, hail; HW, hail wet; HLWS, hail large wet spongy; HLW, hail large wet; RLH, rain/large-hail mixture) (from Meischner et al., 2004)

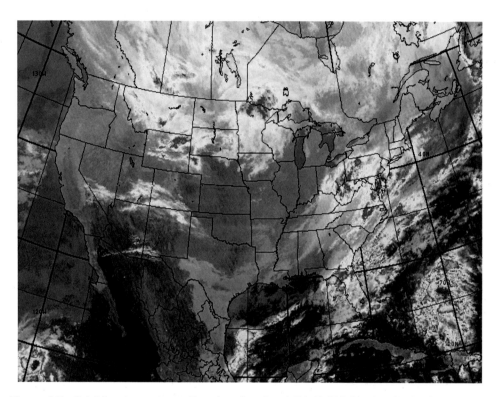

Figure 6.7 Rainfall estimates from infrared satellite data, 1630 GMT 8 October 2012: the US satellite images displayed are infrared (IR) images. Warmest (lowest) clouds are shown in white; coldest (highest) clouds are displayed in shades of yellow, red and purple. Imagery is obtained from the GOES and METEOSAT geostationary satellites, and the two US Polar Orbiter (POES) satellites. POES satellites orbit the Earth 14 times each day at an altitude of approximately 520 miles (870 km) (courtesy WSI Corporation)

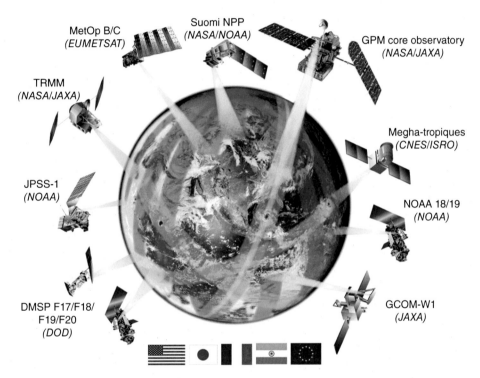

Figure 6.11 Global Precipitation Mission (GPM) Constellation. Suomi NPP, National Polar-orbiting Partnership (2011–); JPSS-1, NOAA Joint Polar Satellite System (2017–); GCOM-W1, Global Change Observation Mission (2012–); Megha-Tropiques, French Water Cycle and Energy Budget in the Tropics (2011–) (see also Table 6.1) (from pmm.nasa.gov/GPM)

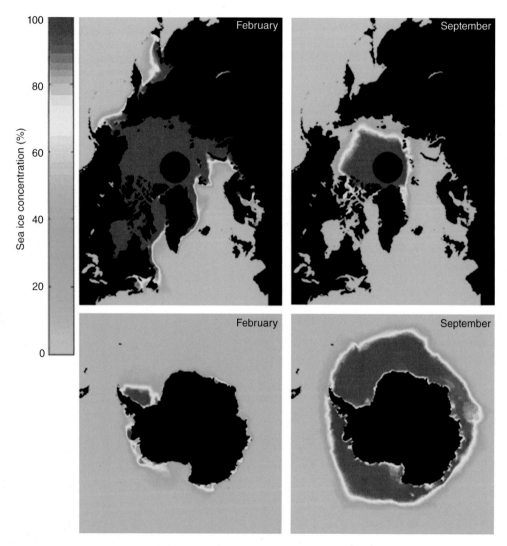

Figure 6.13 Arctic (top panels) and Antarctic (bottom panels) sea ice concentration climatologies from 1978 to 2002: approximate seasonal winter maximum (February in the Arctic and September in the Antarctic) and minimum (September in the Arctic and February in the Antarctic) levels (image provided by National Snow and Ice Data Center, University of Colorado, Boulder, CO)

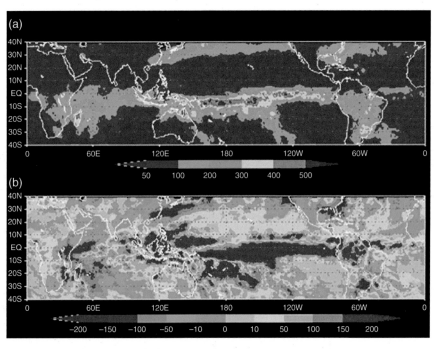

Figure 6.16 (a) TRMM monthly mean rainfall (mm) January and February 1998 and (b) the difference in two-month mean rainfall between 1998 and 1999 (GSFC, NASA)

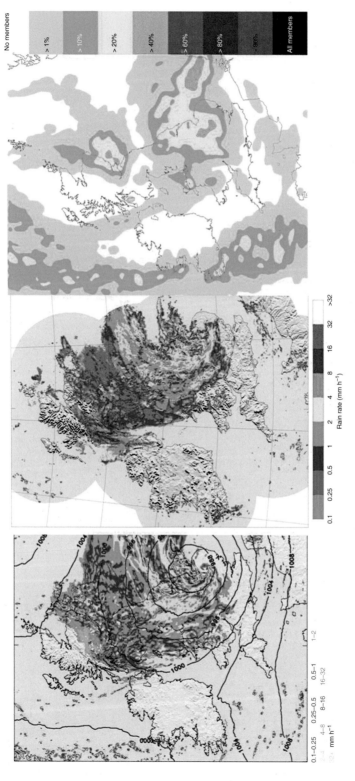

Figure 8.6 Sample of corresponding precipitation forecast products from UKV (left) and MOGREPS-UK (right), along with the verifying radar image (centre), for 0400Z on 25 November 2012. The MOGREPS-UK chart illustrates the 'chance of heavy rain in the hour', which is defined as 4 mm h⁻¹; members are the individual forecasts within the ensemble (from Mylne, 2013)

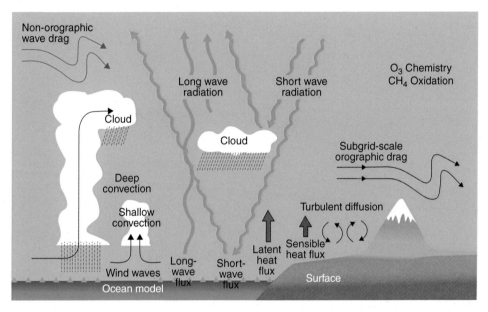

Figure 8.8 Schematic diagram of the different physical processes represented in the ECMWF IFS model (from ECMWF IFS Documentation Cy38r1)

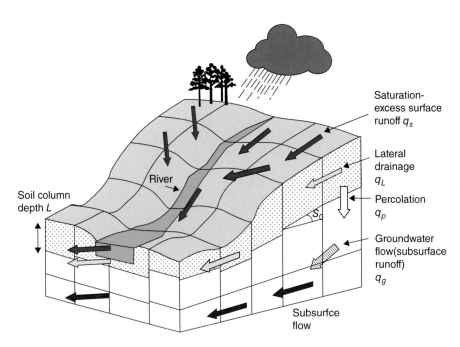

Saturation-
excess surface
runoff q_s

Lateral
drainage
q_L

Percolation
q_p

Groundwater
flow(subsurface
runoff)
q_g

River

Soil column
depth L

S_o

Subsurfce
flow

Figure 9.5   The grid-to-grid model structure (source CEH)

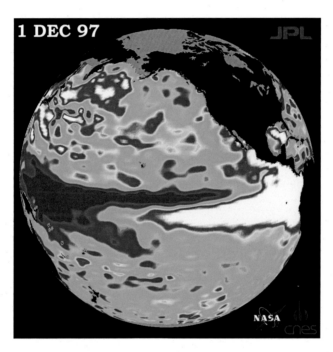

Figure 11.1   The 1997 El Niño observed by TOPEX/Poseidon. The white areas off the tropical coasts of South and North America indicate the pool of warm water (NASA and CNES)

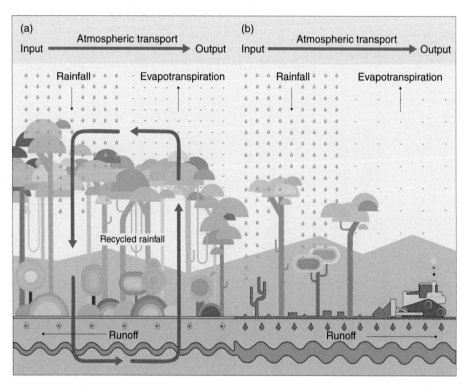

Figure 13.11 Effects of deforestation on rainfall in the tropics. (a) Much of the rainfall over tropical forests comes from water vapour that is carried by the atmosphere from elsewhere. However, a large component is 'recycled' rain, that is water pumped by trees from soil into the atmosphere through evapotranspiration. Water exits from forests either as runoff into streams and rivers, or as vapour that is carried away by the atmosphere. The atmospheric transport of water vapour into the forest is balanced by exit of water in the form of vapour and runoff. (b) The analysis of Spracklen et al. (2012) suggests that deforestation reduces evapotranspiration and so inhibits water recycling. This decreases the amount of moisture carried away by the atmosphere, reducing rainfall in regions to which the moisture is transported. Decreasing evapotranspiration may also increase localized runoff and raise river levels (from Aragao, 2012)

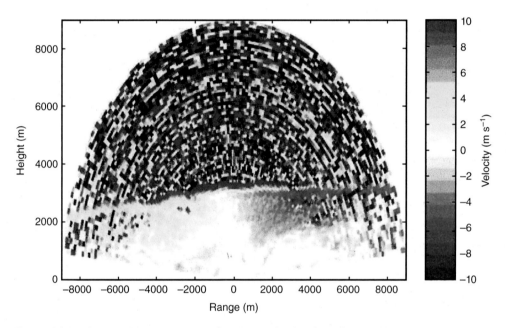

Figure 14.1 A vertical cross-section across the rural–urban boundary in West London, illustrating variation of the cloud base at the top of the boundary layer measured by Doppler lidar. The left-hand side of the image is over the rural area to the west, and the right-hand side of the image is over the urban area to the east. The colours represent the strength of Doppler radial velocities in m s^{-1} towards (positive) and away from (negative) the lidar. Above the boundary layer the colours represent noise in the measurements (from Collier et al., 2005)

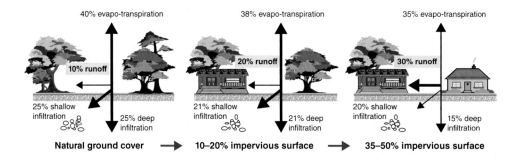

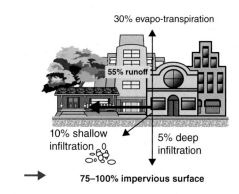

Figure 14.2 How impervious cover affects the water cycle (from California Water Land Use Partnership, WALUP: www.coastal.ca.gov/nps/watercyclefacts.pdf)

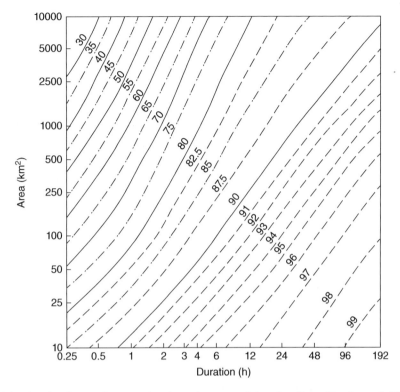

Figure 7.9   Areal reduction factors related to rain area and duration (from Shuttleworth, 2012, after Sumner, 1988 and Rodda et al., 1976)

Generally confidence in the flood rarity calculations can be derived from use of the different procedures. We begin first with an approach for estimating extreme rainfall and floods which has been used all over the world with some success, namely the *probable maximum precipitation* (PMP) and *probable maximum flood* (PMF) approach. We then move on to the statistical approach represented by the *Flood Estimation Handbook* which is used for all flood occurrences.

7.6   Probable maximum precipitation (PMP)

Probable maximum precipitation (PMP) is defined as 'the theoretical greatest depth of precipitation for a given duration that is physically possible over a given size of storm area at a particular geographic location at a certain time of the year' (Hansen et al., 1982). Structures to retain or deliver water must be designed to pass extreme flood events without significant damage or failure. Design criteria may be based upon probable maximum flood (PMF) arising from the PMP.

Empirical formulae represent local and world maximum values of precipitation (WMO, 1986). However these values usually apply to tropical or semi-tropical regions, and do not apply to colder regions. The values of PMP may be given by

$$R = 422D^{0.475} \tag{7.18}$$

where D is storm duration in hours and R is total point rainfall accumulated in millimetres. In evaluating PMP for specific catchments the following steps are involved (WMO, 1973):

1. moisture maximization of observed extreme precipitation amounts
2. transposition to the location of interest
3. envelopment of the maximized transposed depth–duration and depth–area amounts.

The transposition of storms requires that storm records have similar meteorological characteristics, similar orography, and mountains which act as barriers to inflowing moisture. Transposition works best in flat areas where orographic effects can be neglected. The process of adjusting rainfall totals for orography requires the separation of orographic and non-orographic components of rainfall, as described in WMO (1986). Regionalization may be required which involves setting limits to which storms can be transposed.

A stochastic storm transposition technique has been proposed by Foufoula-Georgiou (1989) in which an elliptical storm shape and spread function are used to define extreme events in the US Midwest. This approach has been used in many countries and has led to the derivation of the following formula:

$$PMP = FAFP\left[\frac{T}{C} - m^2\left(\frac{T}{C} - 1\right)\right] \tag{7.19}$$

where FAFP is the free atmosphere forced precipitation; m is the storm intensification factor, which is the ratio of the storm core rainfall to the total rainfall in the storm (this is a measure of the concentration of the rainfall in the core phase of the storm, and usually has a value of around 0.6); T is the 24 hour rainfall of the 100 year return period (or other duration) rainfall; and C is the convergence component of T. Factors which contribute to the FAFP are convergence, forcing from mesoscale weather systems, fronts and baroclinic waves, and condensation efficiency.

An alternative method of computing PMP was proposed by Chow (1951) in which PMP is related to the statistical characteristics of a series of annual maxima of rainfall as follows:

$$X_m = X_n + K_m S_n \tag{7.20}$$

where X_m is the maximum observed rainfall; X_n is the mean of a series of n annual maxima; S_n is the standard deviation of the series; and K_m is the number of standard deviations needed to obtain X_m, and is a function of the length of observational record. The greatest value of K_m to obtain X_m was 15; however the drier the area, the higher the value of K.

Collier and Hardaker (1996) described a method of estimating PMP based upon the use of a storm model The model uses a parameter storm efficiency E, which is the ratio of rainfall to the amount of precipitable water in the representative air column during the storm:

$$E = \frac{\text{total rainfall at ground (cell relative)}}{\text{total cloud water condensed}} \tag{7.21}$$

The total cloud water condensed (precipitable water) is evaluated from radiosonde data, assuming that the air mass is saturated. One problem with estimating E is that a good estimate of the dew point temperature is needed. A small error in the estimate leads to a rather large error in converting to the total condensed cloud water and hence the storm efficiency. The storm model described by Collier and Hardaker (1996) requires as input, among other variables, the maximum storm efficiency for the duration of the event. The model generates PMP values very similar to the *FSR* for storm durations less than 11 hours, but for durations between 11 and around 24 hours, the *FSR* PMP values are exceeded.

7.7   Probable maximum flood (PMF)

The derivation of the probable maximum flood (PMF) requires that the shape of the whole flood hydrograph is used, and therefore the unit hydrograph approach is often adopted. However, extreme floods may be caused by a combination of hydrological factors and therefore the procedure may require some adjustment as follows:

1. A storm profile of the greatest peakiness should be used, allowing for the maximum value of percentage runoff for the particular catchment.
2. Snowmelt at a uniform rate should be added to the precipitation.
3. The time to peak of the unit hydrograph should be reduced by one third, causing the peak flow rate to rise by 50%. Any significant base flow should be added to the final flood hydrograph.

The recommended method for producing a maximum design flood hydrograph remains as described in the *Flood Studies Report*. However, the *Flood Estimation Handbook* does not support this approach as it relies on supposition of what unfavourable conditions might conceivably occur. Unless there is an exceptionally rich source of historical or paleoflood data then statistical analysis of peak flows should play a minor supporting role in estimating PMF. Nevertheless, a statistically derived flood frequency curve may imply an upper bound as shown in Figure 7.10.

7.8   *Flood Studies Report* (FSR)

The *Flood Studies Report* (NERC, 1975) comprises five volumes: I *Hydrological Studies*, II *Meteorological Studies*, III *Flood Routing Studies*, IV *Hydrological Data*, and V *Maps*. Designs may be based upon rainfall events with a frequency of between once in 2 years and once in 100 years. A notation is used throughout to denote return period, e.g. 2M means twice in 1 year, M2 means once in 2 years, M100 means once in 100 years. The extreme rainfall events are analysed by considering only the annual maximum values. Ordered sets of annual maxima are divided into four quartiles and identified by the quartile means in a way which is in good agreement with extreme value theory. The first quartile mean is identified with 2M, and the fourth quartile mean with M10. The four highest values of the ordered data set of annual maxima are used to estimate with less confidence the rarer events up to Mn, where n is the number of years of record.

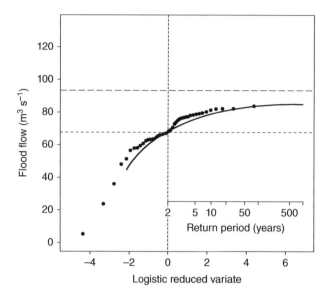

Figure 7.10   Single site flood frequency curve for the River Thames at Eynsham, fitted by the method of L-moments (a robust way of summarizing a distribution: see for example Hosking and Wallis, 1997) using a generalized logistic (GL) distribution; the broken line is the upper bound of the fitted distribution (from Institute of Hydrology, 1999)

Taking the rainfall totals for 2 hours to 25 days at stations in England and Wales with M5 between 30 and 300 mm, and plotting the logarithm of the return period against the logarithm of the rainfall amount, a series of smooth curves can be drawn as shown in Figure 7.11. The ratio of MT to M5, where T may have any value, is referred to as the *growth factor*.

The report contains maps of the geographical distribution of M5 for 60 minute, 2 day and 25 day rainfall, and indicates how these amounts may be related to other durations. Areal reduction factors and depth–duration relationships are derived, and storm intensity profiles with time for summer and winter are given.

The hydrological studies involved the building up of flood region curves by combining sets of stations in regions and countrywide. The approach of using a series of peaks over a threshold with an assumed distribution was adopted as an alternative to the estimation of floods from the annual maximum series. A 'shot noise' model was used for a time series approach to flood frequency analysis.

Where no records exist at a site, the mean annual flood Q may be derived from the following equation:

$$Q = 0.0201\,\mathrm{AREA}^{0.94}\,\mathrm{STMFRQ}^{0.27}\,\mathrm{S1085}^{0.16}\,\mathrm{SOIL}^{1.23}\,\mathrm{RSMD}^{1.03}\,(1+\mathrm{LAKE})^{-0.85}\qquad(7.22)$$

where AREA is catchment area (km²); STMFRQ is stream frequency (junctions km⁻²); S1085 is 10–85% stream slope (m km⁻¹); SOIL is soil index; RSMD is 1 day rainfall of 5 year return period less effective mean soil moisture deficit; and LAKE is fraction of catchment draining through a lake or reservoir. Q may also be estimated from a *peaks over threshold* (POT) series and from the annual maxima series or using unit hydrograph methods.

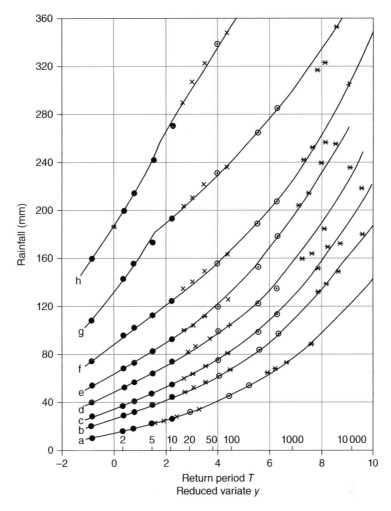

Figure 7.11   Rainfall growth curves, which provide a multiplier of the index rainfall. Designs are commonly based upon rainfall events with a frequency of between once in 2 years and once in 100 years, denoted by a return period of M2 for once in 2 years, M100 for once in 100 years etc. The curves show the growth curves for particular rainfall amounts as follows: for England and Wales: (a) 2 hour rainfall, (b) 1 day rainfall for M5 30–40 mm, (c) 2 day rainfall for M5 40–50 mm, (d) 1 day rainfall for M5 60–75 mm, (e) 4 day rainfall for M5 75–100 mm, (f) 8 day rainfall for M5 100–150 mm, (g) 25 day rainfall for M5 150–200 mm, (h) 8 day rainfall for M5 200–300 mm (NERC, 1975)

The time to peak in hours of the unit hydrograph may be estimated from

$$T_{\mathrm{p}} = 46.6\mathrm{MSL}^{0.14}\mathrm{S1085}^{-0.38}\left(1+\mathrm{URBAN}\right)^{-1.99}\mathrm{RSMD}^{-0.4} \qquad (7.23)$$

where MSL is mean stream length and URBAN is the fraction of catchment in urban development. The duration D of the design storm depends on T_{p} and the mean annual rainfall SAAR:

$$D = \left(1+\mathrm{SAAR}/1000\right)T_{\mathrm{p}} \qquad (7.24)$$

Extreme value distributions are discussed to describe on purely theoretical grounds how annual maximum floods are distributed.

7.9   *Flood Estimation Handbook* (FEH)

The *Flood Estimation Handbook* (Institute of Hydrology, 1999) comprises five volumes: I *Overview*, II *Rainfall Frequency Estimation*, III *Statistical Procedures for Flood Frequency Estimation*, IV *Restatement and Application of the Flood Studies Report Rainfall–Runoff Method*, V *Catchment Descriptors*. Figures 7.12 and 7.13 compare the rainfall frequency analyses carried out in the *FEH* and the *FSR* respectively. The *FEH* rainfall frequency analysis is based on annual maximum rainfalls, i.e. the largest rainfalls observed in each year at a site. The two key parts of the analysis are the mapping of an *index variable* and the derivation of *growth curves*. The index variable was chosen to be the median of annual maximum rainfalls at a site, RMED, whereas the index variable used in the *FSR* was the rainfall with a 5 year return period, M5.

Rainfall growth curves in the *FEH* are derived using a procedure called the *focused rainfall growth curve extension* (FORGEX). This was applied on a 1 kilometre grid, and the resulting grids of growth rates multiplied by the grids of RMED give design rainfalls. A six-parameter depth–duration–frequency model was developed and fitted to all rainfall durations. An example is shown in Figure 7.14. Maps may

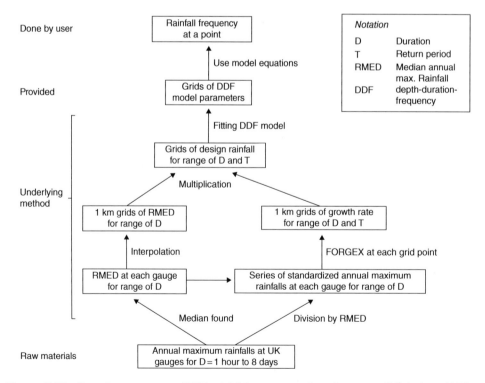

Figure 7.12   Flow chart summary of *FEH* rainfall frequency analysis (Institute of Hydrology, 1999)

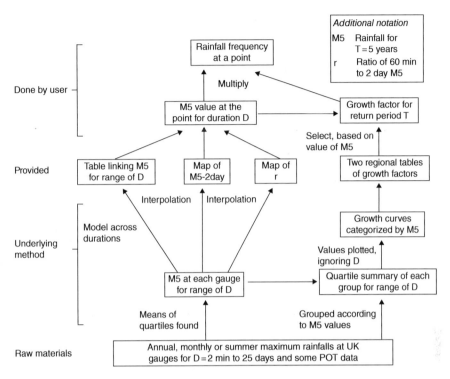

Figure 7.13   Flow chart summary of *FSR* rainfall frequency analysis (NERC, 1975)

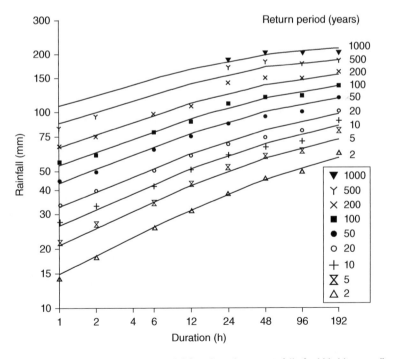

Figure 7.14   Depth–duration–frequency model fitted to design rainfalls for Waddington (Institute of Hydrology, 1999)

then be produced of rainfall frequency on a 1 kilometre grid covering the United Kingdom.

Flood frequencies (QMED) are estimated from flood annual maxima if the record length is 14 years or longer. For shorter record lengths QMED is estimated from flood data abstracted in peaks over threshold (POT) form. However problems arise when there are no flood peak data. In this case QMED is estimated from catchment descriptors:

$$\text{QMED}_{\text{rural}}$$
$$= 1.172 \text{AREA}^{\text{AE}} \left(\text{SAAR}/1000\right)^{1.560} \text{FARL}^{2.642} \left(\text{SPRHOST}/100\right)^{1.211} 0.0198^{\text{RESHOST}} \qquad (7.25)$$

Here AE is the area exponent, given by

$$\text{AE} = 1 - 0.015 \ln\left(\text{AREA}/0.5\right) \qquad (7.26)$$

where AREA is the catchment drainage area (km²); SAAR is the standard average annual rainfall 1961–90 (mm); FARL is the index of flood attenuation due to reservoirs and lakes; SPRHOST is the standard percentage runoff derived from HOST soils data; and RESHOST is a residual soils term (linked to soil responsiveness). It must be stressed that this model only applies to rural UK catchments.

Appendix 7.1   Three-dimensional description of a rainfall surface

Given a network of rainfall observations, the (x, y, z) surface can be described using polynomials, Fourier series or multi-quadratics. The equation for the multi-quadratics rainfall surface (Shaw and Lynn, 1972) is given by

$$z = \sum_{i=1}^{n} c_i \left[\left(x - x_i\right)^2 + \left(y - y_i\right)^2 \right]^{1/2} = \sum_{i=1}^{n} c_i a_{(x_i, y_i)} \qquad (7.27)$$

where n is the number of data points (x_i, y_i). There are n coefficients that need estimation; they are obtained from the n equations for z_1 to z_n, which in matrix form are

$$\mathbf{Z} = \mathbf{CA} \qquad (7.28)$$

where $\mathbf{Z}$ and $\mathbf{C}$ are row vectors and $\mathbf{A}$ is a square matrix with elements $a_{(x_i, y_i)}$. Multiplying by the inverse of $\mathbf{A}$ gives the coefficients

$$\mathbf{C} = \mathbf{ZA}^{-1} \qquad (7.29)$$

The equation for z conditioned by these c_i fits all the data points exactly. The volume of rainfall is obtained by integrating over the area of interest.

Appendix 7.2   Gumbel distribution

The probability density function for the Gumbel distribution (section 7.4) is

$$f(x) = \frac{1}{\beta} \exp\left[-z - \exp(-z)\right] \tag{7.30}$$

where $z = (x - \mu)/\sigma$; μ is the location parameter, σ is the distribution scale ($\sigma > 0$), and β is a scale parameter. The density function is skewed to the right and has its maximum at $x = \xi$. The cumulative Gumbel distribution probabilities can be obtained from

$$F(x) = \exp\left[-\exp\left(-\frac{x - \xi}{\beta}\right)\right] \tag{7.31}$$

where $\beta = \sqrt{(6)}/\pi$ and $\xi = x - \gamma\beta$. Here $\gamma = 0.57721...$ is Euler's constant.

Summary of key points in this chapter

1. The production of a gridded field of precipitation is a difficult problem because of the wide range of spatial and temporal scales present.
2. The quality of a precipitation analysis can vary significantly according to data availability.
3. Point measurements are often a poor representation of area-average precipitation.
4. The isohyetal method of estimating areal precipitation involves constructing isohyets across a catchment based upon measured values. Depth–area relationships for different durations may be derived.
5. The triangle method involves constructing a network of triangles as near equilateral as possible, with one of the gauges in the network at each apex of each triangle.
6. The Thiessen method comprises constructing a network of polygons by drawing the perpendicular bisector of the line joining each pair of gauges.

   Statistical interpolation, also known as optimal interpolation, is often used. Given a set of N precipitation observations y_a^i and a background field x_b^i available on a grid of M points, an analysis x_a^i at the model grid point j is then $x_a^i = x_b^i + \Sigma w^{ij}(y_a^i - y_b^i)$, where y_b^i is the background field spatially interpolated to the observation point i and the summation is over $i = 1$ to N.

   The matrix of observation errors is assumed to be diagonal (uncorrelated observation errors), $\mathbf{O}^{ij} = \sigma_o^2 \delta_{ij}$, where σ_o is the standard deviation of the observation error.
7. The length of records restricts direct statistical analysis of likely extremes.

8. Taylor's hypothesis is the conversion of time into space through the velocity of a storm system.

9. The fractal dimension of unsmoothed shapes shows scaling behaviour, i.e. if one changes the scale of an observation, one is able to detect fluctuations which look statistically similar to those of more erratic periods except for a factor of scale.

10. A plot of correlation coefficient can reveal evidence of the prevailing direction of rain-bearing winds and the location of features in the landscape that are associated with atmospheric activity that generates precipitation.

11. A double mass curve reveals long term sustained trends.

12. Precipitation time series may include periodicity at several frequencies.

13. Kriging is a statistical method that uses the variogram of the rainfall field, i.e. the variance between pairs of points at different distances apart, to optimize the raingauge weightings to minimize the estimation error.

14. It is possible to assess the relative quality of three independent precipitation estimates in log-transformed space.

15. Design storms are precipitation patterns defined for use in the design of a hydrologic system. They are defined by a hyetograph (time distribution of rainfall) or an isohyetal map (spatial distribution of rainfall).

16. Flood frequency estimation may be based upon statistical analysis of flood peak data or upon rainfall–runoff procedures. Extreme value analysis may also be used.

17. Extreme value analysis curves are known as Fisher–Tippett distribution types I, II and III; type I is often known as a Gumbel distribution.

18. A risk-based approach has largely replaced 'return period' analyses.

19. Point–area relationships may be extended by deriving average areal reduction factors.

20. Probable maximum precipitation (PMP) is defined as the theoretical greatest depth of precipitation for a given duration that is physically possible over a given size of storm area at a particular geographic location at a certain time of the year.

21. The storm efficiency parameter is the total rainfall at the ground (cell relative) divided by the total cloud water condensed (precipitable water).

22. Probable maximum flood (PMF) involves using the shape of the whole flood hydrograph with the PMP.

23. Rainfall growth curves may be derived by taking the ratio of rainfall frequency events at different return periods, e.g. once in 2 years to once in 100 years.

24. Flood frequencies (QMED) may be estimated from flood annual maxima if the record length is 14 years or longer.

Problems

1. Explain the isohyetal and triangle methods of estimating areal precipitation.
2. Reproduce a graph of areal rainfall (mm) versus area (km^2) for 6 hour and 24 hour durations.
3. Give examples of statistical interpolation.
4. Given a set of N precipitation observations, what is the statistical interpolation technique correction at a given grid point?
5. What limits the statistical analysis of likely extremes?
6. What is Taylor's hypothesis?
7. Define fractal dimension.
8. What is a double mass curve and what is it used for?
9. Give a formula for a mathematical function illustrating oscillations in a time series.
10. What are conditional probabilities?
11. Describe kriging.
12. Describe how it is possible to assess the relative quality of three independent precipitation estimates.
13. Define a design storm and describe how it is used.
14. Outline extreme value analysis, noting the Fisher–Tippett and Gumbel distributions.
15. What are three principal methods of estimating flood rarity?
16. Why is 'return period' a misleading term?
17. Define probable maximum precipitation (PMP) and probable maximum flood (PMF).
18. Give a relationship between total point rainfall and storm duration, and give an expression for PMP as a function of atmospheric forced precipitation and storm intensification factor.
19. What are rainfall growth curves?
20. Give an expression for flood frequency (QMED).

References

Bastin, G., Lovert, B., Duque, C. and Gevers, M. (1984) Optimum estimation of the average areal rainfall and optimal selection of raingauge locations. *Water Resour. Res.*, 20, 463–470.

Bhargava, M. and Danard, M. (1994) Application of optimum interpolation to the analysis of precipitation in complex terrain. *J. Appl. Meteorol.*, 33, 508–518.

Bradley, A.A. and Smith, J.A. (1994) The hydrometeorological environment of extreme rainstorms in the Southern Plains of the United States. *J. Appl. Meteorol.*, 33, 1418–1431.

Brasnett, B. (1999) A global analysis of snow depth for numerical weather prediction. *J. Appl. Meteorol.*, 38, 726–740.

Bussières, N. and Hogg, W. (1989) The objective analysis of daily rainfall by distance weighting schemes on a mesoscale grid. *Atmos. Ocean*, 27, 521–541.

Chow, V.T. (1951) A general formula for hydrologic frequency analysis. *Trans. Am. Geophys. Union*, 32, 231–237.

Clark, C. (1995) New estimates of probable maximum precipitation in South West England. *Meteorol. Appl.*, 2, 307–312.

Clark, C. (2007) Flood risk assessment. *Int. Water Power Dam Constr.*, 59, 22–30.

Collier, C.G. and Hardaker, P.J. (1996) Estimating probable maximum precipitation using a storm model approach. *J. Hydrol.*, 183, 277–306.

Cowpertwait, P.S. (1994) A generalized point process model for rainfall. *Proc. R. Soc. Lond. A*, 447, 23–37.

Cox, D.R. and Isham, V. (1988) A simple spatial-temporal model of rainfall. *Proc. R. Soc. Lond. A*, 415, 317–328.

Creutin, J.D., Delrieu, G. and Lebel, T. (1988) Rain measurement by raingauge radar combination: a geostatistical approach. *J. Atmos. Oceanic Technol.*, 5, 102–115.

Crossley, A.L. (2001) Estimation of daily rainfall accumulations by merging radar and rain gauge data for near real time use. Forecasting Research Technical Report 365, Met Office, Bracknell.

Daley, R. (1991) *Atmospheric Data Analysis*. Cambridge University Press, Cambridge.

Finkelstein, P.L. (1984) The spatial analysis of acid precipitation data. *J. Clim. Appl. Meteorol.*, 23, 52–62.

Fisher, R.A. and Tippett, L.H.C. (1928) Limiting forms of the frequency distribution of the largest or smallest members of a sample. *Proc. Cam. Phil. Soc.*, 24, 180–190.

Foufoula-Georgiou, E. (1989) A probabilistic storm transposition approach for estimating exceedance probabilities of extreme precipitation depths. *Water Resour. Res.*, 25, 799–815.

Garand, L. and Grassotti, C. (1995) Toward an objective analysis of rainfall rate combining observations and short-term forecast model estimates. *J. Appl. Meteorol.*, 34, 1962–1977.

Gumbel, E.J. (1958) *Statistics of Extremes*. Columbia University Press, New York.

Hansen, E.M., Schreiner, L.C. and Miller, J.F. (1982) Application of probable maximum precipitation estimates, United States east of the 105th meridian. Hydrometeorological Report 52, National Weather Service, US Department of Commerce, Silver Spring, MD.

Heath, M.T. (2005) *Scientific Computing: An Introductory Survey*. McGraw-Hill, Columbus, OH.

Hosking, J.G. and Stow, C.D. (1987) Ground-based high resolution measurements of the spatial and temporal distribution of rainfall. *J. Clim. Appl. Meteorol.*, 26, 1530–1539.

Hosking, J.R.M. and Wallis, J.R. (1997) *Regional Frequency Analysis: An Approach Based on L-Moments*. Cambridge University Press, Cambridge.

Institute of Hydrology (1999) *Flood Estimation Handbook*, vols. 1–5. Institute of Hydrology, Wallingford.

Jenkinson, A.F. (1955) The frequency distribution of the annual maximum (or minimum) values of meteorological elements. *Q. J. R. Meteorol. Soc.*, 81, 158–171.

Kavvas, M.L. and Herd, K.R. (1985) A radar-based stochastic model for short-time increment rainfall. *Water Resour. Res.*, 21, 235–315.

Krajewski, W.F. and Georgakakos, K.P. (1985) Synthesis of radar rainfall data. *Water Resour. Res.*, 21, 764–768.

Lovejoy, S. (1982) The area–perimeter relation for rain and cloud areas. *Science*, 216, 185–187.

Lovejoy, S. and Mandelbrot, B.B. (1985) Fractal properties of rain and a fractal model. *Tellus*, 37A, 209–232.

Mahfouf, J.-F., Brasnett, B. and Gagnon, S. (2007) A Canadian precipitation analysis (CaPA) project: description and preliminary results. *Atmos. Ocean*, 45(1), 1–17.

Mitchell, R., Charette, C., Chouinard, C. and Brasnett, B. (1990) Revised interpolation statistics for the Canadian data assimilation procedure: their derivation and application. *Mon. Weather Rev.*, 18, 1591–1614.

NERC (1975) *Flood Studies Report*, vols. 1–5. Natural Environment Research Council, London.

Niemczynowicz, J. (1982) Areal intensity–duration–frequency curves for short term rainfall events in Lund. *Nord. Hydrol.*, 13, 193–204.

Rodda, J.C., Downey, R.A. and Law, F.M. (1976) *Systematic Hydrology*. Newnes-Butterworth, London.

Rudolf, B., Hauschild, H., Rueth, W. and Schneider, U. (1994) Terrestrial precipitation analysis: operational method and required density of point measurements. In *Global Precipitation and Climate Change*, ed. M. Desbois and F. Desalmand, NATO ASI Series vol. I 26, pp. 173–186. Springer, Berlin.

Shaw, E.M. (1994) *Hydrology in Practice*, 3rd edn. Chapman & Hall, London.

Shaw, E.M. and Lynn, P.P. (1972) Area rainfall evaluation using two surface fitting techniques. *Bull. IAHS*, XVII(4), 419–433.

Shuttleworth, W.J. (2012) *Terrestrial Hydrometeorology*. Wiley-Blackwell, Chichester.

Smith, J.A., Bradley, A.A. and Baeck, M.L. (1994) The space–time structure of extreme storm rainfall in the Southern Plains. *J. Appl. Meteorol.*, 33, 1402–1417.

Sumner, G.N. (1988) *Precipitation Process and Analysis*. Wiley, Chichester.

Taylor, G.I. (1938) The spectrum of turbulence. *Proc. R. Soc. Lond. A*, 164, 476–490.

Weymouth, G., Mills, G.A., Jones, D., Ebert, E.E. and Manton, M.J. (1999) A continental-scale daily rainfall analysis system. *Aust. Meteorol. Mag.*, 48, 169–179.

WMO (1973) *Manual for Estimation of Probable Maximum Precipitation*. Operational Hydrology Report 1. World Meteorological Organization, Geneva.

WMO (1986) *Manual for Estimation of Probable Maximum Precipitation*, 2nd edn. Operational Hydrology Report 1. World Meteorological Organization, Geneva.

Zawadzki, I. (1973) Statistical properties of precipitation patterns. *J. Appl. Meteorol.*, 12, 459–472.

8

Precipitation Forecasting

8.1 Introduction

A tremendous effort by both meteorologists and hydrologists over the last 50 years or so has resulted in *quantitative precipitation forecasts* (QPFs) which are potentially fit for purpose in operational hydrological forecast systems. However, further research is necessary. Considerable improvements have been brought about in recent years through increased understanding of the physical and statistical properties of precipitation processes, the deployment of weather radar networks and satellite systems, and improvements in numerical weather prediction models. However, precipitation forecasts from mesoscale numerical weather prediction (NWP) models often contain features that are not deterministically predictable and require a probabilistic forecast approach.

In this chapter we consider forecast lead times as follows: 0–6 hours (nowcasting), 3–24 hours (short term forecasting), and 3–24 months (seasonal forecasting). The quality of weather forecasts changes as the lead time increases, as shown in Figure 8.1. Whilst this figure illustrates the situation at the end of the 20th century, the levels of performance are constantly changing and the performance of NWP mesoscale models is improving as computers become more and more powerful.

Manual forecasting relies upon human forecasters applying their knowledge of atmospheric systems to output from procedures using observations from ground-based (automatic weather stations), airborne (radiosondes, aircraft, satellites) and remotely sensed (radar, satellites) instrumentation. Nowcasting is largely based upon the use of radar data; as the lead time increases the forecasts are based upon numerical weather models, although these may assimilate observations of all types.

8.2 Nowcasting

8.2.1 Definition

Nowcasts are defined as very short range forecasts usually developed from extrapolation of observations for up to 3 (sometimes up to 6) hours ahead. The usefulness of such forecasts depends very much on whether the precipitation field develops rapidly.

Hydrometeorology, First Edition. Christopher G. Collier.
© 2016 John Wiley & Sons, Ltd. Published 2016 by John Wiley & Sons, Ltd.
Companion website: www.wiley.com/go/collierhydrometerology

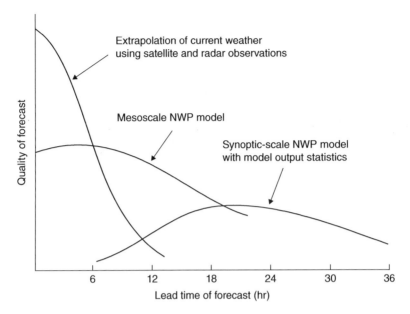

Figure 8.1   The quality of weather forecasts, defined as the product of the accuracy and detail achievable, shown as a function of lead time for three different forecasting methods. The figure is highly schematic, and the stage at which the quality of one technique becomes superior to another will change (from Browning, 1980)

8.2.2   Impact of errors in precipitation measurements

There are five main sources of precipitation measurements, namely manual observations from qualitative descriptions of precipitation (drizzle, light rain etc.), rain-gauges, radar, lightning observations which may be interpreted as precipitation, and observations derived from satellites. All these measurement types have their advantages and disadvantages. Figure 8.2 illustrates the advantage of radar observations compared to those from automatic weather stations (AWSs). Although the distribution of AWSs can be improved, it becomes uneconomic to have a very dense network. Both radar networks and satellite observations are subject to errors (see Chapters 5 and 6), and therefore any forecast system based purely upon observations must embody quality control procedures which are activated before forecasts are made.

Many of the uncertainties in radar measurements of precipitation may be removed by real-time processing at the radar sites and/or at a central location, as described by Harrison et al. (2009). Satellite data may be used in combination with radar data, the FRONTIERS system being one of the first examples of such an operational system (Browning, 1979; Browning and Collier, 1982).

8.2.3   Extrapolation of radar data

When quality controlled radar data are available, radar echoes may be objectively described using a variety of interrelated techniques, as described by Pankiewicz (1995). Collier (1996) summarizes several approaches including,

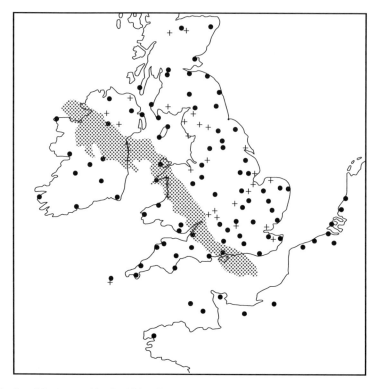

Figure 8.2   Rainfall observed by the UK radar network at 1400 UTC on 15 April 1986. Meteorological stations providing conventional observations are shown (•). The locations of AWS observations are also indicated (+) (1985). Note that the full extent of the rain band is not observed by the conventional observations (© Crown Copyright from Collier, 1996)

- contouring: the most conceptually simple method using selectable intensity thresholding
- Fourier analysis: a description of the contours derived from a Fourier series expansion
- bivariate normal distribution: using an unnormalized bivariate normal distribution, the contours of which in three dimensions are ellipses
- clustering: isolating data into homogeneous subcategories based upon the sequential lowering of an initially high intensity threshold, referred to as hierarchical clustering.

The simplest form of extrapolation is based upon using regression relationships between echo movement and wind velocity (see for example Tatehira et al., 1976) (see Figure 8.3). However more sophisticated approaches to linear extrapolation may be based upon echo centroid tracking (Wilk and Gray, 1970), or cross-correlation techniques (for example Austin and Bellon, 1974) for matching portions of one image with portions of subsequent images. Table 8.1 shows examples of typical extrapolation timescales for types of precipitation fields. When linear extrapolation ceases to produce a good forecast, which may be after only a few minutes or tens of minutes as

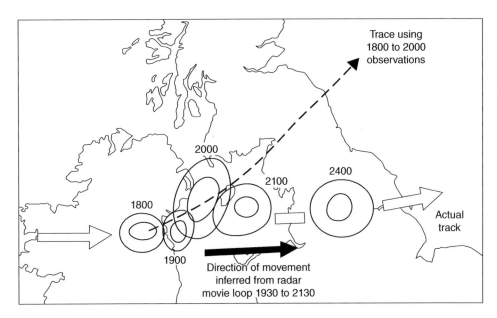

Figure 8.3 Positions of low between 1800 and 2400 UTC on 22 December 1983 according to the analyses of the Central Forecasting Office at the Met Office, Bracknell, UK. Each position is defined by the innermost two isobars (2 hPa intervals) irrespective of the actual pressure values. The possible track inferred from analyses between 1800 and 2000 is shown by the broken arrow, whilst the direction of movement suggested by radar data up to 2130 is shown by the bold full arrow. The 2130 radar picture (available within 2 min) was available before the 2100 surface chart (from Monk et al., 1987)

Table 8.1 Examples of typical linear extrapolation timescales for precipitation fields associated with various weather events (from Collier, 1996, partly from Doswell, 1986)

Weather event	Timescale for linear extrapolation validity	Non-linear predictive capability
Downburst or microburst	≈1 to a few minutes	Very limited
Tornado	≈1 to a few minutes	Currently very limited
Individual thunderstorm or heavy shower	5–20 min	Very limited
Severe thunderstorm	10 min to 1 h	Very limited
Thunderstorm organized on the mesoscale	≈1–2 h	Some
Flash-flood rainfall	≈1 to a few hours	Very limited
Orographically triggered showers	≈1 h	Very limited
Lake-effect snowstorms	A few hours	Very limited
Heavy snow, winter storm or blizzard	A few hours	Some
High wind gusts accompanying shallow showers	5–30 min	Very limited
Hurricane	Many hours	Fair
Frontal passage	Many hours	Fair to good

shown in Table 8.1, then a non-linear forecasting procedure or subjective procedure must be adopted.

It is difficult to assess objectively the relative performance of different methods of extrapolating echo movement. Elvander (1976) described a series of experiments using different techniques. It was concluded that the simple cross-correlation approach was the most effective when used with zero-tilt reflectivity data, but the linear least squares interpolation of echo centroids was the most effective method when vertically integrated liquid water content data were used.

The improvement that can be achieved using subjective techniques was reported by Browning et al. (1982) who compared forecasts for a 400 km² area using a network of four radars in the UK. The forecasts derived during a total of 29 frontal events between November 1979 and June 1980 were prepared using both a totally objective echo centroid technique (Collier, 1981) and subjective linear extrapolation. Figure 8.4 shows the percentage error, regardless of sign, in the forecast hourly rainfall for the 400 km² centred on Malvern (located in the middle of the radar network, plotted as a function of lead time. Various errors in the forecasts were identified and partially corrected such that the subjective forecasts were considered better than the objective forecasts. However, as indicated on the figure, objective procedures developed by Bellon and Austin (1984) may approach the performance of subjective techniques.

So far in this section we have discussed simple forms of artificial intelligence. Improvements in forecasting may be possible using knowledge-based expert systems. For example, Elio et al. (1987) propose the automatic implementation of a decision tree, whereas Dixon and Weiner (1993) use a procedure for recognizing particular types of storms employing contiguous regions exceeding thresholds of intensity and size in their TITAN system.

A different use of artificial intelligence is to attempt to analyse meteorological situations using a decision tree approach (for example Colquhoun, 1987). A more advanced approach is to use object oriented programming (see for example Winston, 1984; Hand and Conway, 1995). Here an application domain is defined as a population of objects of various sorts. For example convective cells are given attributes which are identified in images and then allowed to develop in different ways. One example is the GANDOLF system described by Collier et al. (1995). However, the application of a non-linear model can produce a worse result than linear techniques, as shown in Figure 8.5, and therefore the best approach is to use a mesoscale numerical model as suggested by Pielke (1981) (see section 8.4).

8.3   Probabilistic radar nowcasting

A probabilistic model for radar forecasts of rain was developed by Zuckerberg (1976). However using a Poisson distribution it only provided the probability of rain or no rain. Andersson (1991) developed this further to give probabilities for different rain amounts. The assessment results are presented on reliability diagrams, as discussed by Andersson and Ivansson (1993).

An ensemble-based probabilistic precipitation forecasting scheme has been developed by the UK Met Office and the Australian Bureau of Meteorology; it blends an extrapolation nowcast with a downscaled NWP forecast, and is known as a *short term ensemble prediction system* (STEPS) (Bowler et al., 2006). The uncertainties in the motion

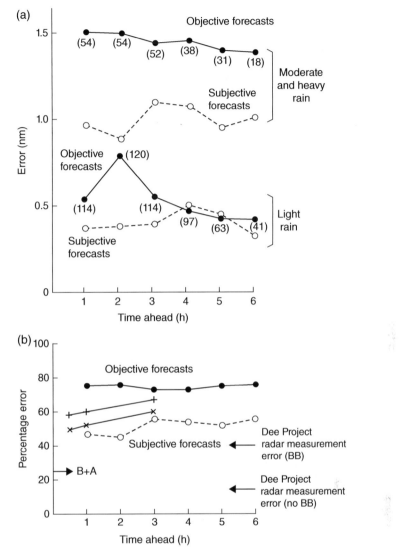

Figure 8.4   (a) The mean error regardless of sign in the forecast hourly rainfall for the 20 km square (400 km²) centred on Malvern plotted as a function of lead time of the forecast for light rain (trace, 1 mm h⁻¹) and moderate–heavy rain (more than 1 mm h⁻¹). Solid lines are objective forecasts without manual modification; dashed lines are subjective forecasts; numbers in parentheses denote number of cases. (b) The percentage error of the cases of moderate–heavy rain. Radar measurement errors found in the Dee Weather Radar Project (Harrold et al., 1974) are indicated, where BB stands for bright band. The performance of the Bellon and Austin (1984) cross-correlation technique is indicated by B + A (partly from Browning et al. 1982; from Collier, 1996)

and evolution of radar-inferred precipitation fields are quantified, and the uncertainty in the evolution of the precipitation pattern is shown to be the more important. The use of ensembles allows the scheme to be used for applications that require forecasts of the probability density function of areal and temporal averages of precipitation, such as fluvial flood forecasting. The output from an NWP forecast model

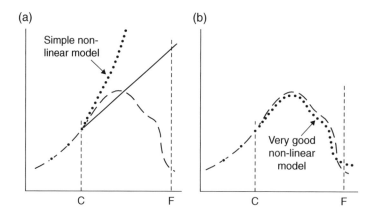

Figure 8.5   (a) An example of a simple non-linear model that is much worse than a linear one. (b) An example of a very good non-linear model. Point C is the time a forecast is made for a period ending at point F. The dashed lines represent the actual behaviour of the rainfall; the dotted lines represent the model predictions of the rainfall; the solid line on the left-hand graph shows a simple linear prediction from point C (after Doswell, 1986 from Collier, 1996)

is downscaled so that the small scales not represented accurately by the model are injected into the forecast using stochastic noise. This allows the scheme to better represent the distribution of precipitation rate at spatial scales finer than those adequately resolved by operational NWP. Performance evaluation statistics show that the scheme possesses predictive skill at lead times in excess of 6 hours.

Foresti and Seed (2014) investigated the spatial distribution and scale dependence of the very short term predictability of precipitation by Lagrangian persistence of composite radar images under different flow regimes in connection with the presence of orographic features. Data from the weather radar composite of eastern Victoria, Australia, a $500 \times 500\,\text{km}^2$ domain at 10 min temporal and $2 \times 2\,\text{km}^2$ spatial resolutions, covering the period from February 2011 to October 2012, were used for the analyses. The predictability of precipitation is measured by its lifetime, which is derived by integrating the Lagrangian autocorrelation function. The lifetimes were found to depend on the scale of the feature as a power law, and also exhibited significant spatial variability and were approximately a factor of two longer on the upwind compared with the downwind slopes of terrain features. These findings provide opportunities to carry out spatially inhomogeneous stochastic simulations of space–time precipitation to account for the presence of orography, which may be integrated into design storm simulations and stochastic precipitation nowcasting systems.

8.4   Numerical models: structure, data requirements, data assimilation

8.4.1   Probabilistic quantitative precipitation forecasting

A number of existing methods generate probabilistic precipitation forecasts based on deterministic forecasts. Regression techniques such as *model output statistics* (MOS) can be used to generate probabilities of exceeding thresholds (see for example Glahn and Lowry, 1972), or to generate quantiles of expected precipitation (see for example Bremnes, 2004). Applequist et al. (2002) found that logistic regression can outperform standard regression, and Hamill et al. (2004) found that this can be further refined by using logistic regression on power-transformed forecasts. These methods, however, do not yield a full predictive probability density function (PDF), giving only probabilities for certain specific events.

An *ensemble forecasting* system samples the uncertainty inherent in weather prediction to provide more information about possible future weather conditions. Rather than producing a single forecast with an NWP model, multiple forecasts are produced by making small alterations either to the starting conditions or to the forecast model itself or both.

The ensemble forecast system is designed so that each individual forecast or *member* should be equally likely, so that the ensemble can be used to forecast the probabilities of different possible outcomes. Where all the forecasts in an ensemble are similar there is more confidence in the forecast. However, where they differ more account must be taken of uncertainty.

Two ensemble forecasting systems are used at the UK Met Office: the Met Office Global and Regional Ensemble Prediction Systems (MOGREPS) (Bowler et al., 2008) is used for short range prediction, and the ECMWF ensemble prediction system is used for the medium range (up to 15 days ahead). Probabilistic forecasts of heavy precipitation over the UK have been compared from two configurations of MOGREPS. These are a high resolution convection permitting UK ensemble (MOGREPS-UK) and a coarser resolution global ensemble (MOGREPS-G). Case study analysis and objective verification provide an insight into the benefits of higher resolution forecasts. Neighbourhood processing of the raw MOGREPS-UK ensemble members is used to provide probabilities of a range of thresholds, either at specific times or within time windows for rain, wind and temperature (Figure 8.6).

Hamill and Colucci (1998) have established errors and ensemble spread for several ensemble systems. Buizza et al. (2005) proposed fitting gamma distributions with parameters based on corrected ensembles or transformations of the ensemble mean. However it is not obvious how to obtain calibrated probability of precipitation (PoP) forecasts using this approach. Bayesian model averaging (BMA) was introduced by Raftery et al. (2005) and Sloughter et al. (2007) as a statistical post-processing method for producing probabilistic forecasts from ensembles in the form of predictive PDFs. The BMA predictive PDF of any future weather quantity of interest is a weighted average of PDFs centred on the individual forecasts.

For a deterministic hydrologic forecast system, it is known that the uncertainty in the quantitative precipitation forecast (QPF) is a primary source of forecast error. Wu et al. (2011) have developed statistical procedures to identify this uncertainty, and the resulting precipitation ensemble forecasts are used to drive a hydrologic model to produce hydrologic ensemble forecasts.

Mascaro et al. (2010) noted that the propagation of errors associated with ensemble QPFs into the ensemble stream flow response is important to reduce uncertainty in operational flow forecasting. A multifractal rainfall downscaling model was coupled with a fully distributed hydrological model to create, under controlled conditions, an extensive set of synthetic hydrometeorological events, taken as observations. For each event, flood hindcasts were simulated by the hydrological model using three ensembles of QPFs, namely one reliable and the other two affected by different

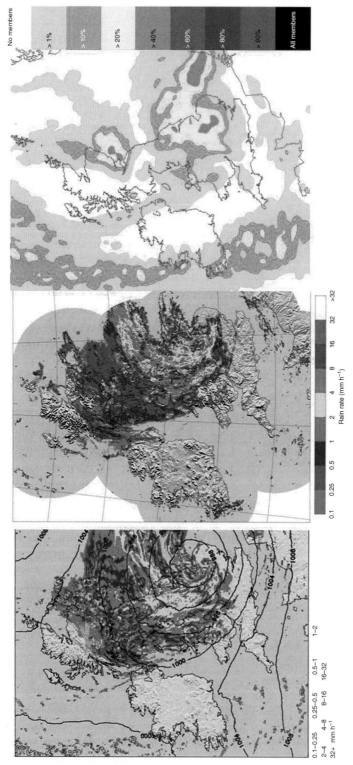

Figure 8.6 Sample of corresponding precipitation forecast products from UKV (left) and MOGREPS-UK (right), along with the verifying radar image (centre), for 0400Z on 25 November 2012. The MOGREPS-UK chart illustrates the 'chance of heavy rain in the hour', which is defined as 4 mm h⁻¹; members are the individual forecasts within the ensemble (from Mylne, 2013) (*see plate section for colour representation of this figure*)

kinds of precipitation forecast errors generated by the downscaling model. The analyses indicated that the best forecast accuracy of the ensemble stream flows was obtained when the reliable ensemble QPFs were used. These results underlined (i) the importance of hindcasting to create an adequate set of data that span a wide range of hydrometeorological conditions, and (ii) the sensitivity of the ensemble stream flow verification to the effects of basin initial conditions and the properties of the ensemble precipitation distributions.

8.4.2   Mesoscale models

The use of high resolution NWP has the potential to improve forecasts of deep convection and the associated heavy rain, though precisely how high the resolution needs to be remains an open question. Clark (2006) reports that experience so far suggests that an important step is allowing explicit convection to form in flows which adequately resolve the mesoscale features which lead to triggering and organization of storms, rather than fully resolving the detail of convective cells. In the UK a great deal of work has been done using configurations of the unified model (UM) at 4 km (driven by the UK mesoscale model at 12 km) and 1 km horizontal resolution. A number of 1 km domains have been used to study different cases.

The initiation of convection is particularly important in mesoscale models for a number of reasons. Different initiation mechanisms, such as surface forced sea breeze convergence or mid-level initiation by gravity waves, may have very different inherent predictability – important to understand when designing and assessing the model. If the model does not correctly initiate convection, then subsequent evolution will be in error.

The triggering, number and intensity of convective cells is strongly controlled by the treatment of turbulence. Also the details of the cloud and precipitation microphysical processes that are parameterized in the model play a vital role in the development of deep convective cells in the forecast. These aspects of the model remain uncertain and the degree of complexity of the system needs to be adequately represented.

8.4.3   Data assimilation

NWP is considered as an initial boundary value problem: that is, given an estimate of the present state of the atmosphere, the model simulates its evolution. However, specification of correct initial conditions and boundary conditions is essential to have a well posed problem and then for the model to make a good forecast. The goal of data assimilation is to construct the best possible initial and boundary conditions from which to integrate forward the model.

The Met Office 3D-Var system uses an incremental formulation (for a review see for example Rihan et al., 2005) producing a multivariate incremental analysis for pressure, density, horizontal and vertical wind components, potential temperature and specific humidity in the model space, consistent with a non-hydrostatic, height coordinate model. Cloud water and fraction are neglected in the basic system. Under the assumption that the background and observation errors are Gaussian, random and independent of each other, and that the increments and model are at the same resolution (as they were in the 12 km model and these experiments), the optimal estimate in the analysis space is given by an incremental cost function. Xu and Gong (2003) assumed that the observation error covariance matrix is diagonal, with constant diagonal elements given by the estimated observation error. An alternative approach was discussed by Rihan et al. (2008).

The current high resolution system in the UK has implemented data assimilation based upon that used in the 12 km UK mesoscale model and subsequently the UK 4 km model (unified model). This makes use of 3D-Var together with latent heat nudging (LHN) and the moisture observation processing system (MOPS) to nudge in rainfall and cloud data. Dixon et al. (2009) tested the data assimilation scheme on five convection-dominated case studies from the Convective Storm Initiation Project (CSIP). Model skill was assessed statistically using radar-derived surface precipitation hourly accumulations via a scale-dependent verification scheme. Data assimilation was shown to have a dramatic impact on skill during both the assimilation and subsequent forecast periods on nowcasting timescales. The resulting forecasts are also shown to be much more skilful than those produced using either a 12 km grid length version of the UM with data assimilation in place, or a 4 km grid length UM version run using a 12 km state as initial conditions.

8.4.4   Performance of high resolution mesoscale model-based nowcasting systems

Since March 2012 and over the London Olympic Games period, the Met Office has run a demonstration trial of the hourly cycling numerical weather prediction system for the southern UK (Ballard et al., 2013). This system combined a 1.5 km resolution version of the unified model and 3 km resolution 4D-variational (4D-Var) data assimilation. This system, referred to as the Nowcasting Demonstration Project, every hour produced analyses and forecasts for the period 0 to 12 hours ahead. The following data sources were assimilated: Doppler radial winds, data from wind profilers, satellite radiances, hourly 3D moisture derived from cloud observations, METEOSAT second-generation cloud and humidity tracked winds, and hourly surface temperature, relative humidity, wind, pressure and visibility. This will form the basis of a future operational forecast system (Sun et al., 2013). Figure 8.7 indicates the level of performance achieved.

8.5   Medium range forecasting

The European Centre for Medium Range Weather Forecasting (ECMWF) integrated forecasting system (IFS) is a global NWP model that runs every 12 hours. Forecasts out to 10 days in 1 day intervals are produced as both a deterministic forecast and an ensemble.

The physical processes represented in the model are shown in Figure 8.8.

ECMWF has extended the range of weather forecast products that are available freely and with no restrictions from its website. The new products are from the ECMWF ensemble prediction system (EPS) that provides guidance on the day-to-day predictability of the atmosphere. This gives an important complement to the information that is already available from the single deterministic model.

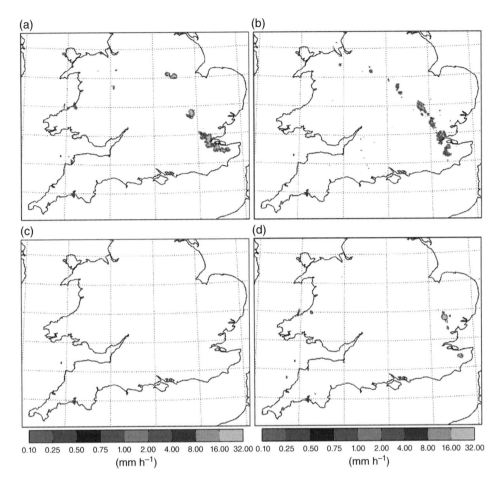

(a)　　　　　　　　　　　　　　　　(b)

(c)　　　　　　　　　　　　　　　　(d)

0.10   0.25   0.50   0.75   1.00   2.00   4.00   8.00   16.00   32.00   0.10   0.25   0.50   0.75   1.00   2.00   4.00   8.00   16.00   32.00

(mm h^{-1})　　　　　　　　　　　　　　　(mm h^{-1})

Figure 8.7   Comparisons of: (a) the observed UK radar-derived surface precipitation rate at 15UTC 28 May 2012, with (b) the 5-hour forecast from the 10UTC Met Office 4D-Var hourly cycling 1.5 km NWP system, (c) the current operational nowcast from 10UTC using extrapolated radar derived rain rates blended with a 4 km resolution UK forecast from 03UTC, and (d) the latest available real-time forecast from the 3 hourly cycling with 3D-Var 1.5 km UK-wide system, which is a 12 hour forecast from 03UTC. The figure shows the comparison on the domain of the hourly cycling Met Office NWP system; the coastline is shown by the black contour (from Ballard et al., 2013)

The ensemble is a set of 51 separate forecasts made by the same computer model, all started from the same initial time. The starting conditions for each member of the ensemble are slightly different. The differences between these ensemble members tend to grow as the forecasts progress. This dispersion varies from day to day, depending on the initial atmospheric conditions, and therefore gives an indication of the uncertainty in the current forecast (Buizza et al., 1999).

The new graphical products show the ensemble mean (average over all 51 members) and ensemble standard deviation (spread) corresponding to each parameter available from the deterministic model: 500 hPa geopotential, 850 hPa wind and temperature and mean sea level pressure. An example of the performance of the model in forecasting heavy rainfall during May and June 2013 over Central Europe was

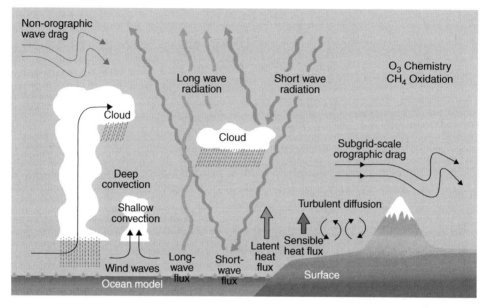

Figure 8.8 Schematic diagram of the different physical processes represented in the ECMWF IFS model (from ECMWF IFS Documentation Cy38r1) (*see plate section for colour representation of this figure*)

described by Pappenberger et al. (2013). Orographic enhancement of precipitation along the northern side of the Alps played an important role in this event. The ECMWF high resolution model captured the spatial extent and location of the rainfall maximum up to 84 hours ahead of its occurrence, but underestimated amounts in the area of heaviest upslope precipitation.

Figure 8.9 illustrates measures of the overall performance of raw ensemble forecasts via the implied probabilities to exceed chosen rainfall thresholds. The plots show the decomposition of the Brier skill score (BSS: Wilks, 2006) into its 'resolution' and 'reliability penalty' components: the latter measures the difference between each forecast probability and the observed event frequency when that probability is forecast, whilst the former measures the ability to discriminate between situations in which the event is very likely to occur and situations in which it is very unlikely (see also Appendix 8.1). The reliability penalty can in principle be eliminated by remapping the forecast probability values given sufficient representative training data. The resolution score is more tied to the quality of the underlying forecast system, although it can be affected by corrections to the bias or spread of the underlying ensemble, since these will change the distribution of members with respect to thresholds. The overall BSS is the resolution score minus the reliability penalty. A perfect forecast would have a resolution score of one and a reliability penalty of zero, whilst BSS less than zero indicates a forecast no better than the climatological probability of the event.

The ECMWF temperature and precipitation ensemble forecasts have been evaluated by Verkade et al. (2013) for biases in the mean, spread and forecast probabilities, and how these biases propagate through to stream flow ensemble forecasts. The forcing ensembles were subsequently post-processed to reduce bias and increase skill, and to investigate whether this leads to improved stream flow ensemble forecasts. Multiple

(a)   (b)   (c)

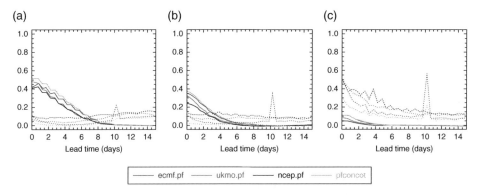

Figure 8.9 Resolution (solid) and reliability (dotted) components of the Brier skill score (vertical axis) for selected quantiles of local climatology: (a) > 50% (b) > 95% (c) > 99%. Forecasts use the perturbed members from the ECMWF, Met Office and NCEP 15 day ensemble systems, and the simple aggregation of all these members (right panel). The spike in ECMWF reliability penalty at $T+10$ days is associated with a coarsening of the model grid, and may be an artefact of the data interpolation code used (from Flowerdew, 2012)

post-processing techniques were used, namely quantile-to-quantile transform, linear regression with an assumption of bivariate normality, and logistic regression. Both the raw and post-processed ensembles are input to a hydrologic model of the river Rhine to create stream flow ensembles. The results are compared using multiple verification metrics and skill scores: relative mean error, Brier skill score and its decompositions, mean continuous ranked probability skill score and its decomposition, and the relative operating characteristic (ROC) score. The derivation of ROC is based on contingency tables giving the hit rate and false alarm rate for deterministic or probabilistic forecasts (see Stanski et al., 1989). Verification of the stream flow ensembles was performed at multiple spatial scales: relatively small headwater basins, large tributaries and the Rhine outlet at Lobith. The stream flow ensembles were verified against simulated stream flow, in order to isolate the effects of biases in the forcing ensembles and any improvements therein. The results indicated that the forcing ensembles contain significant biases, and that these cascade to the stream flow ensembles.

8.6   Seasonal forecasting

Although individual weather systems might be chaotic and unpredictable beyond a week or so, the statistics describing them may be perturbed in a deterministic and predictable way, particularly by the ocean. In the past, seasonal forecasts of atmospheric variables have largely been based on empirical relationships, which were not reliable. However, recently, atmosphere models forced by ocean conditions have been used. Stockdale et al. (1998) used the ECMWF fully coupled global ocean–atmosphere general circulation model to make seasonal forecasts of the climate system with a lead time of up to 6 months. Although reliable verification of probabilistic forecasts is difficult, the results achieved in this study compared to observations were encouraging for the prospects for seasonal forecasting. Rainfall predictions for 1997 and the first half of 1998 show a marked increase in the spatial extent of statistically significant anomalies during the present El Niño, and include strong signals over Europe.

The US National Weather Service began routinely issuing long lead forecasts of 3 month mean US temperature and precipitation by the beginning of 1995. The ability to produce useful forecasts for certain seasons and regions at projection times of up to 1 year is attributed to advances in data observing and processing, computer capability, and physical understanding – particularly for tropical ocean–atmosphere phenomena. Much of the skill of the forecasts comes from anomalies of tropical sea surface temperatures (SSTs) related to El Niño/Southern Oscillation (ENSO). Sea surface temperature in the equatorial Pacific Ocean may be characterized by unusually warm (El Niño) or cool (La Niña) temperatures.

Barnston et al. (1994) highlighted long lead forecasts of the tropical Pacific SST itself, which have higher skill than the US forecasts that are made largely on their basis. The performance of five ENSO prediction systems were examined: two are dynamical (the Cane–Zebiak simple coupled model of Lamont–Doherty Earth Observatory and the non-simple coupled model of the National Centers for Environmental Prediction, NCEP); one is a hybrid coupled model (the Scripps Institution for Oceanography–Max Planck Institute for Meteorology system with a full ocean general circulation model and a statistical atmosphere); and two are statistical (canonical correlation analysis and constructed analogues, used at the Climate Prediction Center of NCEP). With increasing physical understanding, dynamically based forecasts have the potential to become more skilful than purely statistical ones. The simplest model performed about as well as the more comprehensive models. At a lead time of 6 months (defined here as the time between the end of the latest observed period and the beginning of the predictand period), the SST forecasts had an overall correlation skill in the range 0.60–0.69 for 1982–93, which easily outperforms persistence and is regarded as useful. Skill for extratropical surface climate is this high only in limited regions for certain seasons. Both types of forecast are not much better than local higher order autoregressive controls. However, continual progress is being made in understanding relations among global oceanic and atmospheric climate-scale anomaly fields.

Understanding future changes in the frequency, intensity and duration of extreme events in response to increased greenhouse gas forcing is important for formulating adaptation and mitigation strategies that minimize damage to natural and human systems. Singh et al. (2013) quantified transient changes in daily-scale seasonal extreme precipitation events over the United States using a five-member ensemble of nested, high resolution climate model simulations covering the 21st century in the Intergovernmental Panel on Climate Change (IPCC) Special Report on Emission Scenarios (SRES) A1B scenario. They found a strong drying trend in annual and seasonal precipitation over the south-west United States in autumn, winter and spring and over the central United States in summer. These changes are accompanied by statistically significant increases in dry-day frequency and dry-spell lengths. The results also showed substantial increases in the frequency of extreme wet events over the north-western United States in autumn, winter and spring and the eastern United States in spring and summer. In addition, the average precipitation intensity increased relative to the extreme precipitation intensity in all seasons and most regions, with the exception of the south-east.

Appendix 8.1   Brier skill score

The Brier skill score (BSS) is used for the verification of probabilistic events and is the mean squared error of the probability forecasts. It can be expressed as the sum of three terms, reliability, resolution and uncertainty:

$$BSS = \frac{1}{n}\sum_{i=1}^{I}N_i\left(y_i-\bar{o}_i\right)^2 - \frac{1}{n}\sum_{i=1}^{I}N_i\left(\bar{o}_i-\bar{o}\right)^2 + \bar{o}\left(1-\bar{o}\right)$$

$$\underbrace{\qquad\qquad}_{(reliability)} \quad \underbrace{\qquad\qquad}_{(resolution)} \quad \underbrace{\qquad}_{(uncertainty)}$$

(8.1)

where i is the index denoting a numbering of the n forecast/event pairs, o is the observation, and I is the discrete number of forecast values y. More accurate forecasts are represented by smaller values of BS. The reliability term summarizes the calibration or conditional bias of the forecasts, whereas the resolution term summarizes the ability of the forecasts to discriminate subsample forecast periods with different relative frequencies of the event. The uncertainty term depends upon the variability of the observations. This equation can be illustrated by an attributes diagram as developed by Hsu and Murphy (1986).

Summary of key points in this chapter

1. Considerable improvements in quantitative precipitation forecasts (QPFs) have been brought about in recent years through increased understanding of the physical and statistical properties of precipitation processes, the deployment of weather radar networks and satellite systems, and improvements in numerical weather prediction models.

2. Nowcasting refers to forecasts with lead times of 0–6 hours, whereas short term forecasts refer to lead times of 3–24 hours and seasonal forecasts to lead times of 3–24 months.

3. The levels of performance are constantly changing and the performance of NWP mesoscale models is improving as computers become more and more powerful.

4. Manual forecasting relies upon human forecasters applying their knowledge of atmospheric systems.

5. There are five main sources of precipitation measurements, namely manual observations, raingauges, radar, lightning observations, and observations from satellites.

6. Many of the uncertainties in radar measurements of precipitation may be removed by real-time processing.

7. Several approaches to the extrapolation of radar data include contouring, Fourier analysis, bivariate normal distribution, and clustering. The simplified form of extrapolation is based upon using regression relationships between echo movement and wind velocity.

8. The simple cross-correlation approach is generally the most effective when used with zero-tilt reflectivity data, but the linear least squares interpolation

of echo centroids can be the most effective with integrated liquid water content data.

9. Objective procedures may approach the performance of subjective techniques.

10. Improvements in forecasting may be possible using knowledge-based expert systems.

11. A short term ensemble prediction system (STEPS) blends an extrapolation nowcast with a downscaled NWP forecast. The use of ensembles allows the scheme to be used for applications that require forecasts of the probability density function of areal and temporal averages of precipitation such as fluvial flood forecasting.

12. The spatial distribution and scale dependence of the very short term predictability of precipitation depend on the Lagrangian persistence of composite radar images under different flow regimes.

13. Regression techniques such as model output statistics (MOS) can be used to generate probabilities of exceeding thresholds, or to generate quantiles of expected precipitation.

14. An ensemble forecasting system samples the uncertainty inherent in weather prediction.

15. Bayesian model averaging is a statistical post-processing method for producing probabilistic forecasts from ensembles in the form of predictive probability density functions.

16. The propagation of errors may lead to a reduction in uncertainty in operational flow forecasting.

17. The use of high resolution NWP has the potential to improve forecasts of deep convection and the associated heavy rain.

18. NWP is considered as an initial boundary value problem.

19. The goal of data assimilation is to construct the best possible initial and boundary conditions from which to integrate forward the model.

20. Data assimilation has been shown to have a dramatic impact on skill during the assimilation and subsequent forecast periods.

21. New products from the European Centre for Medium Range Weather Forecasting (ECMWF) ensemble system provide guidance on the day-to-day predictability of the atmosphere.

22. Numerical model verification metrics include the Brier skill score and the relative operating characteristic (ROC) score.

23. Although individual weather systems might be chaotic and unpredictable beyond a week or so, the statistics describing them may be perturbed in a deterministic and predictable way.

24. Much of the skill of seasonal forecasts comes from anomalies of tropical sea surface temperatures related to the El Niño/Southern Oscillation (ENSO).

25. Understanding the future changes in the frequency, intensity and duration of extreme events in response to increased greenhouse gas forcing is important for formulating adaptation and mitigation strategies.

Problems

1. What has led to improvements in quantitative precipitation forecasts over the last 50 years or so?
2. Define the lead times for nowcasting, short term forecasting and seasonal forecasting.
3. What is necessary before forecasts are made based upon observations?
4. Generally, by what may uncertainties in radar measurements be removed?
5. Name four approaches to extrapolating radar data to make forecasts.
6. Give the timescale for linear extrapolation validity for individual thunderstorms, hurricanes and fronts.
7. Give an example of the percentage error, regardless of sign, in the forecast hourly rainfall over an area of 400 km^2.
8. Outline the STEPS system.
9. What is the Lagrangian persistence of composite radar images under different flow regimes?
10. What are model output statistics (MOS)?
11. Explain what comprises an ensemble forecasting system.
12. What is Bayesian model averaging?
13. How may flood hindcasts be simulated?
14. What do ensemble stream flows indicate?
15. Name why the initiation of convection is important in mesoscale models.
16. Given that NWP is an initial boundary value problem, why is data assimilation important?
17. What are the assumptions made in 3D-Var data assimilation?
18. What does the ECMWF ensemble prediction system provide?
19. Name two skill scores used as NWP verification metrics.
20. How can the statistics of chaotic and unpredictable weather systems impact seasonal forecasts?
21. What provides much of the skill in seasonal forecasts?
22. What changes in forecasting extreme events are important, and why?

References

Andersson, T. (1991) An advection model for probability nowcasts of accumulated precipitation using radar. In *Hydrological Applications of Weather Radar*, ed. I.D. Cluckie and C.G. Collier, pp. 325–330. Ellis Horwood, Chichester.

Andersson, T. and Ivansson, K.-I. (1993) A model for probability nowcasts of accumulated precipitation using radar. *J. Appl. Meteorol.*, 30, 135–141.

Applequist, S., Gahrs, G.E., Pfeffer, R.L. and Niu, X.-F. (2002) Comparison of methodologies for probabilistic quantitative precipitation forecasting. *Weather Forecast.*, 17, 783–799.

Austin, G.L. and Bellon, A. (1974) The use of digital weather records for short-term precipitation forecasting. *Q. J. R. Meteorol. Soc.*, 100, 658–664.

Ballard, S.P., Li, Z., Simonin, D., Tubbs, R., Kelly, G. and Caron, J.-F. (2013) The Met Office NWP-based Nowcasting Demonstration Project for the summer 2012 floods and London Olympics 2012. *Geophys. Res. Abstr.*, 15, EGU2013-9631, EGU General Assembly 2013.

Barnston, A.G., Van den Dool, H.M., Rodenhuis, D.R., Ropelewski, C.R., Kousky, V.E., O'Lenic, E.A., Livezey, R.E., Zebiak, S.E., Cane, M.A., Barrett, T.P., Graham, N.E., Ji, M. and Leetmaa, A. (1994) Long-lead seasonal forecasts: where do we stand? *Bull. Am. Meteorol. Soc.*, 75, 2097–2114.

Bellon, A. and Austin, G.L. (1984) The accuracy of short-term rainfall forecasts. *J. Hydrol.*, 70, 1778–1787.

Bowler, N.E., Pierce, C. and Seed, A.W. (2006) STEPS: a probabilistic precipitation forecasting scheme which merges an extrapolation nowcast with downscaled NWP. *Q. J. R. Meteorol. Soc.*, 132, 2127–2155.

Bowler, N.E., Arribas, A., Mylne, K.R., Robertson, K.B. and Beare, S.E. (2008) The MOGREPS short-range ensemble prediction system. *Q. J. R. Meteorol. Soc.*, 134, 703–722.

Bremnes, J.B. (2004) Probabilistic forecasts of precipitation in terms of quantiles using NWP model output. *Mon. Weather Rev.*, 132, 338–347.

Browning, K.A. (1979) The FRONTIERS plan: a strategy for using radar and satellite imagery for very-short-range precipitation forecasting. *Meteorol. Mag.*, 108, 161–184.

Browning, K.A. (1980) Local weather forecasting, *Proc. Roy. Soc. Lond. A*, 371, 179–211.

Browning, K.A. and Collier, C.G. (1982) An integrated radar–satellite nowcasting system in the UK. Chapter 1.5 in *Nowcasting*, ed. K.A. Browning, pp. 47–61. Academic Press, London.

Browning, K.A., Collier, C.G., Larke, P.R., Menmuir, P., Monk, G.A. and Owens, R.G. (1982) On the forecasting of frontal rain using a weather radar network. *Mon. Weather Rev.*, 110, 534–552.

Buizza, R., Miller, M. and Palmer, T.N. (1999) Stochastic representation of model uncertainties in the ECMWF Ensemble Prediction System. *Q. J. R. Meteorol. Soc.*, 125, 2887–2908.

Buizza, R., Houtekamer, P.L., Toth, Z., Pellerin, G., Wei, M. and Zhu, Y. (2005) A comparison of the ECMWF, MSC and NCEP global ensemble prediction systems. *Mon. Weather Rev.*, 133, 1076–1097.

Clark, P. (2006) High resolution model progress and plans. MOSAC-11, 17–18 November, Paper 11.9.

Colquhoun, J.R. (1987) A decision tree method of forecasting thunderstorms, severe thunderstorms and tornadoes. *Weather Forecast.*, 2, 337–345.

Collier, C.G. (1981) Objective rainfall forecasting using data from the United Kingdom weather radar network. Proceedings IAMAP Symposium, Hamburg, 25–28 August 1981, Special Publication ESA SP-165, pp. 201–206.

Collier, C.G. (1996) *Applications of Weather Radar Systems: A Guide to Uses of Radar Data in Meteorology and Hydrology*, 2nd edn. Wiley, Chichester.

Collier, C.G., Hardaker, P.J. and Pierce, C.E. (1995) The GANDOLF system: an automatic procedure for forecasting convection. Preprints 27th Conference on Radar Meteorology, 9–13 October, Vail, CO, pp. 606–608. American Meteorological Society, Boston.

Dixon, M. and Weiner, E. (1993) TITAN: Thunderstorm Identification Tracking, Analysis and Nowcasting: a radar-based methodology. *J. Atmos. Oceanic Technol.*, 10, 785–797.

Dixon, M., Zhihong, L., Lean, H., Roberts, N. and Ballard, S. (2009) Impact of data assimilation on forecasting convection over the United Kingdom using a high-resolution version of the Met Office Unified Model. *Mon. Weather Rev.*, 137, 1562–1584.

Doswell, C.A. III (1986) Short-range forecasting. Chapter 29 in *Mesoscale Meteorology and Forecasting*, ed. P. Ray, pp. 689–719. American Meteorological Society, Boston.

Elio, R., DeHaan, J. and Strong, G.S. (1987) METEOR: an artificial intelligence system for convective storm forecasting. *J. Atmos. Oceanic Technol.*, 4, 19–28.

Elvander, R.C. (1976) An evaluation of the relative performance of three weather radar echo forecasting techniques. Preprints 17th Conference on Radar Meteorology, Seattle, WA, pp. 526–532. American Meteorological Society, Boston.

Flowerdew, J. (2012) Calibration and combination of medium-range ensemble precipitation forecasts. Met Office Forecasting Research, Technical Report 567.

Foresti, L. and Seed, A. (2014) The effect of flow and orography on the spatial distribution of the very short-term predictability of rainfall from composite radar images. *Hydrol. Earth Syst. Sci.*, 18, 4671–4686.

Glahn, H.R. and Lowry, D.A. (1972) The use of model output statistics (MOS) in objective weather forecasting. *J. Appl. Meteorol.*, 11, 1203–1211.

Hamill, T.M. and Colucci, S.J. (1998) Evaluation of Eta-RSM ensemble probabilistic precipitation forecasts. *Mon. Weather Rev.*, 126, 711–724.

Hamill, T.M., Whitaker, J.S. and Wei, X. (2004) Ensemble re-forecasting: improving medium-range forecast skill using retrospective forecasts. *Mon. Weather Rev.*, 132, 1434–1447.

Hand, W.H. and Conway, B.J. (1995) An object oriented approach to nowcasting showers. *Weather Forecast.*, 10, 327–341.

Harrison, D.L., Scovell, A.W. and Kitchen, M. (2009) High-resolution precipitation estimates for hydrological users. *Proc. Inst. Civil Eng. Water Manag.*, 162, 125–135.

Harrold, T.W., English, E.J. and Nicholass, C.A. (1974) The accuracy of radar-derived rainfall measurements in hilly terrain. *Q. J. R. Meteorol. Soc.*, 100, 331–350.

Hsu, W.-R. and Murphy, A.H. (1986) The attributes diagram: a geometrical framework for assessing the quality of probability forecasts. *Int. J. Forecast.*, 2, 285–293.

Mascaro, G., Vivoni, E.R., and Deidda, R. (2010) Implications of ensemble quantitative precipitation forecast errors on distributed streamflow forecasting. *J. Hydrometeorol.*, 11, 69–86.

Monk, G.A., Browning, K.A. and Jonas, P.R. (1987) Forecasting applications of radar derived precipitation echo velocities in the vicinity of polar lows. *Tellus*, 39A, 426–433.

Mylne, K. (2013) Scientific framework for the ensemble prediction system for the UKV. MOSAC and SRG Meetings 2013, 13–15 November, MOSAC Paper 18.6. Met Office, Exeter.

Pankiewicz, G. (1995) The application of pattern recognition techniques to cloud detection. *Meteorol. Appl.*, 2, 257–271.

Pappenberger, F., Wetterhall, F., Albergel, C., Alfieri, L., Balsamo, G., Bogner, K., Haiden, T., Hewson, T., Magnusson, L., DeRosnay, P., Munoz-Sabater, J. and Tsonevsky, I. (2013) Floods in Central Europe in June 2013. *ECMWF Newsletter*, 136 (Summer), 9–11.

Pielke, R.A. (1981) Mesoscale dynamic modelling. *Adv. Geophys.*, 23, 186–344.

Raftery, A.E., Gneiting, T., Balabdaoui, F. and Polakowski, M. (2005) Using Bayesian model averaging to calibrate forecast ensembles. *Mon. Weather Rev.*, 133, 1155–1174.

Rihan, F.A., Collier, C.G. and Roulstone, I. (2005) Four-dimensional variational data assimilation for Doppler radar wind data. *J. Comput. Appl. Math.*, 176, 15–34.

Rihan, F.A., Collier, C.G., Ballard, S.P. and Swarbrick, S.J. (2008) Assimilation of Doppler radial winds into a 3D-Var system: errors and impact of radial velocities on the variational analysis and model forecasts. *Q. J. R. Meteorol. Soc.*, 134, 1701–1716.

Singh, D., Tsiang, M., Rajaratnam, B. and Diffenbaugh, N.S. (2013) Precipitation extremes over the continental United States in a transient, high-resolution, ensemble climate model experiment. *J. Geophys. Res. Atmos.*, 13, 7063–7086.

Sloughter, J.M., Raftery, A.E., Gneiting, T. and Fraley, C. (2007) Probabilistic quantitative precipitation forecasting using Bayesian model averaging. *Mon. Weather Rev.*, 135, 3209–3220.

Stanski, H.R., Wilson, L.J. and Burrows, W.R. (1989) Survey of common verification methods in meteorology. World Weather Watch Technical Report 8, WMO/TD 358. World Meteorological Organization, Geneva.

Stockdale, T.N., Andesson, D.L.T., Alves, J.O.S. and Balmaseda, M.A. (1998) Global seasonal rainfall forecasts using a coupled ocean–atmosphere model. *Nature*, 392, 370–373.

Sun, J., Xue, M., Wilson, J.W., Zawadzki, I., Ballard, S.P., Onvlee-Hooimeyer, J., Joe, P., Barker, D., Li, R.W., Golding, B., Xu, M. and Pinto, J. (2013) Use of NWP for nowcasting convective precipitation: recent progress and challenges. *Bull. Am. Meteorol. Soc.*, early online release, journals.ametsoc.org/toc/bams/0/0.

Tatehira, R., Sato, H. and Makino, Y. (1976) Short-term forecasting of digitised echo pattern. *J. Meteorol. Res. Japan Met. Agency*, 28, 61–70.

Verkade, J.S., Brown, J.D., Reggoani, P. and Weerts, A.H. (2013) Precipitation extremes over the continental United States in a transient, high-resolution, ensemble climate model experiment. *J. Hydrol.*, 501, 73–91.

Wilk, K.E. and Gray, K.C. (1970) Processing and analysis techniques used with the NSSL weather radar system. Preprints 14th Conference on Radar Meteorology, pp. 369–374. American Meteorological Society, Boston.

Wilks, D.S. (2006) *Statistical Methods in the Atmospheric Sciences*, 2nd edn. Academic Press, San Diego.

Winston, P.H. (1984) *Artificial Intelligence*. Addison Wesley, New York.

Wu, L., Seo, D.-J., Demargne, J., Brown, J.D., Cong, S. and Schaake, J. (2011) Generation of ensemble precipitation forecast from single-valued quantitative precipitation forecast for hydrologic ensemble prediction. *J. Hydrol.*, 399, 281–298.

Xu, Q. and Gong, J. (2003) Background error covariance functions for Doppler radial-wind analyses. *Q. J. R. Meteorol. Soc.*, 129, 1703–1720.

Zuckerberg, F.L. (1976) An application of the binomial distribution to radar observations for short-range precipitation forecasting, Preprints 6th Conference on Weather Forecasting and Analysis, Albany, NY, pp. 228–231. American Meteorological Society, Boston.

9

Flow Forecasting

9.1 Basic flood forecasting techniques

Beven (2001) and Moore and Bell (2001) have provided a comparison of rainfall–runoff models for flood forecasting. Collier (1985) and others have suggested two basic approaches to hydrological forecasting:

- *The reactive approach: upstream and downstream river level correlation.* In this approach, real-time forecasting and river control are accomplished by using *a priori* derived operating rules. These may be based upon river stage-to-stage correlations relating present values from telemetering raingauges, river-flow sensors or storage measurements indicative of upstream-to-downstream conditions. This method can be used as a forecasting technique where the relationship between the two sites is always the same and can easily be calculated. The method is often used on larger rivers like the Nile or in the United Kingdom the Thames. Also included here is extrapolation forecasting (also known as rate-of-rise extrapolation) and rainfall depth-duration thresholds.
- *The adaptive approach: rainfall–runoff modelling.* Forecasts and control decisions in the current real-time period are based upon future anticipated inflows as well as on current conditions, so that available storage and flow capacities may be utilized in the best way, and warnings issued as quickly as possible. This is a more complex forecasting tool and there are many different methods to choose from, for example, probability distributed model (PDM) and physically realizable transfer function (PRTF) model. Real-time rainfall and river data are put into the model to predict the timing and magnitude of maximum river levels or flows. The model can use rainfall data either as actual recorded rainfall or as forecast rainfall taken from radar predictions. Rainfall–runoff models can provide extra lead time, but rely on good quality rainfall forecasts.

In addition to these two basic approaches, other types of model include routing models and hydrodynamic models. (The basic equations for describing flow are outlined in Appendix 9.1.) Routing models route the flow from upstream to downstream river monitoring stations, incorporating inflows from other tributary streams.

Hydrometeorology, First Edition. Christopher G. Collier.
© 2016 John Wiley & Sons, Ltd. Published 2016 by John Wiley & Sons, Ltd.
Companion website: www.wiley.com/go/collierhydrometerology

Hydrodynamic models were originally developed for flood mapping purposes; however, in some places these have now been adapted for use as a real-time forecasting tool. These models rely on detailed topographic survey data and use standard software packages, such as Mike 11 or Flood Modeller (formerly ISIS), to forecast levels at set locations. Another package is HEC-RAS, developed by the US Army Corps of Engineers, which is freely available. Good calibration is essential for reliable, confident forecasts.

9.2 Model calibration and equifinality

Generally it is not possible to estimate the parameters of a model by either measurement or prior estimation (see for example Beven et al., 1984; Loague and Kyriakidis, 1997). As pointed out by Beven (2001) there are two main reasons for difficulties with model calibration. First, the scale of the measurement techniques available is generally much less than the scale at which parameter values are required. For example hydraulic conductivity is measured over areas of around 1 m^2, whereas models often use grid boxes of around 100 m^2. Second, parameter values are optimized by comparing the results of repeated simulations with observations of the catchment response such that 'best fit' parameter sets are found.

Finding algorithms that are robust with respect to the complexity of the parameter sets can be very difficult. Optimization may not involve enough information in the data and is therefore flawed. Also optimization generally assumes that the observations are error free and the model is a true representation of the data; this is often not true. Consequently the parameter values may only be valid inside the model structure used in the calibration. Also there may be many optimum parameter sets that are nearly as good as others.

Given the limitations of both observed data and model structures, *equifinality* – the concept that there are many possible parameter sets – has been introduced. Multiple acceptable parameter sets will result in a range of predictions, so allowing for a measure of uncertainty. We discuss this procedure in the next section.

9.3 Flood forecasting model development

Rainfall–runoff model types include input–storage–output (ISO), event driven, distributed and hydrodynamic. However, there are three generic types of rainfall–runoff model, namely *black box* (or *metric*), *conceptual* (or *parametric*) and *physics based* (or *mechanical*). Generally these models are characterized by increasing complexity.

Black box models are usually linear, and may use time series analysis techniques. The component parts are taken as input, storage, output (Lambert, 1972) and slow flow (base flow). Initial conditions are modified according to observed records, and model parameters are estimated using regression equations with stream flow records. Burn and Boorman (1993) describe a method involving the assignment of ungauged catchments to a group based upon the physical characteristics of the catchment and the use of similarity measures.

Grey box models are combinations of several linear models which may work reasonably well over limited areas and time periods. However since the rainfall–runoff process is non-linear and time variant, and the catchment characteristics are inhomogeneous, the results are not always reliable.

Conceptual models are based upon parameters which are meaningful, but which are not usually measurable, and therefore must be optimized. Autoregressive models (Box and Jenkins, 1970) are of this type, and the parameters are usually estimated by least square methods, the method of moments or the maximum likelihood method (see for example Wang and Singh, 1994). This process is referred to as *model calibration*, this being a process of systematically adjusting model parameter values to obtain a set of parameters which provides the best estimate of the observed stream flow (see for example Vidal et al., 2007). Problems may arise when the optimization process identifies different parameter sets, and it is unclear which set is optimum (Duan et al., 1992; Sorooshian et al., 1993).

Generalized likelihood uncertainty estimation (GLUE) is a statistical method for quantifying the uncertainty of model predictions (Beven and Binley, 1992). The basic idea of GLUE is that given our inability to represent physical processes exactly in a mathematical model, there will always be several different models that mimic equally well an observed natural process (such as river discharge). Such equally acceptable or behavioural models are therefore called equifinal (Beven, 2006). The methodology deals with models whose results are expressed as probability distributions of possible outcomes, often in the form of Monte Carlo simulations, and the problem can be viewed as assessing, and comparing between models, how good are these representations of uncertainty. Hunter et al. (2005) describe a calibration framework based on GLUE that can be used to condition hydrological model parameter distributions in scarcely gauged river basins, where data are uncertain, intermittent or non-concomitant.

Physics-based models use quantities which are, in principle, measurable (see for example Beven and Kirby, 1979). Model parameters have physical significance with homogeneous systems. However, it is necessary to associate the small scale physics with the model grid scale in meteorological models. This is the case in the Système Hydrologique Européen (SHE) model (Abbot et al., 1986a; 1986b) (Figure 9.1). Bathurst (1986) describes how this is done in applying the SHE model in the Wye catchment in the United Kingdom. The flow of water over and through the soil surface is described in Appendix 9.2.

Hydrodynamic models were originally developed for flood mapping purposes; however, in some places they have now been adapted for use as a real-time forecasting tool. These models rely on detailed topographic survey data and use standard commercial model software packages such as the Halcrow (now CH2M HILL) ISIS package or the DHI Mike 11 to forecast levels at set locations. Good calibration is essential for confident forecasts.

The LISFLOOD-FP is a two-dimensional hydrodynamic model specifically designed to simulate floodplain inundation in a computationally efficient manner over complex topography (Bates et al., 1992). It is capable of simulating grids (Figure 9.2) up to 10^6 cells for dynamic flood events and can take advantage of new sources of terrain information from remote sensing techniques such as airborne laser altimetry and synthetic aperture radar (SAR). The model predicts water depths in each grid cell at each time step, and hence can simulate the dynamic propagation of flood waves over fluvial, coastal and estuarine floodplains. It is a non-commercial research code (Bates and De Roo, 2000) developed as part of an effort to improve our

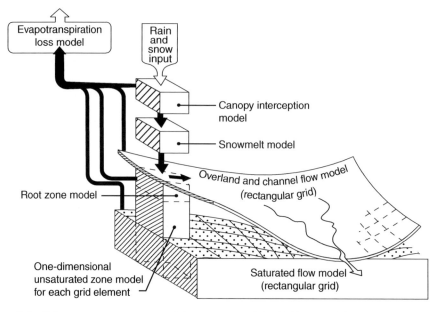

Figure 9.1   Schematic representation of the structure of the SHE model (after Abbott et al., 1986b; from Collier, 1996)

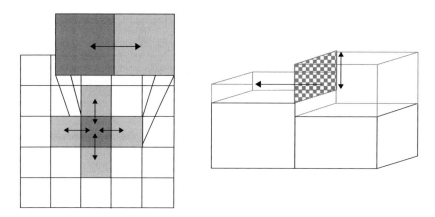

Figure 9.2   Grid cells and the transfer of water between grid cells

fundamental understanding of flood hydraulics, flood inundation prediction and flood risk assessment.

LISFLOOD-FP has been used as a research and operational tool, for example within the European Flood Awareness System (EFAS) discussed in section 9.8, and as part of the Risk Assessment for Flood and Coastal Defence for Strategic Planning (RASP) tiered methodology for flood risk assessment developed for the UK Environment Agency and DEFRA. It has also been used for research studies at a number of institutions, including Ohio State and the University of Washington in the United States and the University of Messina in Italy.

9.4   Conversion of detailed hydrodynamic models to simplified models suitable for real-time flood forecasting

Flood risk mapping and scheme design models tend to be calibrated primarily for historical flood events, whereas a real-time model should be able to operate year round in a range of flow conditions. However, it should be noted that the improved flood forecasting model is optimized for flood forecasting, and not for the more complex task of full flow forecasting, which would require modelling of abstractions, discharges and other complicating factors. An example of a hydrodynamic model developed for real-time use is the Eden/Carlisle model built using the ISIS (v2.3) software (Chen, 2005).

The overall objective of the hydraulic model improvement for real-time flood forecasting was to predict flood levels at the defined flood forecast points and to improve the model runtime, convergence and stability for real-time use, thus allowing a 'slimmer' model to run on a higher time step and deliver a shorter model runtime whilst preserving model accuracy at the flood forecast points. In general, different simplification and modification approaches are required for different features of the models, as follows:

- Significant hydraulic controls for local water levels needed to be retained within the model to achieve the accuracy specifications.
- Explicitly modelled areas that are not specifically required to deliver accurate simulations at flood forecast points can be removed from the model, or represented in a lumped fashion (typically these are sections of the model upstream of the reach for which forecasts are required).
- Areas known to be potential causes of instability (and therefore drivers for low time steps and more numerical iterations) needed to be modified in order to increase the model time step.

A reduction in the number, and hence an increase in the spacing, of channel cross-sections can also often reduce model runtimes. However, it is sometimes necessary to add cross-sections and spills to improve model stability and convergence and hence reduce model runtime.

Xiong and O'Connor (2002) investigated four different updating models, namely the most widely used single autoregressive (AR) model, the autoregressive threshold (AR-TS) updating model, the fuzzy autoregressive threshold (FU-AR-TS) updating model and the artificial neural network (ANN) model. As a baseline updating procedure, against which the performance of models can be compared, the results of fitting an AR model to the observed discharge series were used.

The calibration procedure of the standard AR model in updating the simulation errors of a rainfall–runoff model is as follows. First, the simulation errors of the rainfall–runoff (substantive) model are obtained as:

$$e_i = Q_i - \hat{Q}_i \tag{9.1}$$

where e_i denotes the simulation error of the selected substantive model, Q_i is the observed discharge, and $\hat{Q}_i$ is the estimated discharge. If the mean value $\bar{e}$ of the simulation error series of the calibration period is not equal to zero, then that mean should be subtracted from the simulated errors to produce a corresponding zero-mean time series ε_i:

$$\varepsilon_i = e_i - \bar{e} \tag{9.2}$$

Hence, the AR updating model at the current time step i is:

$$\hat{e}_i = a_1 \varepsilon_{i-1} + a_2 \varepsilon_{i-2} + \cdots + a_p \varepsilon_{i-p} \tag{9.3}$$

where $a_1, a_2, \ldots, a_p$ are the coefficients used in the AR model, $\hat{e}_i$ is the estimate of ε_i, and p is the number of AR coefficients. Recalling that the autocorrelation function of the ε_i time series satisfies a form of linear difference equation analogous to that of equation 9.3, one can obtain the Yule–Walker estimates of the parameters by replacing the theoretical autocorrelations by their respective estimates (obtained from the ε_i time series) (Box and Jenkins, 1970). In practice, however, the AR parameters $a_1, a_2, \ldots, a_p$ are generally estimated by the ordinary least squares method, for the chosen value of order p, by treating equation 9.3 as a linear regression.

To prevent, mitigate and manage urban pluvial flooding, it is necessary to accurately model and predict the spatial and temporal distribution of both rainfall and surface flooding. Schellart et al. (2011) describe a number of cases of urban pluvial flood modelling.

The first case study explores the use of STEPS (Chapter 8) as input to the commercial hydrodynamic sewer network model InfoWorks CS v10.0 to predict rainfall runoff from the urban area as well as flows through the sewer network conduits. The hydrodynamic sewer model was obtained from the sewer operators and had been calibrated following current industrial standards (WaPUG, 2002).

In the second case study, two types of physically based surface flooding models are employed: a 1D–1D model and a 1D–2D model. In the 1D–2D approach, the surface network is modelled as a 2D mesh of triangular elements based on a digital terrain model (DTM). The 2D model of the surface network is coupled with the 1D model of the sewer network to obtain a 1D–2D model. Although the 2D models of the surface network are detailed and accurate, modelling 2D surface flows is computationally intensive and the simulations take a long time, making it unsuitable for real-time applications. In the 1D–1D approach, the Automatic Overland Flow Delineation (AOFD) tool (Maksimović et al., 2009) is used to create the 1D model of the overland network, which is coupled with the 1D model of the sewer network. The AOFD tool uses a high resolution DTM, obtained from 1 m resolution lidar data, for the creation of a network of ponds (modelled as storage nodes) connected by preferential pathways (modelled as conduits with associated geometry derived from the DTM). The output of the AOFD tool is a 1D model of the overland network which can be imported into InfoWorks CS and is coupled with the sewer network model (the connection between these two systems takes place at the manholes). The 1D–1D dual drainage models can reproduce the behaviour of the system, while keeping computational time reasonably short, enabling the use of these models for real-time forecasting of pluvial flooding. However, in order to further decrease simulation time, techniques for simplifying the 1D–1D physically based models have been implemented (Simões et al., 2010) and hybrid models that combine 1D–1D and 1D–2D are currently under development.

In the third case study the application of AI techniques for flood modelling was explored. The RAdar Pluvial flooding Identification for Drainage System (RAPIDS) is an artificial neural networks (ANNs) technique. RAPIDS1 is a feedforward multi-layer perceptron (MLP), which provides a fast surrogate data driven model (DDM) for a conventional hydrodynamic simulator (such as InfoWorks). It rapidly relates incoming rainstorm data to the extent of flooding present at each manhole in the sewer network. A moving time window approach is implemented: rainfall data (intensity, cumulative total, elapsed time) spanning recent history (of 10 time steps, i.e. 30 minutes) provide input to the ANN. Output target signals for training and evaluation of ANN performance are provided by the corresponding flood level hydrographs generated by a hydrodynamic simulator for each manhole (Duncan et al., 2011). By time-advancing the target data used for training, the ANN can provide prediction of flooding for up to 60 minutes ahead. The framework of the ANN is such that it can be trained using the flow information from either hydraulic modelling results or field measurements whenever the observations are available. The flooding assessment of RAPIDS1 relies on rainfall input, for which UK Met Office NIMROD 1 km composite radar data were used.

9.5   Probabilistic flood forecasting and decision support methods

Probabilistic flood forecasting has a number of benefits. It represents the inherent uncertainties associated with flood forecasts and can improve the utility of forecasts for flood warning in situations of greater uncertainty, such as in convective storms or for longer lead times. Overall, it allows certain actions to be taken earlier in a more informed way and provides a more complete picture of potential flood risk as an event develops.

Laegar et al. (2010) suggest the following categories of uncertainty techniques:

1. techniques for forward uncertainty propagation in which the uncertainty is assumed to be known or specified in advance, and is used to determine the likely range or distribution of model outputs.
2. data assimilation and conditioning approaches in which the model outputs are conditioned on current and historical observations and/or real-time observed data are assimilated to update and improve a forecast.

They reported some experiences with the following techniques:

- gain updating and Kalman filter approaches, including the database mechanistic (DBM) methodology (Young, 2006), based upon a stochastic approach to data assimilation
- quantile regression (Koenkert, 2005), a method for estimating conditional quantiles
- autoregressive moving average (ARMA) error prediction with uncertainty estimation (Moore et al., 2010), the PDM rainfall–runoff model.

Flood forecast uncertainty deriving from both model uncertainty and rainfall forecast uncertainty can be jointly assessed if ensemble rainfall forecasts are available. To illustrate this, the parametric ARMA approach was applied using ensemble rainfall forecasts – produced out to 6 hours ahead by STEPS (Bowler et al., 2006; Chapter 8) – as input to the rainfall–runoff model for future times, emulating the real-time forecasting situation. The probability bands of model uncertainty for selected ensemble forecast percentiles can then be calculated using the parametric ARMA method.

Due to the relatively small spatial scale and rapid response of urban drainage systems, the use of quantitative rainfall forecasts for providing quantitative flow and depth predictions is very challenging. These predictions are important when consideration is given to urban pluvial flooding and receiving water quality. Schellart et al. (2014) considered three quantitative precipitation forecast methods of increasing complexity, and used them to create quantitative forecasts of sewer flows 0–3 hours ahead in the centre of a small town in the north of England. The HyRaTrac radar nowcast model was employed, as well as two different versions of the more complex STEPS model (Chapter 8). The STEPS model was used as a deterministic nowcasting system, and was also blended with the numerical weather prediction (NWP) model MM5 to investigate the potential of increasing forecast lead times (LTs) using high resolution NWP. Predictive LTs between 15 and 90 min gave acceptable results, but were a function of the event type. Schellart et al. concluded that higher resolution rainfall estimation as well as nowcasts are needed for prediction of both local pluvial flooding and combined sewer overflow spill events.

9.6   Derivation of station rating (stage-discharge) curves

A stage-discharge rating curve represents the relation of water level at a given point in a stream to a corresponding volumetric rate of flow. The shape of a curve can be derived by conducting synchronized measurements of stage and discharge and investigating the pattern of points on a scatter plot. Any physical change to the stream channel will alter this relation and these changes must be accounted for in the derivation of a discharge hydrograph from a time series of stage data.

There are often only enough measurements to hint at what the true shape of the curve is and how it changes over time. A poorly conceived rating curve can produce discharge data that do not pass the test of hydrologic reason. Extreme values from curve extrapolation may be too high (e.g. a rainfall–runoff ratio > 1), or too low. The early years of record may have peaks defined by a badly extrapolated curve relative to 'better', more mature, extrapolations resulting in an apparent (but not real) trend in peak flows. The seasonal water balance may be skewed by poorly modelled backwater effects.

9.7   Performance testing of forecasting models and updating procedures

A basic method of calculating the predictive uncertainty in flood forecasts involves computation of the exceedance probability of alert and alarm levels. Such results allow far more complete information to be provided to decision makers (in comparison with deterministic model-based forecasts). The uncertainty estimation is based on the statistical analysis of historical flood forecasting results. The simplest methods of estimating performance from training, validation and testing are evaluated by the following measures for goodness-of-fit: root mean square error (RMSE) and coefficient of correlation (CC). RMSE furnishes a quantitative indication of the model error in units of the variable, with the characteristic that larger errors receive greater attention than smaller ones. The qualitative evaluation of the model performance is made in terms of the coefficient of correlation between the measured and simulated data. Another important criterion of the model performance for flood forecasting is to predict the peak flood as accurately as possible.

The forecast residuals (differences between predictions and measurements at river gauging stations) have been analysed by Van Steenbergen et al. (2012) using a non-parametric technique. Because the residuals are correlated with the value of the simulated water level and time horizon, the residuals are split up into discrete classes of simulated water levels and time horizons. For each class, percentile values of the residuals are calculated and stored in a so-called 'three-dimensional error matrix'. Based on 3D interpolation in the error matrix, confidence intervals on forecasted water levels may be calculated and visualized.

Although a summary of flood forecasting models has been provided by the World Meteorological Organization (WMO, 2011), in recent years many non-linear approaches, such as the artificial neural networks, fuzzy logic and genetic algorithm approaches, have been used in solving flood forecasting problems including assessing performance. The computational side of artificial neural networks is very useful in real-time operations.

When conceptual rainfall–runoff models are intended for use in an actual real-time forecasting situation they are generally associated with an updating procedure whereby, at the time of making the forecast, errors already observed in the recent non-updated forecasts are used to modify the forecast about to be made. These simulation errors of the rainfall–runoff models can be due to, for example, inadequacies in the model structure, incorrect estimation of the model parameters, errors in the data applied to the model, or indeed the absence of any substantial, consistent or coherent relationship in the data used to calibrate the models (Kachroo and Liang, 1992). Through the analysis of the persistence structure of these errors, an updating procedure can be established. Although the objective of the updating procedure is to improve the flow forecast by using the recent values of the observed discharges, other factors can be used, such as rainfall (including that over the lead time of the forecast), soil moisture content, model parameter estimates, and simulated (non-updated) flow values of a rainfall–runoff model (Georgakakos and Smith, 1990; WMO, 1992; Suebjakla, 1996).

The most popular updating scheme is that achieved through the indirect updating of the outputs (i.e. the simulated discharges) of the rainfall–runoff model by a time series model of the forecast error series, usually an autoregressive (AR) model (WMO, 1992). A real-time forecasting procedure, using the AR model for updating, consists of three steps (WMO, 1992; Kachroo and Liang, 1992). The first step is to calibrate the selected, substantive rainfall–runoff model in order to obtain its estimates of the observed river flows. The second step is to calibrate and use the corrective AR model to forecast the simulation errors of the substantive model. The third and final step is to combine the discharge forecast from the substantive model and the corresponding error forecast from the error model, in order to obtain the updated discharge forecast, i.e. the final real-time output of the forecasting procedure. Xiong and O'Connor (2002) compared four updating models for real-time river flow forecasting, and recommended the use of the AR model.

Xiong and O'Connor (2002) tested the potential of the four updating models discussed in section 9.4 on 11 catchments using daily data. The lumped conceptual soil moisture accounting model (SMAR) of Kachroo (1992) was used in a non-updating mode. It was concluded that all four updating methods can effectively improve the flow forecasting efficiency by means of simulating the non-updated forecast errors of the SMAR model.

9.8 Configuration of models on to national and international forecasting platforms

Recent large floods in Europe, such as those that occurred in the Meuse and Rhine basins in 1993 and 1995, over large areas of the United Kingdom in 1998 and 2000, in the Elbe Basin in summer 2002, and in Switzerland, Austria and Romania in August 2005, have increased investments in research, development and upgrading of flood forecasting systems at the scale of medium and large river basins. One of the initiatives in Europe has been the development of the European Flood Forecasting System (EFFS) (De Roo et al., 2003), based upon WL | Delft Hydraulics (now Deltares) Delft-FEWS flood forecasting platform, and developed around the LISFLOOD suite of raster-based hydrology and hydraulic codes (De Roo et al., 2003; Bates and De Roo, 2000; Bergstrom et al., 2002). Its development took place over the period 2000–3 with the objective of providing a flood forecasting platform that could be applied at European scale (which led to the launch of the European Flood Awareness System, EFAS). The EFFS system was developed by a consortium of 19 European research institutes, universities and state agencies, with WL | Delft Hydraulics as a leading partner (Gouweleeuw et al., 2004). In the past, most flood forecasting systems have been built around the application of hydrological and/or hydraulic routing models. Contrary to this, EFFS has been developed as a data management platform. On the one hand, this platform is open to various data sources supplying measured or forecasted weather state variables, such as precipitation and temperature. On the other hand, the platform enables the generic coupling of a variety of hydrological/hydraulic flood routing models. The data management platform of the EFFS has been equipped with generic tools providing a variety of data handling tasks, such as data validation, interpolation, aggregation and error correction in forecasts, including a variety of visualization and forecast dissemination options. An example of the output of EFFS using the LISFLOOD-FF model is shown in Figure 9.3.

At the heart of the UK National Flood Forecasting System is Delft-FEWS (Flood Early Warning System) by Deltares. The National Flood Forecasting System (NFFS) comprises the complete fluvial and coastal flood forecasting capability of the Environment Agency for England and Wales. The NFFS was developed by the former WL | Delft Hydraulics (now Deltares) with subcontractor Tessella Scientific Software Solutions (UK). Since 2006, NFFS has been fully operational in all Environment Agency regions as the primary flood forecasting system.

The NFFS is set up as a client–server application. The main system servers are hosted centrally, together with dedicated 'number crunching' forecasting shell servers. Forecasters from the regional forecasting centres connect to these central servers using fully functional client software (Figure 9.4), while forecast results are

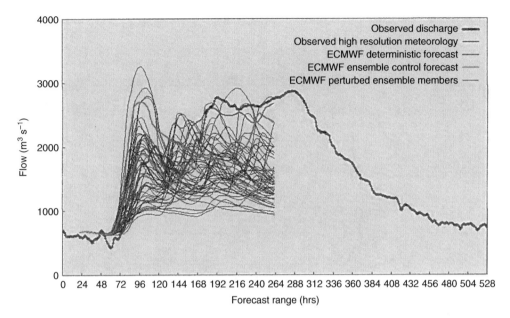

Figure 9.3   Example of the 10 day flow forecasts from the LISFLOOD-FF model for 1200UTC on 22 January 1995 for the Borgharen gauging station on the River Meuse, The Netherlands. The observed discharge is shown as a thick black line. The simulation is driven by the ECMWF TL511L60 deterministic forecast (shown as thick grey line) and the ensemble control forecasts. The ensemble forecast members are shown black (from Gouweleeuw et al., 2005 and De Roo et al., 2003)

disseminated to relevant users in the Environment Agency through a dedicated webserver connected to the Agency's intranet.

Particular emphasis has been given to the resilience of the forecasting system. This is found not only in the generic shell, where extensive functionality is included to ensure the continuity of the flood forecasting process, even where external data sources are incomplete or inconsistent, but also in the duplication of the central servers in two locations to run in duty/standby or duty/duty mode. The system has been fully developed in Java™, and resilient communication in the J2EE compliant client–server application is assured through the use of state-of-the-art standards such as JMS.

Configuration of the system was undertaken in close cooperation with flood forecasters from the regions. During the development of the system, workshops held at 2 month intervals gave detailed attention both to exactly how the system is configured to provide the regional forecasting requirements, and to forecasting functionality being added.

In each of the regions for which the NFFS is being developed, a range of forecasting modules are in use, including a number of hydrological and hydraulic models. To allow these to interconnect with the published interface to NFFS, module adapters have been developed as a part of the project; adapters have also been created by Halcrow, the Centre for Ecology and Hydrology (CEH), Wallingford, and DHI in cooperation with Deltares.

The Delft-FEWS team continues to provide courses and seminars on the latest improvements to the software. Users are invited to the annual FEWS user days

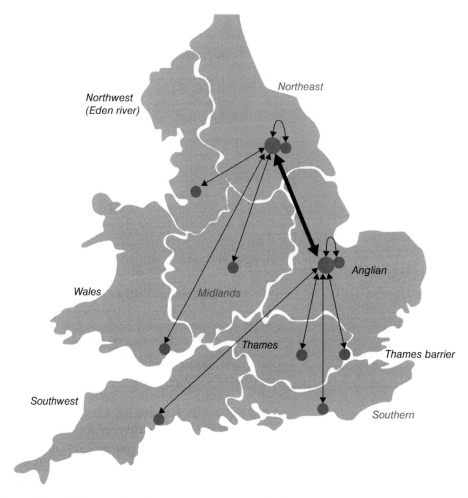

Figure 9.4   NFFS regional configuration and interaction (Deltares)

which provide insight into a range of operational and research projects using Delft-FEWS technology.

The grid-to-grid hydrological model is a tool developed by the Centre for Ecology and Hydrology (CEH) for translating rainfall into river flows to predict potential river flooding (Bell et al., 2007). Intensive work on the grid-to-grid (G2G) project began in 2002 at CEH and continues today in collaboration with the Environment Agency (EA). Existing regional hydrological models did not have the capability to make maximum use of high resolution rainfall forecasts since they are based on large geographical areas draining to a location where river flow is measured. The grid-to-grid model is configured on a 1 km² grid, and so can take better account of the location and intensity of rainfall forecast by high resolution weather models, allowing for very targeted river flood warnings. The model calculates the runoff from each grid square, which is then routed into the next downhill grid square, and so on, down to the bottom of the catchment: the model converts rainfall into river runoff and river

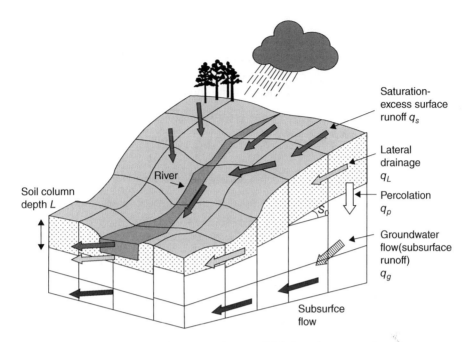

Saturation-
excess surface
runoff q_s

Lateral
drainage
q_L

Percolation
q_p

Groundwater
flow(subsurface
runoff)
q_g

River

Soil column
depth L

S_d

Subsurfce
flow

Figure 9.5   The grid-to-grid model structure (source CEH) (*see plate section for colour representation of this figure*)

flows. Surface and subsurface runoff and lateral water movements in the soil are accounted for (see Figure 9.5).

The US National Oceanic and Atmospheric Administration (NOAA) National Weather Service (NWS) is responsible for the production of the nation's river flood forecasts and warnings in support of the protection of life and property and enhancement of national commerce. For the past 30 years, NWS hydrologists have used the NWS River Forecast System (NWSRFS) as the core infrastructure for their hydrologic operations. NWSRFS is remarkable in that it has met most NWS needs for such a long time. With increasing operational needs and escalating support costs, the NWSRFS will be retired and replaced by the Community Hydrologic Prediction System (CHPS). CHPS has been developed by the NWS in collaboration with Deltares in The Netherlands. Delft-FEWS serves as the infrastructure for CHPS, with NWS hydrologic models and US Army Corps of Engineers (USACE) hydraulic models providing the forecasting core. The CHPS migration project realizes significant improvements for the NWS and its hydrologic forecast offices. First, CHPS provides all of the 'standard' benefits of a legacy system replacement, having lower support costs, better use of modern computing technology, and increased flexibility for adding service improvements. Second, the plug-and-play nature of CHPS expands the ability to partner with federal, state, local and university researchers. Third, CHPS allows the NWS to re-examine its hydrologic forecast concept of operations, and consider new ways to meet its mission requirements. A fourth benefit is the immediate connection between the NWS and the international flood forecasting community through the Delft-FEWS user community. The NWS anticipates being able to exchange forecast techniques with many others very easily.

9.9   Flood warnings and levels of service

The type of flood warning service available varies greatly from country to country, and a location may receive warnings from more than one service. There follow outlines of procedures in the United Kingdom and the United States. Flood warnings are issued in a number of ways:

- direct to the population at risk
- via an internet website
- from a recorded telephone message
- through the media
- from designated flood wardens calling at individual properties
- activation of sirens or other similar means.

9.9.1   United Kingdom

Arrangements for flood warnings vary across the United Kingdom, with several agencies leading on warnings for emergency responders and the public. The Environment Agency, Natural Resources Wales and the Scottish Environment Protection Agency all undertake location specific flood warning activities for communities at risk depending upon the scale of flood risk, technical challenges and investment needed to deliver a reliable service. Table 9.1 gives the meanings of the flood warnings that are issued.

Prior to issuing a flood warning consideration is given to:

- the needs of communities to activate emergency response plans
- the nature of the catchment or coastline and the lead time that may be provided
- meteorological observations and forecast information on rainfall and coastal water levels
- hydrological observations and flood forecasts
- reference to thresholds of historic or forecast flood levels.

Dissemination of flood warnings has moved towards a service whereby those at risk can pre-register to receive warnings by phone, email or text message from an automatic system, Floodline. Both warnings and updates about current conditions are also carried by local radio stations. In addition live updates are carried by the Environment Agency website, showing which locations have flood warnings in place and the severity of those warnings.

There is currently no flood warning system in Northern Ireland but the Met Office does issue weather warnings. Flood risk management is the responsibility of the Rivers Agency in Northern Ireland. Consideration will be given to the introduction of a warning system as part of the implementation of the EU Floods Directive.

9.9.2   United States and Canada

In the US, the National Weather Service issues flood watches and warnings for large scale, gradual river flooding. Watches are issued when flooding is possible or expected within 12–48 hours, and warnings are issued when flooding over a large

Table 9.1 The meaning of the Environment Agency flood warnings

Online flood risk forecast	FLOOD ALERT 	FLOOD WARNING	SEVERE FLOOD WARNING	Warning no longer in force
What it means Be aware. Keep an eye on the weather situation	*What it means* Flooding is possible. Be prepared	*What it means* Flooding is expected. Immediate action required	*What it means* Severe flooding. Danger to life	*What it means* No further flooding is currently expected in your area
When it's used Forecasts of flooding on our website are updated at least once a day	*When it's used* Two hours to two days in advance of flooding	*When it's used* Half an hour to one day in advance of flooding	*When it's used* When flooding poses a significant threat to life	*When it's used* When river or sea conditions begin to return to normal
What to do • Check weather conditions • Check for updated flood forecasts on our website	*What to do* • Be prepared to act on your flood plan • Prepare a flood kit of essential items • Monitor local water levels and the flood forecast on our website	*What to do* • Move family, pets and valuables to a safe place • Turn off gas, electricity and water supplies if safe to do so • Put flood protection equipment in place	*What to do* • Stay in a safe place with a means of escape • Be ready should you need to evacuate from your home • Cooperate with the emergency services • Call 999 if you are in immediate danger	*What to do* • Be careful. Flood water may still be around for several days • If you've been flooded, ring your insurance company as soon as possible

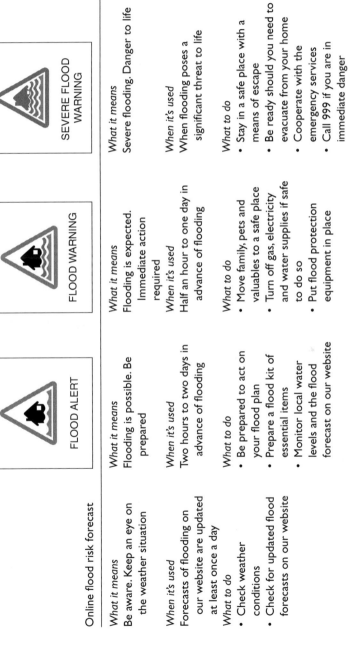

area or river flooding is imminent or occurring. Both can be issued on a county-by-county basis or for specific rivers or points along a river. When rapid flooding from heavy rain or a dam failure is expected, flash flood watches and warnings are issued.

In the United States and Canada dissemination of flood warnings is covered by Specific Area Message Encoding (SAME) code FLW, which is used in the US Emergency Alert System and NOAA Weatheradio network and in Canada's Weatheradio Canada network.

'Flood statements' are issued by the National Weather Service to inform the public of flooding along major streams in which there is not a serious threat to life or property. They may also follow a flood warning to give later information.

9.10 Case studies worldwide: river and urban

There are a great number of case studies of flood modelling in both rural and urban areas throughout the world. Einfalt et al. (2007) describe the URBAS project, which considers the urban occurrence of flash floods in Germany with a view to increasing the preparedness and the range of possible actions of communities and public organizations. Radar data are used to characterize convective cells, and this information is used to identify the frequency of occurrence of heavy precipitation pixels to help develop communal measures of precaution.

On a larger scale, Verwey et al. (2006) provide an overview of developments of the Delft-FEWS modelling system and its application in the Mekong basin. Although the system was developed for the Rhine and Po river catchments, it could be successfully applied to the wider Mekong area including Vietnam, Thailand, Cambodia and Laos.

The use of the ECMWF ensemble forecasts of precipitation as input to the Precipitation–Runoff–Evapotranspiration–Hydrological response units (PREVAH) model having a resolution of $500 \times 500 \, \mathrm{m}^2$ and a time step of 1 hour for the alpine tributaries of the Rhine has been described by Verbunt et al. (2007). Probabilistic forecast guidance on the early indications of extreme floods are successfully demonstrated, as shown in Figure 9.6.

Appendix 9.1 St Venant equations

The St Venant equations assume that flow may be described in terms of average cross-sectional velocities and depths from the balances of both mass and momentum in the flow (see for example Beven, 2001). For a flow of average velocity v and depth h in a cross-sectional area A, with wetted perimeter P, a bed slope S_0 and lateral inflow per unit length of slope or channel i, the *mass balance equation* is

$$\frac{\partial A}{\partial t} = -A\frac{\partial v}{\partial x} - v\frac{\partial A}{\partial t} + i \tag{9.4}$$

and the *momentum balance equation*, assuming the water is incompressible, is

$$\frac{\partial Av}{\partial t} + \frac{\partial Av^2}{\partial x} + \frac{\partial Agh}{\partial x} = gAS_0 - \frac{Pfv^2}{2} \tag{9.5}$$

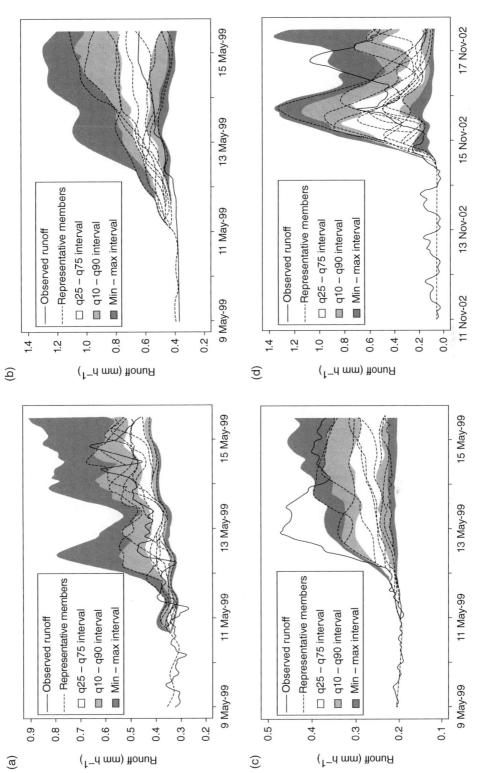

Figure 9.6 Several interquantile ranges together with 10 representative members for different catchment areas and different case studies: (a) Aare/Hagneck (5128 km²), (b) Reuss/Luzern (2251 km²), (c) Rhine/Rheinfelden (3455 km²), and (d) Rhine/Domat-Ems (3229 km²) (from Verbunt et al., 2007)

where f is the Darcy–Weisbach uniform roughness coefficient. One or two terms in the momentum equation can be neglected to give the *diffusion wave approximation*

$$\frac{\partial A \rho g h}{\partial x} = \rho g A (S_0 - S_i) \tag{9.6}$$

and the *kinematic wave approximation*

$$\rho g A (S_0 - S_i) = 0 \tag{9.7}$$

so that when $S_0 = S_i$ it is assumed that the water surface is always parallel to the bed. The local hydrostatic pressure p at the bed is given by $p = \rho g h$. This gives for the *transient flow* case

$$v = \left[(2g/f) S_0 R_h \right]^{0.5} \tag{9.8}$$

where R_h is the hydraulic radius of the flow ($R_h = A/p$). For a channel that is wide relative to its depth, or for an overland flow on a relatively smooth slope, $R_h \approx h$.

Appendix 9.2   Flow in unsaturated and saturated zones

Water accumulated on the soil surface responds to gravity by flowing down the gradient over the land surface to the channel system where it discharges through the stream channel to the outlet of the catchment. This is described by the equations of unsteady free surface flow represented by the conservation of mass and momentum.

The flow in the unsaturated zone can be estimated using Richard's equation

$$C = \frac{\partial \psi}{\partial t} = \frac{\partial}{\partial Z}\left(K \frac{\partial \psi}{\partial Z} \right) + \frac{\partial K}{\partial Z} + S \tag{9.9}$$

where ψ is the pressure head, t is the time variable, Z is the vertical coordinate (positive upwards), $C = \partial \theta / \partial \psi$ is the soil-water capacity, θ is the volumetric moisture content, K is the hydraulic conductivity and S is a source/sink term.

The governing equation describing the flow in the saturated zone is the non-linear Boussinesq equation

$$S \frac{\partial h}{\partial t} = \frac{\partial}{\partial x}\left(K_x H \frac{\partial h}{\partial x} \right) + \frac{\partial}{\partial y}\left(K_y H \frac{\partial h}{\partial y} \right) + R \tag{9.10}$$

where S is the specific yield, h is the phreatic surface level, K_x and K_y are the saturated hydraulic conductivities in the x and y directions respectively, H is the saturated thickness, t is the time variable, x and y are the horizontal space coordinates, and R is an instantaneous recharge/discharge term. This equation combines Darcy's law and the mass conservation of two-dimensional laminar flow in an anisotropic, heterogeneous aquifer. The recharge/discharge term R is

$$R = \Sigma q - \frac{\partial}{\partial t} \int_{h}^{q_s} \theta(Z,t) dz \qquad (9.11)$$

where Σq includes the transpiration q_R, the soil evaporation q_s, the infiltration q_1, the stream/aquifer exchange q_0, the external boundary flows q_e and the soil-moisture content in the unsaturated zone θ (see Abbott, 1986a; 1986b; WMO, 1994; Van Steenbergen and Willems, 2013).

Summary of key points in this chapter

1. There are two basic approaches to hydrological forecasting: the reactive approach based upon upstream and downstream river level correlation using *a priori* derived operating rules, and the adaptive approach based upon rainfall–runoff modelling.

2. Other types of model include routing models, which route the flow from upstream to downstream river monitoring stations; and hydrodynamic models, originally used for flood mapping and more recently as a real-time forecasting tool.

3. There are two difficulties with model calibration: the scale of the measurement techniques is generally much less than the scale at which parameter values are required, and parameter values are optimized to find 'best fit' parameter sets.

4. Equifinality is the concept that there are many possible parameter sets that are acceptable to give a range of predictions.

5. Rainfall–runoff model types are input–output–storage (ISO), event driven, distributed and hydrodynamic.

6. There are three generic types of model: black box (metric), conceptual (parametric) and physics based (mechanical).

7. Grey box models are combinations of several linear models, whereas conceptual models are based upon parameters which are meaningful but which are not usually measurable and therefore must be optimized.

8. Model calibration is a process of systematically adjusting model parameters to obtain a set of parameters which provides the best estimate of the observed stream flow.

9. Generalized likelihood uncertainty estimation (GLUE) is a statistical method for quantifying the uncertainty of model predictions.

10. Physics-based models use quantities which are, in principle, measured.

11. Hydrodynamic models rely on detailed topographic survey data and use standard commercial model software packages.

12. LISFLOOD-FP is a two-dimensional hydrodynamic model designed to simulate floodplain inundation. It has been used within the European Flood Awareness System (EFAS).

13. The overall objective of hydraulic model improvement for real-time flood forecasting is to predict flood levels at defined flood forecast points, and to improve model runtime, convergence and stability.

14. Updating models include single autoregressive models and artificial neural network models.

15. Autoregressive parameters are generally estimated by least square methods.

16. To prevent, mitigate and manage urban pluvial flooding, it is necessary to accurately model and predict the spatial and temporal distribution of both rainfall and surface flooding.

17. A 1D–2D surface flooding model is developed as a 2D mesh of triangular elements based on a digital terrain model.

18. Artificial intelligence techniques are used for flood modelling.

19. Probabilistic flood forecasting represents the inherent uncertainties associated with flood forecasts, and can improve the utility of forecasts for flood warning.

20. Categories of uncertainty techniques include techniques for forward uncertainty propagation and techniques in which model outputs are conditioned on current and historical observations.

21. A stage-discharge rating curve represents the relation of water level at a given point in a stream to a corresponding volumetric rate of flow.

22. The simplest methods of estimating performance involve deriving goodness-of-fit using root mean square error and coefficient of correlation.

23. The UK National Flood Forecasting System (NFFS) is the Delft-FEWS flood forecasting system by Deltares. It is set up as a client–server application.

24. The grid-to-grid hydrological model has been developed by the UK Centre for Ecology and Hydrology using a 1 km^2 grid and calculates the runoff from each grid square.

25. The ECMWF ensemble forecasts of precipitation are input to the Precipitation–Runoff–Evapotranspiration–Hydrological response units (PREVAH) model for the alpine tributaries of the Rhine.

Problems

1. Describe two basic approaches to hydrological forecasting.
2. What is the physically realizable transfer function model?
3. What are the main difficulties with model calibration?
4. Define equifinality.

5. What are the three generic types of rainfall–runoff model?
6. Describe what is referred to as model calibration.
7. Outline the GLUE procedure.
8. What are physics-based models? Give an example.
9. What are hydrodynamic models? Give an example.
10. Outline three different approaches for improving features of hydraulic models.
11. Give an equation for an autoregressive model.
12. Give two types of physically based surface flooding model.
13. Outline categories of uncertainty techniques used in probabilistic flood forecasting.
14. Explain what is a stage-discharge rating curve.
15. Give a basic method of calculating the predictive uncertainty in flood forecasts.
16. What are forecast residuals?
17. How can an updating procedure be established?
18. Name three steps in an autoregressive model for updating.
19. Outline the European Flood Forecasting System (EFFS).
20. Draw a schematic of the grid-to-grid model structure, showing how water moves through the system.
21. What factors need to be considered in issuing a flood warning?
22. Give an example of how ECMWF ensemble forecasts are input to a hydrological model.

References

Abbott, M.B., Bathurst, J.C., Cunge, C.A., O'Connell, P.E. and Rasmussen, J. (1986a) An introduction to the European Hydrological System–Système Hydrologique Européen 'SHE'. 1: History and philosophy of a physically-based, distributing modelling system. *J. Hydrol.*, 87, 45–59.

Abbott, M.B., Bathurst, J.C., Cunge, C.A., O'Connell, P.E. and Rasmussen, J. (1986b) An introduction to the European Hydrological System–Système Hydrologique Européen 'SHE'. 2: Structure of a physically-based, distributed modelling system. *J. Hydrol.*, 87, 61–77.

Bates, P.D. and De Roo, A.P.J. (2000) A simple raster-based model for flood plain inundation. *J. Hydrol.*, 236, 54–77.

Bates, P.D., Anderson, M.G., Baird, L., Walling, D.E. and Simm, D. (1992) Modelling flood plain flows using a two-dimensional finite element model. *Earth Surf. Proc. Land.*, 17, 575–588.

Bathurst, J.C. (1986) Physically-based distributed modelling of an upland catchment using the Système Hydrologique Européen. *J. Hydrol.*, 87, 79–102.

Bell, V.A., Kay, A.I., Jones, R.G. and Moore, R.J. (2007) Development of a high resolution grid-based river flow model for use with regional climate model output. *Hydrol. Earth Syst. Sci.*, 11, 532–549.

Bergstrom, S., Lindstrom, G. and Pettersson, A. (2002) Multi-variable parameter estimation to increase confidence in hydrological modelling. *Hydrol. Proc.*, 16, 413–421.

Beven, K.J. (2001) *Rainfall–Runoff Modelling: The Primer.* Wiley, Chichester.

Beven, K.J. (2006) A manifesto for the equifinality thesis. *J. Hydrol.*, 320, 18–36.

Beven, K.J. and Binley, R.M. (1992) The future of distributed models: model calibration and uncertainty prediction. *Hydrol. Proc.*, 6, 279–298.

Beven, K.J. and Kirby, M.J. (1979) A physically based variable contributing area model of basin hydrology. *Hydrol. Sci. Bull.*, 24, 43–69.

Beven, K.J., Kirby, M.J., Schoffield, N. and Tagg, A. (1984) Testing a physically-based flood forecasting model (TOPMODEL) for three UK catchments. *J. Hydrol.*, 69, 119–143.

Bowler, N.E., Pierce, C.E. and Seed, A.W. (2006) STEPS: a probabilistic precipitation forecasting scheme which merges an extrapolation nowcast with downscaled NWP. *Q. J. R. Meteorol. Soc.*, 132, 2127–2155.

Burn, D.G.H. and Boorman, D.B. (1993) Estimation of hydrological parameters at ungauged catchments. *J. Hydrol.*, 143, 429–454.

Box, G.E.P. and Jenkins, G.M. (1970) *Time Series Analysis and Control.* Holden-Day, San Francisco.

Chen, Y. (2005) River Eden/Carlisle flood forecast modelling. Atkins, Warrington.

Collier, C.G. (1985) Remote sensing for hydrological forecasting. In *Facets of Hydrology*, vol. II, ed. J.C. Rodda, pp. 1–24. Wiley, Chichester.

Collier, C.G. (1996) *Applications of Weather Radar Systems: A Guide to Uses of Radar Data in Meteorology and Hydrology.* Wiley-Praxis, Chichester.

De Roo, P.J. Ad, Gouweleeuw, B., Thielen, J., Bartholmes, J., Bonngioannini-Cerlini, P., Todini, E., Bates, P.D., Horritt, M., Hunter, N., Beven, K., Pappenberger, F., Heise, E., Rivin, G., Hollingsworth, A., Holst, B., Kwadijk, J., Reggiani, P., Van Dijk, M., Sattler, K. and Sprokkercef, E. (2003) Development of a European flood forecasting system. *Int. J. River Basin Manag.*, 1, 49–59.

Duan, Q., Sorooshian, S. and Gupta, V.K. (1992) Effective and efficient global optimisation for conceptual rainfall–runoff models. *Water Resour. Res.*, 28, 1015–1031.

Duncan, A.P., Chen, A.S., Keedwell, E.C., Djordjevic, S. and Savic, D.F. (2011) Urban flood prediction in real-time from weather radar and rainfall data using artificial neural networks. Weather Radar and Hydrology, Proceedings of a Symposium held in Exeter, UK, April 2011, IAHS Publication 351.

Einfalt, T., Hatzfeld, F., Wagner, A., Seltmann, J., Castro, D. and Frerichs, S. (2007) URBAS: forecasting and management of flash floods in urban areas. NOVATECT 2007, Session 1.2, pp. 75–82.

Georgakakos, K.P. and Smith, G.F. (1990) On improved operational hydrologic forecasting results from a WMO real-time forecasting experiment. *J. Hydrol.*, 114, 17–45.

Gouweleeuw, B.T., Thielen, J., Franchello, G., De Roo, A.P.J. and Buizza, R. (2005) Flood forecasting using medium-range probabilistic weather prediction. *Hydrol. Earth System Sci.*, 9(4), 365–380.

Gouweleeuw, W.B., Reggiani, P. and De Roo, A.J.P. (2004) A European Flood Forecasting System (EFFS). Full Report, Project EVG1-CT-1999-00011, EC Report EUR 21208 EN.

Hunter, N.M., Bates, P.D., Horritt, M., DeRoo, P.J. and Werner, M.G.F. (2005) Utility of different data types for calibrating flood inundation models within a GLUE framework. *Hydrol. Earth System Sci.*, 9, 412–430.

Kachroo, R.K. (1992) River flow forecasting. 5: Application of a conceptual model. *J. Hydrol.*, 133, 141–178.

Kachroo, R.K. and Liang, G.C. (1992) Non-linear modelling of the rainfall–runoff transformation. *J. Hydrol.*, 133, 17–40.

Koenkert, R. (2005) *Quantile Regression.* Cambridge University Press, Cambridge.

Laeger, S., Cross, R., Sene, K., Weerts, A., Beven, K., Leedal, D., Moore, R.J., Vaughn, M., Harrison, T. and Whitlow, C. (2010) Probabilistic flood forecasting in England and Wales: can it give us what we crave? BHS 3rd International Symposium on Managing Consequences of a Changing Global Environment, Newcastle, pp. 1–7.

Lambert, A.O. (1972) Catchment models based on ISO functions. *J. Inst. Water Eng. Sci.*, 26, 413–422.

Loague, K.M. and Kyriakidis, P.C. (1997) Spatial and temporal variability in the R-5 infiltration data set: déjà vu and rainfall–runoff simulations. *Water Resour. Res.*, 2883–2896.

Maksimović, C., Prodanović, D., Boonya-Aroonnet, S., Leitão, J.P., Djordjević, S. and Allitt, R. (2009) Overland flow and pathway analysis for modelling of urban pluvial flooding. *J. Hydraulic Res.*, 47, 512–523.

Moore, R.J. and Bell, V.A. (2001) Comparison of rainfall–runoff models for flood forecasting. 1: Literature review of models. R&D Technical Report W241, CEH Project C00260. Environment Agency, Bristol.

Moore, R.J., Robson, A.J., Cole, S.J., Howard, P.J., Weerts, A. and Sene, K. (2010) Sources of uncertainty and probability bands for flood forecasts: an upland catchment case study. *Geophys. Res. Abs.*, 12, EGU 2010.

Schellart, A., Ochoa, S., Simões, N., Wang, L.-P., Rico-Ramirez, M., Liguori, S., Zhu, D., Duncan, A., Chen, A.S., Keedwell, E., Djordjević, S., Savić, D.A., Saul, A. and Maksimović, C. (2011) Urban pluvial flood modelling with real time rainfall information: UK case studies. Proceedings 12th International Conference on Urban Drainage, 10–15 September, Porto Alegre, Brazil.

Schellart, A., Liguori, S., Krämer, S., Saul, A., and Rico-Ramirez, M.A. (2014) Comparing quantitative precipitation forecast methods for prediction of sewer flows in a small urban area. *Hydrol. Sci. J.*, 59, 1418–1436.

Simões, N.E., Leitão, J.P., Maksimović, C., Sa Marques, A. and Pina, R. (2010) Sensitivity analysis of surface runoff generation in urban flood forecasting. *Water Sci. Technol.*, 61(10), 2595–2601.

Sorooshian, S., Duan, Q. and Gupta, V.K. (1993) Calibration of rainfall–runoff models: application of global optimisation to the Sacramento soil moisture accounting model. *Water Resour. Res.*, 29, 1185–1194.

Suebjakla, W. (1996) Modification of the autoregressive (AR) error updating model for real-time streamflow forecasting, Unpublished MSc thesis, Department of Engineering Hydrology, National University of Ireland, Galway.

Van Steenbergen, N. and Willems, P. (2013) Increasing river flood preparedness by real-time warning based on wetness state conditions. *J. Hydrol.*, 489, 227–237.

Van Steenbergen, N., Ronsyn, J. and Willems, P. (2012) A non-parametric data-based approach for probabilistic flood forecasting in support of uncertainty communication. *Environ. Model. Software*, 33, 92–105.

Verbunt, M., Walser, A., Gurtz, J., Montani, A. and Schar, C. (2007) Probabilistic flood forecasting with a limited-area ensemble prediction system: selected case studies. *J. Hydrometeorol.*, 8, 897–909.

Verwey, A., Heynert, F., Werner, M., Reggiani, P., VonKappel, B. and Brinkman, J.-J. (2006) The potential of the Delft-FEWS flood forecasting platform for application in the Mekong Basin. 4th Annual Mekong Flood Forum, Siem Reap, Cambodia, 18–19 May, Chapter 2, pp. 110–115.

Vidal, J.-P., Moisan, S., Faure, J.-B. and Dartus, D. (2007) River model calibration from guidelines to operational support tools. *Environ. Model. Software*, 22, 1628–1640.

Wang, G.T. and Singh, V.P. (1994) An autocorrelation function method for estimation of parameters of autoregressive models. *Water Resour. Manag.*, 8, 33–56.

WaPUG (2002) Code of practice for the hydraulic modelling of sewer systems. www.ciwem. org/media/44426/Modelling_Ver_03.pdf.

WMO (1992) Simulated real-time intercomparison of hydrological models. Operational Hydrology Report 38, WMO 779. World Meteorological Organization, Geneva.

WMO (1994) Guide to hydrological practices: data acquisition and processing, analysis, forecasting and other applications, 5th edn. WMO 168. World Meteorological Organization, Geneva.

WMO (2011) Manual on flood forecasting and warning: quality management framework. WMO 1072. World Meteorological Organization, Geneva.

Xiong, L. and O'Connor, K.M. (2002) Comparison of four updating models for real-time river flow forecasting. *Hydrol. Sci. J.*, 47, 621–639.

Young, P.C. (2006) Updating algorithms in flood forecasting. FRMRC Research Report UR5. www.floodrisk.org.uk.

10

Coastal Flood Forecasting

10.1 Types of coastal flooding

Coastal flooding occurs when normally dry, low lying land is inundated by sea water. Wind, waves and elevated water levels either singularly or together may cause such flooding. Several different pathways are possible: *direct inundation*, where the sea height exceeds the elevation of the land; *overtopping sea defences* (Figure 10.1), which may be natural or human engineered (see for example Allsop et al., 2005); or the *breaching of a barrier*. Generally coastal flooding may be caused by tides and storm surges together with waves. Waves and storm surges are caused by storm events with high winds. In addition, tsunamis, resulting from undersea earthquakes, landslides, volcanic eruptions or even meteorites impacting the sea, may also cause massive coastal flooding.

Storms can cause flooding when the waves are significantly larger than normal, and when this coincides with a high astronomical tide. Wind blowing in an onshore direction can cause the water to 'pile up' against the coast. This is referred to as *wind setup*. Low atmospheric pressure tends to increase the surface sea level, referred to as *barometric setup*. Increased wave break height results in a higher water level in the surf zone, referred to as *wave setup*. All these processes may act together. Sea level rise in the future due to climate change (as discussed in Chapter 13), together with other risks arising from extreme events (IPCC, 2012), are expected to result in an increase in the intensity and frequency of storm events. Coastal flooding is usually a natural process, and it is inherently difficult to prevent it. Adapting to the risk is often the best option rather than building defence structures, although such measures may be cost effective in some circumstances.

Appendix 10.1 outlines the mathematics of wave overtopping.

10.2 Models used to predict storm surge flooding

The likelihood of coastal flooding involves the near shore transformation of wave and surge forecasts, transformation of waves in the surf zone, the effect of wind, waves and still water level in causing beach movement, overtopping and breaching, and a probability of damage to people and property. The Environment Agency (2007)

Hydrometeorology, First Edition. Christopher G. Collier.
© 2016 John Wiley & Sons, Ltd. Published 2016 by John Wiley & Sons, Ltd.
Companion website: www.wiley.com/go/collierhydrometerology

Figure 10.1   Overtopping wave at Hartlepool, UK (source: HR Wallingford)

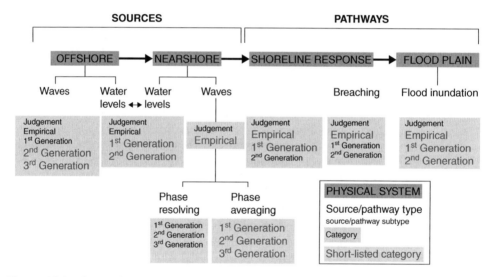

Figure 10.2   Coastal flooding model categories; shortlisted categories are those considered to be practical and cost-effective for use in coastal flood forecasting (from Environment Agency, 2007)

describes model development and model evaluation, and in that report a range of models are discussed as follows. Figure 10.2 provides a summary of model types.

10.2.1   *Empirical models*

These involve the 'projection' of predicted water level across a ground level contour, taking no account of defences. They do not provide good estimates of flood risk in large, low lying or extensive areas where flows occur through a breach, and flood-plain propagation may be critical in determining the extent of the flood.

10.2.2   First-generation models

These are one-dimensional models based upon a grid of flood cells, which are defined with reference to topographical information. The flood cells are linked using spill units placed in areas where flow from one cell to another is most likely to occur. Examples of these models are the ISIS and Mike 21 models.

10.2.3   Second-generation models

These models use the St Venant equations (see Appendix 9.1). The St Venant equation, derived by combining equations 9.4 and 9.5 in Appendix 9.1, when expressed in Cartesian coordinates in the x direction, can be written as:

$$\frac{\partial u}{\partial t} + u\frac{\partial u}{\partial x} + v\frac{\partial u}{\partial y} + w\frac{\partial u}{\partial z} = -\frac{\partial p}{\partial x}\frac{1}{\rho} + v\left(\frac{\partial^2 u}{\partial x^2} + \frac{\partial^2 u}{\partial y^2} + \frac{\partial^2 u}{\partial z^2}\right) + f_x \qquad (10.1)$$

where u is the velocity in the x direction, v is the velocity in the y direction, w is the velocity in the z direction, t is time, p is the pressure, ρ is the density of water, v is the kinematic viscosity, and f_x is the body force in the x direction. The advantage of using this equation is that it can lead to a rapid representation of the two-dimensional flood inundation. Breaches at any location can be simulated, as can the full propagation process. Examples of these models are LISFLOOD-FP and HYDROF.

10.2.4   Third-generation models

These models have two-way coupling with third-generation breaching models such that breaching in three dimensions can be simulated with the flood inundation in two dimensions. Examples of these models are FINEL2D and FINEL3D.

10.2.5   Wave, tide and surge models

To drive wave and current models, a model such as the third-generation spectral Proudman WAve Model (ProWAM) is used which solves a wave action balance equation (see for example Monbaliu et al., 2000). The Proudman Oceanographic Laboratory Coastal Modelling System (POLCOMS) model solves the equations of state and motion, and includes surface forcing, river discharges and turbulence closure. (The Proudman Oceanographic Laboratory is now the National Oceanography Centre, Liverpool.) Three-dimensional flow, turbulence, salinity and a range of other hydro-dynamic parameters at oceanic and coastal scales are generated (see for example Holt and James, 2001; Pan et al., 2009).

The SWAN model is a third-generation spectral wave model that computes random, short crested, wind generated waves in coastal regions and inland water, based on wind, bottom topography, currents and tides (Booij et al., 1999). It is nested in the POLCOMS/ProWAM models using a resolution of 3–5 km to provide higher resolution hydrodynamics close to the surf zone.

The Reynolds-averaged Navier–Stokes volume of fluid (RANS-VOF) model solves the two-dimensional Reynolds-averaged Navier–Stokes equations. This model captures free surface flows (Hirt and Nichols, 1981), wave structure (see for example Garcia et al., 2004) and wave overtopping (Losada et al., 2008; Reeve et al., 2008).

The POLCOMS/ProWAM modelling system has been used (Brown et al., 2010a; 2010b) to model combined tides, surges, waves and wave current interaction in the Irish Sea on a 1.85 km grid. An 11 year hindcast (1996–2006) has been performed to investigate the meteorological conditions that cause extreme surge and/or wave conditions in Liverpool Bay. A one-way nested model approach was used. For waves, a 1° North Atlantic WAM model forces the boundary of the Irish Sea model, driven by ERA-40 wind (1° resolution every 6 hours). To capture the external surge generated outside the Irish Sea, the $(1/9° \times 1/6°)$ Proudman operational surge model extending to the continental shelf edge was run for tide and surge and was forced by Met Office mesoscale winds (12 km resolution every hour). The data implied that the largest surges at Liverpool are generally driven by winds from the south to the west while the largest waves are forced by winds from the west to the north-west. The worst storm conditions in Liverpool Bay result under south-westerly wind conditions that veer to the west. The large tidal range in the region acts to enhance the impact of the surge through tide–surge interaction. Moreover, the highest water levels in Liverpool Bay are in response to south-westerly winds combined with high water spring tide. Even though no significant surge occurs at this time, the flood threat is at its greatest.

In the United States, the NOAA (www.noaa.gov) uses the SLOSH and P-Surge computational models for storm surges from tropical systems, ETSS and ESTOFS models for extratropical systems, and WAVEWATCH III® for modelling waves. These models allow NOAA to simulate many different storms in order to understand the risks involved and forecast storm surge.

The Sea, Lake, and Overland Surges from Hurricanes (SLOSH) model is the computer model developed by the National Weather Service for coastal inundation risk assessment and the prediction of storm surges. It estimates storm surge heights resulting from historical, hypothetical or predicted hurricanes. SLOSH computes storm surge by taking into account a storm's atmospheric pressure, size, forward speed, track and winds. The calculations are applied to a specific locale's shoreline, incorporating the unique bay and river configurations, water depths, bridges, roads, levees and other features.

The model accounts for astronomical tides (which can add significantly to the water height) by specifying a constant tide level to coincide with landfall, but does not include rainfall amounts, river flow, or wind driven waves (those riding on top of the surge). The SLOSH model is the basis for the 'hazard analysis' portion of coastal hurricane evacuation plans. The model does not explicitly model the impacts of waves on top of the surge, and it does not account for normal river flow or rain flooding. Graphical output from the model shows colour-coded storm surge heights for a particular area in either feet above ground level (inundation) or feet above a specific reference level.

Numerical storm surge models depend on an accurate forecast of the hurricane's track, intensity and size. Even the best hurricane forecasts still have considerable uncertainty. The National Hurricane Center's forecast landfall location, for example, can be in error by tens of miles even during the final 12 to 24 hours before the hurricane centre reaches the coast.

To predict the surge accompanying an extratropical storm, the National Weather Service runs the Extratropical Storm Surge Model (ETSS). This model was developed by the Meteorological Development Laboratory and is a variation on the SLOSH model, which is used for tropical (hurricane) storm surge forecasting. ETSS predicts storm surge flooding along US coastlines, but is not able to predict the extent of overland flooding.

Because extratropical storms have larger time and length scales than tropical storms, the ETSS model uses the basic SLOSH model but with winds and pressures generated from the Global Forecast System atmospheric model. The model performs well for both positive and negative surge events (a negative surge occurs when the winds are strong offshore, and the water level is driven below the normal level at that time). The model is particularly useful for forecasting storm surge associated with East Coast north-east winds, and in western Alaska where storm surge associated with extratropical cyclones can devastate low lying coastal communities.

Coverage for the model includes the east, west and Gulf coasts of the continental United States and the Bering Sea and Arctic regions of Alaska. The Ocean Prediction Center website has area forecasts, while the Meteorological Development Laboratory website has station (point) specific forecasts.

The ETSS model runs four times daily, out to 96 hours. The Extratropical Water Level Forecast website combines the model's results with tidal predictions and compares the results to observed water levels. It then combines the model results, the tidal predictions, and an adjustment from recent observations to create a graph of the total predicted water level. At present, the ETSS model does not simulate tides, waves or river effects on storm surge.

The Extratropical Surge and Tide Operational Forecast System (ESTOFS), a new generation hydrodynamic modelling system, was developed for the Atlantic and Gulf coasts. ESTOFS provides real-time forecasts of surges with tides, and provides water levels to WAVEWATCH III® for coastal surge + tide + wave predictions, using Global Forecast System forcing to create a 6 hour nowcast, followed by a 180 hour forecast, four times per day. The ESTOFS provides the National Weather Service with a second extratropical surge system that includes tidal simulations, in addition to the ETSS model.

NOAA uses wind wave modelling to predict the evolution of waves caused by severe coastal storms which can cause flooding. The wave modelling simulations consider multiple factors to describe wave heights, periods and directions. Coastal engineering efforts have led to the development of wind wave models specifically designed for coastal applications. The resulting wave hindcasts and forecasts help predict storm surge.

The National Weather Service computes the influence of wind waves using the WAVEWATCH III® numerical ocean wave prediction computer model, which covers the entire globe. Experts couple the results with high resolution local wave models to increase detail along the coast. WAVEWATCH III® provides important information about potentially dangerous surf conditions resulting from storms hundreds or even thousands of miles away. The model is run four times a day. Each run produces forecasts of every 3 hours from the initial time out to 180 hours (84 hours for the Great Lakes). The wave model suite consists of global and regional nested grids.

10.3   Probabilistic surge forecasting

Zou et al. (2013) present an integrated 'clouds-to-coast' ensemble modelling framework of coastal flood risk due to wave overtopping, as shown in Figure 10.3. The modelling framework consists of four key components: meteorological forecasts, wave/tide/surge predictions, near shore wave models, and surf zone models.

This type of ensemble prediction approach allows us to estimate the probabilities of various outcomes and so improve our understanding of the reliability of results. It also provides a measure of the uncertainty associated with predictions, which can be particularly large for extreme storms, and allows us to assess how the uncertainty propagates from meteorological forecasts to overtopping and subsequent coastal flood risk predictions (Figure 10.4).

The operational Storm Tide Forecasting Service in the United Kingdom is provided by the Met Office and the Environment Agency, working in close partnership. Golding (2009) describes an example of coastal flood warnings for eastern England in 2007, when successive spring high tides on 9 and 23 November were accompanied by the passage of storms that threatened to generate large magnitude storm surges in the North Sea. During this period, an experimental storm surge prediction system was being run, with each Met Office Global and Regional Ensemble Prediction System (MOGREPS: Bowler et al., 2007) regional ensemble member coupled to the Proudman Oceanographic Laboratory continental shelf surge prediction model (Flather, 2000) as part of an Environment Agency project. The results were encouraging and demonstrated the potential of ensemble forecasts.

The limitations of a deterministic SLOSH may result in surge forecasts that are incorrect. To help overcome these limitations, NOAA forecasters use probabilistic storm surge forecasts. The Probabilistic Hurricane Storm Surge (P-Surge) model predicts the likelihood of various storm surge heights above a datum or above ground level based on an ensemble of SLOSH model runs using the official hurricane advisory. These storm surge heights and probabilities are based on the historical accuracy of hurricane track and wind speed forecasts, and an estimate of storm size. P-Surge also computes the probability of surge above ground to more clearly communicate where the surge will occur.

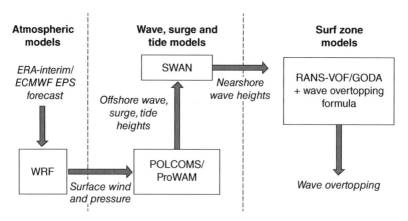

Figure 10.3   Integrated 'cloud to coast' ensemble modelling framework for coastal flood risk arising from overtopping and scour (Zou et al., 2013)

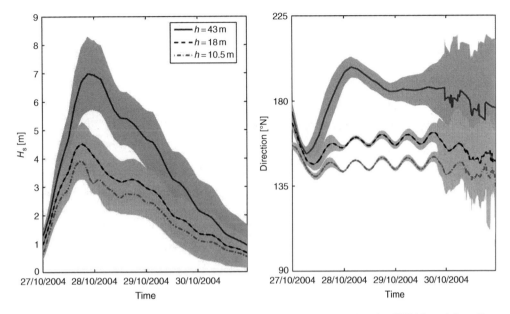

Figure 10.4 Ensemble mean of wave height and direction predicted by the SWAN model at 43 m, 18 m and 10.5 m water depth at Newlyn Harbour, 27–30 October 2004; shaded area indicates one standard deviation about the ensemble mean denoted by lines of same shading; horizontal axis is UTC time (Zou et al., 2013)

10.4 Tsunamis

Tsunamis are a series of enormous waves created by an underwater disturbance such as an earthquake, landslide, volcanic eruption or possibly a meteorite. These disturbances displace a large water mass from its equilibrium position: for example, in the case of earthquake generated tsunamis, the water column is disturbed by the uplift or subsidence of the sea floor (Figure 10.5). A tsunami can move at hundreds of miles per hour in the open ocean and reach land with waves as high as 30 m (100 ft) or more.

From the area where the tsunami originates, waves travel outward in all directions. Once the wave approaches the shore, it builds in height. The topography of the coastline and the ocean floor will influence the size of the wave. There may be more than one wave and the succeeding one may be larger than the one before. That is why a small tsunami at one beach can be a giant wave a few miles away.

On 11 March 2004 a tsunami generated by an earthquake of magnitude up to 9.3 off the west coast of Sumatra caused great devastation around the Indian Ocean (Figure 10.6). The following are the key facts on this tsunami, as established by the Tsunami Evaluation Coalition:

1. A total of 275,000 people were killed in 14 countries across two continents, with the last two fatalities being swept out to sea in South Africa, more than 12 hours after the earthquake.
2. An estimated 40,000 to 45,000 more women than men were killed in the tsunami.

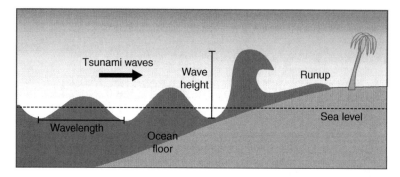

Figure 10.5 Cross-section of the building of the wave during a tsunami (copyright EnchantedLearning.com)

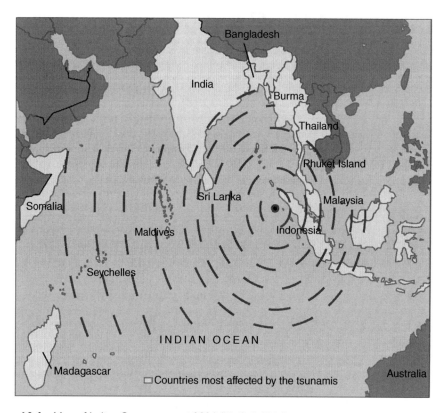

Figure 10.6 Map of Indian Ocean tsunami 2004 (McCall, 2014)

3. The preliminary estimate of the value of economic, infrastructural and human development losses was US$9.9 billion.
4. As many as 141,000 houses were destroyed, which accounts for 47.9% of the total damage (IBRB/World Bank, 2008).
5. Over 600,000 people in Aceh lost their livelihoods (in some cases only for a few months) including all those in the fishery sector and 30% of those in agriculture.
6. A 1200 km section of the Earth's crust shifted beneath the Indian Ocean and the earthquake released stored energy equivalent to more than 23,000 Hiroshima bombs.

7. Speeds of 500 km h^{-1} were reached as the tsunamis radiated through the Indian Ocean. Tsunamis reached 20 m in height at landfall in parts of Aceh. In other locations they spread 3 km inland, carrying debris and salt water with them. The retreating waters eroded whole shorelines.
8. Within 10 minutes of the earthquake, tsunami waves started to strike the Nicobar and Andaman Islands. Banda Aceh was struck within another 10 minutes.
9. Within 2 hours of the earthquake, both Thailand and Sri Lanka had been hit. The east coast of India was hit shortly afterwards.
10. Three hours after the earthquake, tsunamis rolled over the Maldives; and more than 7 hours after the earthquake they hit the Somali coast.
11. Over 1000 German and Swedish tourists were killed. Germany and Sweden were the worst affected countries outside the region and lost more citizens than all but the four most affected countries.

A further example occurred on 26 December 2011, when a magnitude 9.0 earthquake struck off the coast of Honshu, Japan, resulting in a tsunami that not only caused great devastation there but also caused destruction and fatalities in other parts of the world, including Pacific islands and the US West Coast.

10.5   Examples of coastal flooding in the United Kingdom

10.5.1   The surge of 1953

More than 2000 people drowned at the end of January 1953 when the greatest surge on record happened in the North Sea. The surge measured nearly 3 metres in Norfolk and even more in The Netherlands. About 160,000 hectares of eastern England were flooded and 307 people died. A further 200,000 hectares were flooded in The Netherlands and 1800 people drowned. The storm that caused this disastrous surge was among the worst the United Kingdom had experienced. This event led to the construction of the Thames Barrier to protect London.

10.5.2   Wirral floods 2013

Widespread flooding along the Wirral coastline occurred on 5 December 2013, causing internal flooding to more than eight properties within 1 km^2. In addition the flooding resulted in disruption due to the closure of promenade highways at West Kirby, Meols, Hoylake and New Brighton and also caused the closure of businesses at Marine Point, New Brighton (Figure 10.7).

10.5.3   Surges along the east coast of England, December 2013

The surge that came on 8–10 December 2013 brought much damage to the Norfolk and Suffolk coast. Such surges usually occur in February, but recently they have occurred in April, November and December.

Figure 10.7   Flooding on 5 December 2013 at the Marine Point development centred on the Caffe Cream in New Brighton, UK (*Liverpool Echo*, Trinity Mirror)

The devastating North Sea surge wrought serious damage to many homes by wave overtopping, breaches to those sea defences that were present, and undermining of the ground that homes stood on where their only protection was the naturally formed marram dunes. Most of the houses at Walcott, Norfolk, were severely damaged due to waves overtopping the sea wall, whilst serious flooding resulted at Cley, Blakeney and Wells. Five bungalows were undermined and destroyed at Hemsby, with more set to descend from their foundations when the undermined, now vertical dunes dropped to the beach below. Altogether 20 km of 840 km of tidal defences along the Norfolk, Suffolk and Essex coast were damaged or breached by the raging sea.

It was by far the worst coastal impact for 60 years, with a close repeat of the meteorological conditions of 1953. As East Anglia has sunk by almost 10 cm since 1953 and the sea has risen by 23 cm through global warming induced glacier melt and thermal expansion, the impact could have been far worse. In 1953 300 people died, but no one died on this occasion mainly because 9000 people were evacuated from their homes and businesses. This was not the case at Hemsby though, which was not a designated flood area, and therefore was initially ignored by the emergency authorities and not evacuated.

10.5.4   *Aberystwyth floods January 2014*

In January 2014 the worst storm conditions occurred along the Welsh coastline in over 20 years. These unusual events were the result of a combination of factors. Deep low pressure systems generated a series of surges of between 0.5 m and 1 m and brought gale force south-westerly winds and waves of over 5 metres. The surge and high winds coincided with the highest of the spring tides. The significantly raised sea levels, in many cases above the existing defences, brought overtopping, flooding to properties and coastal damage in many different locations (Figure 10.8).

Figure 10.8 Damage caused to Aberystwyth (Wales) seafront after 5 January 2014 high tide (from *Daily Post North Wales*)

Tide levels recorded in Milford Haven, Newport and Barmouth were the highest since detailed records began in 1997.

10.6 Some examples of coastal flooding worldwide

At a rough estimate more than 200 million people worldwide live along coastlines less than 5 metres above sea level. By the end of the 21st century this figure is estimated to increase to 400 to 500 million (World Ocean Review, worldoceanreview.com/en/wor-coasts/living-in-coastal-areas/). In terms of the overall cost of damage, the cities most at risk, in order beginning with the greatest, are: Guangzhou, Miami, New York, New Orleans, Mumbai, Nagoya, Tampa, Boston, Shenzen and Osaka. The top four cities alone account for 43% of the forecast total global losses (Hallegatte et al., 2013). If global carbon emissions continue at current levels and sea levels are affected by climate change, then about 2.6% of the global population (about 177 million people) will be living in a place at risk of regular flooding. There follow three examples of major flooding, selected from the large number of examples across many countries.

Bangladesh lies mostly on the delta of the Ganges, Brahmaputra and Meghna, and 80% of the country lies on a floodplain (1 metre above sea level). During the summer of 2004 36 million people were left homeless due to flooding, 800 people died due to the flood and disease from contaminated water, and 1 million acres of crops were destroyed. There were several causes of this flooding. The sources of many rivers are in the Himalayas, and snowmelt in spring adds to discharge. The monsoon season between May and September results in low pressure, bringing winds blowing south-west across the Bay of Bengal and producing heavy rain in coastal regions. In 2004 the wet period generated more rain than normal, and flooding started at the confluence

Figure 10.9 Flooding in 2008 in Venice, Italy (photograph by Andrea Pattero/AFP/Getty Images)

of four rivers. In addition, deforestation of the Himalayas resulted in decreased rainfall interception and evapotranspiration.

On 29 August 2005 the centre of Hurricane Katrina passed south-east of New Orleans. Although the most severe portion of Katrina missed the city, it made its final landfall to the east. The storm surge extended over 6 miles inland and affected all 57 miles (92 km) of the coastline, reaching a maximum of 16 feet (4.9 m), not including wave action. In the city of New Orleans, the storm surge caused more than 50 breaches in drainage canal levees and also in navigational canal levees and led to the worst engineering disaster in the history of the United States. By 31 August, 80% of New Orleans was flooded, with some parts under 15 feet (4.6 m) of water. Most of the city's levees broke somewhere, leading to most of the flooding. Oil refining was stopped in the area, increasing oil prices worldwide.

On 1 December 2008 high water in Venice reached one of the highest ever levels – 156 cm above the average mid-tide measurement (Figure 10.9). The flooding was triggered by a strong low pressure system that passed through Europe, producing rainfall of 1–2 inches (25–50 mm) which helped fill the salt water lagoon surrounding Venice with high levels of river runoff. The counter-clockwise flow of air around the low brought warm, southerly winds up the length of the Adriatic Sea, which piled high levels of ocean water into the lagoon. Sustained winds reached 30 mph (50 km h^{-1}) on the morning of 1 December.

Appendix 10.1 Wave overtopping at the coast

Many kilometres of the coastal infrastructure around Europe are protected against wave overtopping and/or erosion by steep sea walls. Although it is generally appreciated that sea walls can reduce but not wholly prevent wave overtopping, they are now designed to provide protection against acceptable mean overtopping discharges over given return periods. For beaches and gently sloping structures, the physical form of the wave near maximum runup ξ_{op} can be predicted using

$$\xi_{op} = \tan\alpha/S_{op}^{0.5} \qquad (10.2)$$

where α is the beach slope and the wave steepness is S_{op}. Conditions vary from 'spilling' with $\xi_{op} < 0.4$ through 'collapsing' with $2.3 < \xi_{op} < 3.2$ to 'surging' with $\xi_{op} > 3.2$ (see Allsop et al., 2005). On steep walls 'pulsating' conditions occur when waves are relatively small in relation to the local water depth, and of lesser wave steepness. However, 'impulse' conditions occur on steep walls when waves are larger in relation to local water depth.

For overtopping, a wave breaking parameter h, based on the depth h_s at the toe of the wall, is given by

$$h = h_s / H_{si}\left(2\pi h_s/gT_m^2\right) \qquad (10.3)$$

where H_{si}(m) is the inshore incident significant wave height, $g(=9.81\mathrm{ms^{-2}})$ is the acceleration due to gravity and T_m(s) is the average wave period calculated from spectral moments or zero-crossing analysis. The toe is the visible portion of the sea wall where it meets the surface of the beach, although the toe may also lie below this level. It is suggested that pulsating conditions predominate at the wall when $h > 0.3$ and impulse conditions occur when $h \leq 0.3$.

Summary of key points in this chapter

1. Coastal flooding occurs when normally dry, low lying land is inundated by sea water.
2. Several different pathways are possible, namely direct inundation, overtopping sea defences or breaching of a barrier.
3. Waves and storm surges are caused by storm events with high winds.
4. Massive coastal flooding may be caused by tsunamis.
5. Coastal flooding is usually a natural process and it is inherently difficult to prevent it. Adapting to the risk is often the best option.
6. Models used to predict storm surge flooding include empirical models involving projections of water level and first- and second-generation models linked to numerical equations.
7. The St Venant equations are used as the basis of second-generation models.
8. Third-generation models may be spectral models and have two-way coupling with breaching models.

9. The wave action balance equation is used to solve third-generation spectral models.

10. An 11 year hindcast (1996–2006) has been performed to investigate the meteorological conditions that cause extreme surge and/or wave conditions in Liverpool Bay.

11. The largest surges at Liverpool are generally driven by winds from the south and the west.

12. NOAA uses the SLOSH and P-Surge computational models for storm surges from tropical systems.

13. The SLOSH model was developed by the National Weather Service in the United States for coastal inundation risk assessment and the prediction of storm surges.

14. The US Extratropical Storm Surge Model (ETSS) is a variation of the SLOSH model, and runs four times daily out to 96 hours ahead.

15. The National Weather Service uses a numerical ocean wave prediction model, WAVEWATCH III®, covering the entire globe.

16. An ensemble prediction approach allows estimation of the probabilities of various outcomes and the uncertainty associated with predictions.

17. Tsunamis are a series of enormous waves created by an underwater disturbance such as an earthquake, a landslide, a volcanic eruption or possibly a meteorite.

18. From the area where the tsunami originates, waves travel outward in all directions.

19. As the wave approaches the shore it builds in height, but there may be more than one wave.

20. On 11 March 2004 a tsunami generated by an earthquake of magnitude up to 9.3 off the west coast of Sumatra caused severe devastation around the Indian Ocean, killing about 275,000 people.

21. The tsunami on 26 December 2011 resulted from an earthquake of magnitude 9.0 off the coast of Japan. Over 300,000 people were drowned in Japan and around the Pacific.

22. More than 2000 people drowned at the end of January 1953 during a coastal surge over the east coast of England and in The Netherlands. Further flooding occurred along the east coast of England in December 2013.

23. More than 200 million people worldwide live along coastlines less than 5 metres above sea level. This figure is expected to increase to 400 to 500 million.

24. Around the world, the cities most at risk, in order beginning with the greatest, are Guangzhou, Miami, New York, New Orleans and Mumbai.

25. On 29 August 2005 the centre of Hurricane Katrina passed south-east of New Orleans, causing a storm surge extending over 6 miles inland; 80% of New Orleans was flooded.

Problems

1. What causes coastal flooding?
2. What are the different pathways leading to inundation?
3. Describe empirical models used to predict storm surge flooding.
4. Give the St Venant equation expressed in Cartesian coordinates in the x direction, and say what the advantage is of using this equation.
5. Summarize the sources and pathways for the different model types.
6. Give three examples of wave, tide and surge models.
7. Describe the meteorological conditions causing extreme surge and/or wave conditions in Liverpool Bay.
8. What is the SLOSH model, and what organization developed it?
9. What models might be used to forecast storm surges for extratropical and tropical cyclones?
10. Describe an ensemble modelling framework.
11. Outline how one might overcome the limitations in surge forecasts.
12. What is a tsunami? Illustrate how it develops and what causes it.
13. Describe a real example of an extreme tsunami.
14. What was the consequence of the storm surge in 1953 for the southern North Sea area coastline?
15. How many people worldwide live along coastlines less than 5 metres above sea level? Name five major cities at risk from major coastal flooding.
16. What caused the flooding of New Orleans in 2005?

References

Allsop, W., Bruce, T., Pearson, J. and Besley, P. (2005) Wave overtopping at vertical and steep sea walls. *Proc. Inst. Civil Eng. Mar. Eng.*, 158(September), 103–114.

Booij, N., Ris, R.C. and Holthuijsen, J.H. (1999) A third generation wave model for coastal regions. 1: Model description and validation. *J. Geophys. Res.*, 104, 7649–7666.

Bowler, N.E., Arribas, A., Mylne, K., Robertson, K., Beare, S. and Legg, T. (2007) The MOGREBS short-range ensemble prediction system. II: Performance and verification. Met Office Forecasting Research Technical Note 498. www.metoffice.gov.uk/research/nwp/publications/papers/technical.reports/index.html.

Brown, J.M., Souza, A.J. and Wolf, J. (2010a) An investigation of recent decadal-scale storm events in the eastern Irish Sea. *J. Geophys. Res.*, 115, C05018. doi 10.1029/2009JC005662.

Brown, J.M., Souza, A.J. and Wolf, J. (2010a) An 11-year validation of wave-surge modelling in the Irish Sea, using a nested POLCOMS-WAM modelling system. *Ocean Model.*, 33, 118–128.

Environment Agency (2007) Coastal flood forecasting: model development and evaluation. Science Project SC050069/SR1. Environment Agency, Bristol.

Flather, R.A. (2000) Existing operational oceanography. *Coastal Eng.*, 41, 13–40.

Garcia, N., Lara, J.L. and Losada, I.J. (2004) 2-D numerical analysis of near-field flow at low-crested permeable breakwaters. *Coastal Eng.*, 51, 991–1020.

Golding, B.W. (2009) Long lead time flood warnings: reality or fantasy? *Meteorol. Appl.*, 16, 3–12.

Hallegatte, S., Green, C., Nicholls, R.J. and Corfee-Morlot, J. (2013) Future flood losses in major coastal cities. *Nat. Clim. Change Lett.*, 3, 802–806.

Hirt, C.W. and Nichols, B.D. (1981) Volume of fluid (VOF) method for the dynamics of free boundaries, *J. Comput. Phys.*, 39, 201–225.

Holt, J.T. and James, D.J. (2001) An s-coodinate density evolving model of the northwest European continental shelf. 1: Model description and density structure. *J. Geophys. Res.*, 106, 14015–14034.

IBRB/World Bank (2008) *Data against Natural Disasters: Establishing Effective Systems for Relief, Recovery and Reconstruction*, ed. S. Amin and M. Goldstein. World Bank, Washington, DC.

IPCC (2012) Managing the risks of extreme events and disasters to advance climate change adaptation: summary for policy makers. (ed. Field, C.B., Barros, V., Stacker, T.F., Qin, D., Dokken, D.J., Ebi, K.L., Mastrandrea, M.D., Mach, K.J., Plattner, G.-K., Allen, S.K., Tignar, M. and Midgley, P.M.) A special report of Working Groups I and II of the Intergovernmental Panel on Climate Change. World Meteorological Organization, Geneva.

Losada, I.J., Lara, J.L., Guanche, R. and Gonzalez-Ondina, J.M. (2008) Numerical analysis of wave overtopping of rubble mound breakwaters. *Coastal Eng.*, 55, 47–62.

McCall, C. (2014) Remembering the Indian Ocean tsunami. *Lancet*, 384, 2095–2098.

Monbaliu, J., Padilla-Hernandez, R., Hargreaves, J.C., Carretero-Albiach, J.C., Luo, W., Sciavo, M. and Gunther, H. (2000) The spectral wave model WAM adapted for applications with high spatial resolution. *Coastal Eng.*, 41, 41–62.

Pan, S., Chen, Y., Wolf, J. and Du, Y. (2009) Modelling of waves in the Irish Sea: effects of oceanic wave and wind forcings. *Ocean Dynam.*, 59, 827–836.

Reeve, D.E., Soliman, A. and Lin, P.Z. (2008) Numerical study of combined overflow and wave overtopping over a smooth impermeable seawall. *Coastal Eng.*, 55, 155–166.

Zou, Q.-P., Chen, Y., Cluckie, I., Hewston, R., Pan, S., Peny, Z. and Reeve, D. (2013) Ensemble prediction of coastal flood risk arising from overtopping by linking meteorological, ocean, coastal and surf zone models. *Q. J. R. Meteorol. Soc.*, 139, 298–313.

11

Drought

11.1 Definitions

Drought is caused by a prolonged period of abnormally low rainfall, leading to a shortage of water. It impacts many sectors of the economy, and operates on many different time scales. As a result four types of drought have been defined:

1. *Meteorological drought* is brought about when there is a prolonged period with less than average precipitation. Meteorological drought usually precedes the other kinds of drought, and it can begin and end rapidly.
2. *Agricultural drought* affects crop production or the ecology of an area. This condition can arise independently from any change in precipitation levels when soil conditions and erosion triggered by poorly planned agricultural activity cause a shortfall in water available to the crops. However, more traditionally, agricultural drought is caused by an extended period of below average precipitation, and takes a long time to develop and then recover.
3. *Hydrological drought* is brought about when the water reserves available in sources such as aquifers, lakes and reservoirs fall below the statistical average. Hydrological drought tends to become evident more slowly because it involves stored water that is used but not replenished. Like an agricultural drought, this can be triggered by more than just a loss of rainfall, for example by water being diverted from rivers by other nations.
4. *Socioeconomic drought* relates the supply and demand of various commodities to drought.

Most droughts occur when regular weather patterns are interrupted, causing disruption to the water cycle (Chapter 1). Changes in atmospheric circulation patterns can cause storm tracks to be stalled for months or years, dramatically impacting the amount of precipitation a region normally receives and resulting in droughts or floods. Changes in wind patterns (Chapter 12) can also affect how much moisture a region can absorb.

It has been found that there is a link between certain climate patterns and drought. El Niño (Figure 11.1) is a weather event where the surface water in the Pacific Ocean along the central South American coast rises in temperature. El Niño periodically

Hydrometeorology, First Edition. Christopher G. Collier.
© 2016 John Wiley & Sons, Ltd. Published 2016 by John Wiley & Sons, Ltd.
Companion website: www.wiley.com/go/collierhydrometerology

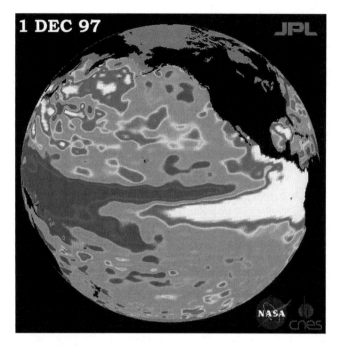

Figure 11.1   The 1997 El Niño observed by TOPEX/Poseidon. The white areas off the tropical coasts of South and North America indicate the pool of warm water (NASA and CNES) (*see plate section for colour representation of this figure*)

develops off the Pacific coast of South America. It is a phase of the El Niño/Southern Oscillation (ENSO), which refers to variations in the temperature of the surface of the tropical eastern Pacific Ocean and in air pressure at the surface in the tropical western Pacific. These warmer waters alter storm patterns and are associated with droughts in Indonesia, Australia, and north-eastern South America. El Niño events are not predictable and can occur every 2 to 7 years. The monitoring and dynamics of droughts are discussed in the following sections.

11.2   Drought indices

Many different indices have been developed to measure drought in these various climate sectors, as shown in Table 11.1. The Palmer drought severity index (PDSI) (also referred to simply as the Palmer index) was created by Palmer (1965) with the intent to measure the cumulative departure (relative to local mean conditions) in atmospheric moisture supply and demand at the surface. It incorporates antecedent precipitation, moisture supply and moisture demand into a hydrological accounting system. Dai et al. (2004) summarized the PDSI, noting that Palmer used a two-layer bucket-type model for soil moisture computations and made certain assumptions relating to field water holding capacity and transfer of moisture to and from the layers based on limited data from the central United States. The Palmer model also computes, as an intermediate term in the computation of the PDSI, the Palmer moisture anomaly index (referred to as the Z index), which is a measure of surface moisture

Table 11.1   Summary of the commonly used drought indices (from Mu et al., 2013)

Indices	Description	Strengths	Weaknesses	Citations
PNP	A simple calculation, by dividing the 30 year average precipitation for the region and multiplying by 100%	Effective for a single region or season	Precipitation does not have a normal distribution. PNP depends on location and season. PNP cannot identify specific drought impacts	Werick et al. (1994)
Deciles	A simple calculation, by grouping precipitation into deciles distributed from 1 to 10. The lowest value indicates conditions drier than normal and the highest value indicates conditions wetter than normal	Accurate statistical measurement of drought response to precipitation, and providing uniformity in drought classifications	Accurate calculations require a long climatology record of precipitation	Gibbs and Maher (1967)
SPI	A simple calculation based on the concept that precipitation deficits over varying periods or timescales influence ground water, reservoir storage, soil moisture, snowpack and stream flow	Computed for flexible multiple timescales, provides early warning of drought and helps in assessing drought severity	Precipitation data are the only input data. SPI values based on long term precipitation may change. The long timescale (up to 24 months) is not reliable	McKee et al. (1993)
PDSI	Calculated using precipitation, temperature, and soil moisture data. Soil moisture algorithm has been calibrated for relatively homogeneous regions	The first comprehensive drought index used widely to detect agricultural drought (see text for details)	PDSI may lag emerging droughts. Not effective for mountainous areas with frequent climatic extremes, or in winter and spring (see text for details)	Palmer (1965); Alley (1984)
PHDI	Derived from PDSI to quantify long term impact from hydrological drought	Same as PDSI, but more effective to determine when a drought ends	PHDI may change more slowly than PDSI	Palmer (1965)
CMI	A derivative of PDSI. CMI reflects moisture supply in the short term	Effective for the detection of short term agricultural drought sooner than PDSI	CMI cannot monitor long term droughts well	Palmer (1968)
SWSI	Developed from the Palmer index by combining hydrological and climatic features	SWSI takes into account reservoir storage, stream flow, snowpack and precipitation. Effective under snowpack conditions	SWSI is difficult to compare between different basins. SWSI cannot detect extreme events effectively. Not a suitable indicator for agricultural drought	Shafer and Dezman (1982); Wilhite and Glantz (1985); Doesken et al. (1991)

(Continued)

Table 11.1   (Continued)

Indices	Description	Strengths	Weaknesses	Citations
RDI	Similar to SPI, based on precipitation and PET	Drought is based on both precipitation and PET. Appropriate for climate change scenarios	Uncertainties in input data for the calculation of PET. RDIs at different basins cannot be compared with each other and have been computed seasonally	Tsakiris and Vangelis (2005); Tsakiris et al. (2007)
USDM	Based on several key physical indicators, such as PDSI, SPI, PNP, soil moisture model percentiles, daily stream flow percentiles, remotely sensed satellite vegetation health index, and many supplementary indicators	Integrating remotely sensed satellite vegetation health index together with other drought indices	USDM is weighted to precipitation and soil moisture in short term. USDM inherits the weaknesses of the other indices it uses	Svoboda et al. (2002)

PNP, percentage of normal precipitation; SPI, standardized precipitation index; PDSI, Palmer drought severity index; PHDI, Palmer hydrological drought index; CMI, crop moisture index; SWSI, surface water supply index; RDI, reconnaissance drought index; USDM, US drought monitor classification.
PET: potential evapotranspiration.

anomaly for the current month without consideration of the antecedent conditions that characterize the PDSI. The Z index can track agricultural drought, as it responds quickly to changes in soil moisture (Karl, 1986). The Z index relates to the PDSI through the following equation (Palmer, 1965):

$$\text{PDSI}(m) = \text{PDSI}\left\{ m-1+\left[\frac{Z(m)}{3} - 0.103\text{PDSI}(m-1)\right]\right\} \tag{11.1}$$

where m is a month index.

There are certain drawbacks of the Palmer index, as indicated by Fuchs (2012) and others, such as that the values quantifying the intensity of drought and indicating the beginning and end of a drought are arbitrarily selected. Also the index is sensitive to the available water content; the use of two soil layers may be too simple for some locations; and the lag between when precipitation falls and the resulting runoff is not considered. Nevertheless, the Palmer index is widely used operationally, with Palmer maps published weekly by the United States Government's National Oceanic and Atmospheric Administration (Figure 11.2).

A drought monitoring and forecasting system for sub-Saharan Africa has been described by Sheffield et al. (2014) for the management of water resources and food security. This system also uses a number of drought indices. On a more local scale Steinemann (2014) discusses the links between state and local drought plans in the United States, pointing out that these are often unrelated. The Palmer index has also has been used by climatologists to standardize global long term drought analysis.

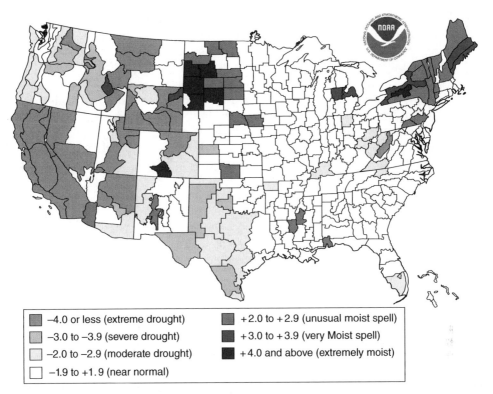

Figure 11.2 Long term Palmer drought severity index for the United States for week ending 9 August 2014, issued by the NOAA Climate Prediction Center (www.cpc.ncep.noaa.gov/products/analysis_monitoring/regional_monitoring/palmer/2014/08-09-2014.gif)

Global Palmer data sets have been developed based on instrumental records beginning in the 19th century (see for example Dai et al., 2004). In addition, dendrochronology has been used to generate estimated Palmer index values for North America for the past 2000 years, allowing analysis of long term drought trends. The index has also been used as a means of explaining the late Bronze Age collapse.

11.3   The physics of drought

Bravar and Kavvas (1991a) developed a conceptual framework for the physics of droughts using the omega equation of atmospheric dynamics (Holton, 1979). The duration and severity of droughts are dependent not only on the atmosphere, but also on the hydrologic processes which feed moisture to the atmosphere. After a dry period caused by a long lasting high pressure, the ability of the Earth's surface to release moisture by evapotranspiration is limited. The piezometric levels in aquifers are lower, which makes evaporative fluxes smaller. In addition, the vegetation conserves its water. When a low pressure system passes with its associated convergence zone, the air rises and cools; however, it may not have a water content high enough to result in a condensation level efficiently low. As evapotranspiration moisture is missing, the disturbance will not be able to release the water it is transporting because the relative humidity level necessary for condensation has not been reached. Note that the moisture released at the ground is immediately available to the atmospheric system because of the presence of the atmospheric boundary layer. This is able to transport the water vapour vertically across its whole depth (1–2 km) in tens of minutes because of the turbulence that characterizes it.

This results in a self-supporting, positive feedback mechanism in which the less water that is present, the less probable rain becomes, and therefore the water content diminishes further. Only disturbances that carry enough moisture from outside the dry area will be able to produce rain and help end the drought period. However, not all of the disturbances have such characteristics. Furthermore, the longer the dry period, the smaller will be the moisture available at the soil surface, and thus the more intense the disturbance will have to be to restore normal conditions.

These are the factors that induce and maintain dry conditions at mid-latitudes. The restriction on the latitude is necessary because the quasi-geostrophic theory, and thereby the omega equation, is valid only in areas such as the United States and Europe.

Bravar and Kavvas (1991b) developed a simple general circulation model (GCM) to investigate the interaction of atmospheric and hydrologic processes during droughts. Only the vertically and latitudinally averaged mean temperature and mean water vapour content over the strip of Earth between latitudes 30°N and 50°N are considered as atmospheric state variables. Only two hydrologic state variables, the water storage (through the water balance equation) and the ground temperature, are considered in the model for describing the behaviour of the hydrologic system. The ocean temperature is specified as a function of season only. The coupling between the atmosphere and the Earth's surface occurs through exchanges of thermal energy (radiation, sensible and latent heat fluxes) and of water (evapotranspiration and rain).

Using this GCM, the mechanism leading to drought conditions described in the earlier paragraphs is investigated. The non-linear positive feedback mechanism is confirmed in the sense that it feeds on itself. Imposing a temperature wave over a geographical region (western United States in the study) for a short period of the order of a few months induced geophysical conditions in that region which tend to reduce the hydrologic water storage for a period of the order of years (called 'drought growth period'). These conditions consisted initially of increased cloudiness over the region, which reduces the net radiative flux of energy available at the surface. This in turn reduces the evapotranspiration which is needed to achieve condensation conditions in the atmosphere, producing a reduction in rainfall. After the end of the drought growth period, the hydrologic system recovered the water storage mainly owing to decreased evapotranspiration (related to reduced water storage) while rainfall resumed its normal levels. The time necessary to return to the climatological average conditions of water storage (called the 'drought recovery period') is proportional to the peak drought severity during the growth period. A definition of aridity is given in Appendix 11.1.

11.4   Frequency analysis: predictability

More than half (52%) of the spatial and temporal variance in multi-decadal drought frequency over the conterminous United States is attributable to the Pacific decadal oscillation (PDO) and the Atlantic multi-decadal oscillation (AMO) (McCabe et al., 2004).

An additional 22% of the variance in drought frequency is related to a complex spatial pattern of positive and negative trends in drought occurrence, possibly related to increasing northern hemisphere temperatures or some other unidirectional climate trend. Recent droughts with broad impacts over the conterminous United States (1996, 1999–2002) were associated with North Atlantic warming (positive AMO) and north-eastern and tropical Pacific cooling (negative PDO). Much of the long term predictability of drought frequency may reside in the multi-decadal behaviour of the North Atlantic Ocean. Should the current positive AMO (warm North Atlantic) conditions persist into the upcoming decade, two drought scenarios are possible which resemble the continental-scale patterns of the 1930s (positive PDO) and 1950s (negative PDO) drought.

Although long considered implausible, there is growing promise for probabilistic climatic forecasts one or two decades into the future based on quasi-periodic variations in sea surface temperatures (SSTs), salinities and dynamic ocean topographies. Such long term forecasts could help water managers plan for persistent drought across the conterminous United States. This need was reinforced by the US droughts in 1996 and 1999–2003, which reminded people of the droughts of the 1930s and 1950s.

Analyses and forecasting of US precipitation have focused primarily on the Pacific Ocean, and more specifically on oceanic indices such as those used to track the El Niño/Southern Oscillation (ENSO) and the Pacific decadal oscillation (PDO). Much of the long term predictability in North American climate, however, may actually reside in both observed and modelled multi-decadal (50–80 years) variations in North Atlantic SSTs. Modelling studies indicate that multi-decadal variability in the North Atlantic climate is dominated by a single mode of SST variability. An important characteristic of this mode of SST variability is that the SST anomalies have the same sign across the entire North Atlantic Ocean and resemble the Atlantic multi-decadal oscillation (AMO). The AMO is an index of detrended SST anomalies averaged over the North Atlantic from 0° to 70°N and has been identified as an important mode of climate variability. Warm phases occurred during 1860–80 and 1930–60, and cold phases occurred during 1905–25 and 1970–90. Since 1995 the AMO has been positive, but it is uncertain whether this condition will persist long enough to be considered a new warm phase. These large swings in North Atlantic SSTs are probably caused by natural internal variations in the strength of ocean thermohaline circulation and the associated meridional heat transport.

Positive AMO conditions (warm North Atlantic SSTs) since 1995, and the cold PDO episode from 1998 to 2002, have together raised concerns among scientists about the potential for an emerging megadrought that could pose serious problems for water planners. Here, drought frequency in the conterminous United States may be decomposed into its primary modes of variability without *a priori* consideration of climate forcing factors. These modes are then related both spatially and temporally to the PDO and AMO to determine the relative influence of the SST patterns tracked by these indices on the probability of drought. Consideration of the potential role of the northern hemisphere (NH) temperature or some other unidirectional trend also proved necessary to understand the variance in drought frequency.

Hannaford et al. (2011) use over 500 river flow time series from 11 European countries with a gridded precipitation data set, to examine the spatial coherence of drought in

Europe. The drought indicators were generated for 24 homogeneous regions and, for selected regions, historical drought characteristics were compared with previous work. The spatial coherence of drought characteristics was examined at a European scale. Historical droughts generally have distinctive signatures in their spatiotemporal development, so there was limited scope for using the evolution of historical events to inform forecasting. Rather, relationships were explored in time series of drought indicators between regions. Correlations were generally low, but multivariate analyses revealed broad continental-scale patterns, which appear to be related to large scale atmospheric circulation indices (in particular, the North Atlantic oscillation and the East Atlantic–West Russia pattern). A methodology for forecasting was developed and demonstrated with reference to the United Kingdom, which predicts drought using the spatial coherence of drought to facilitate early warning of drought in a target region, arising from drought which is developing elsewhere in Europe. At present the skill of the methodology is modest, but it allows prediction for all seasons, and also shows some potential for forecasting the termination of drought conditions.

11.5   Modelling the occurrence of drought

Charney (1975) suggested that the high albedo of a desert contributes to a net radiative heat loss relative to its surroundings, and that the resultant horizontal temperature gradients induce a frictionally controlled circulation which imports heat aloft and maintains thermal equilibrium through sinking motion and adiabatic compression. In the subtropics this sinking motion is superimposed on the descending branch of the mean Hadley circulation but is more intense. Thus the desert feeds back upon itself.

This feedback mechanism could conceivably lead to instabilities in desert border regions. It is argued that a reduction of vegetation, with consequent increase in albedo, in the Sahel region at the southern margin of the Sahara would cause sinking motion and additional drying, and would therefore perpetuate the arid conditions. Numerical integrations with the general circulation model of NASA's Goddard Institute for Space Studies appeared to substantiate this hypothesis. Increasing the albedo north of the Inter-Tropical Convergence Zone (ITCZ) from 14% to 35% had the effect of shifting the ITCZ several degrees of latitude south and decreasing the rainfall in the Sahel about 40% during the rainy season.

Lau et al. (2006) used 19 coupled general circulation models (CGCMs) in 20th century simulations of the Sahel during the 1970s to 1990s. Correlation, regression and cluster analyses were applied to observations and model outputs including Sahel monthly precipitation, evaporation, soil moisture and sea surface temperature (SST). It was found that only eight CGCMs produced a reasonable Sahel drought signal, while seven CGCMs produced excessive rainfall over the Sahel during the observed drought period. Even the model with the highest prediction skill of the Sahel drought could only predict the increasing trend of severe drought events, and not the beginning and duration of the events. From analyses of the statistical characteristics of these models, it was concluded that a good simulation of the Sahel drought requires: (i) a strong coupling between Sahel rainfall and Indian Ocean SST, with warm (cold)

SST identified with Sahel drought (flood); (ii) a significant coupling between Sahel rainfall and the Atlantic Ocean SST, with a warm equatorial Atlantic and cold extratropical North Atlantic coexisting with Sahel drought, and vice versa; and (iii) a robust land surface feedback with strong sensitivity of precipitation and land evaporation to soil moisture.

More recently Meng et al. (2011) used the Weather Research and Forecasting (WRF) regional model to simulate the severe drought that occurred in south-east Australia from 2000 to 2008. The model used the following physics schemes: the WRF single moment 5-class microphysics scheme; the rapid radiative transfer model (RRTM) long wave radiation scheme; the Dudhia short wave radiation scheme; the Monin–Obukhov surface layer similarity; the Noah land surface scheme; the Yonsei University boundary layer scheme; and the Kain–Fritsch cumulus physics scheme. Boundary conditions from the NCEP/NCAR reanalysis were used with an outer 50 km resolution grid and an inner 10 km resolution grid. Both grids used 30 vertical levels spaced closer together in the planetary boundary layer.

WRF was run in control mode with the default climatological surface albedo and vegetation fraction data sets, as well as with these data sets prescribed using satellite data. Comparison of these simulations demonstrated the importance of capturing the dynamic nature of these fields as the climate moves into (and then out of) a persistent multi-year drought. Both simulations capture the drought reasonably well, emphasizing changes in the large scale circulation as a primary cause. Differences in the surface conditions do however provide local influence on the intensity and severity of drought. A similar use of the WRF model to investigate drought in the Yunnan Province of China was reported by Zhao et al. (2013).

Blenkinsop and Fowler (2007) investigated the performance for the British Isles of integrations from six regional climate models (RCMs) driven by four different general circulation models (GCMs). Mean precipitation and drought statistics have been developed for the 1961–90 period, using two drought severity indices based on monthly precipitation anomalies. Spatially averaged statistics are examined in addition to spatial variations in model performance over water resource regions and compared with observations. Estimates of the range and sources of uncertainty in future changes are examined for the SRES A2 2071–2100 emissions scenario.

Results indicated that the RCMs are able to reproduce the spatially averaged annual precipitation cycle over the British Isles, but the spatial anomalies suggest that they may have difficulty in capturing important physical processes responsible for precipitation. The RCMs were unable to simulate the observed frequency of drought events in their control climate, particularly for severe events. This is possibly due to a failure to simulate persistent low precipitation. Future projections suggested an increase in mean precipitation in winter and decrease in summer months. Short term summer drought is projected to increase in most water resource regions except Scotland and Northern Ireland, although the uncertainty associated with such changes is large. Projected changes in longer droughts are influenced by the driving GCM and are highly uncertain, particularly for the south of England, although the longest droughts are projected to become shorter and less severe by most models.

11.6   Major drought worldwide

Droughts occur around every 5–10 years on average in the United Kingdom. These droughts are usually confined to summer, when a blocking high pressure system causes hot, dry weather for an extended period. One of the most major meteorological droughts of recent times occurred in the year 1976, where a dry winter in 1975 was followed by one of the hottest and driest summers since records began. The drought effectively began in October 1975, but with low temperatures and therefore low evaporation rates during the winter, the below average rainfall did not present an immediate problem. Then as the dry winter ended it was followed by the hot and particularly dry summer of 1976.

A significant hydrological drought occurred between 1995 and 1998 in the United Kingdom where the warm, dry summers were followed by dry, cool winters. Over the three years, the lack of winter precipitation followed by the dry summers resulted in a fall in the water table and reservoir water levels. Water levels were only replenished after some exceptionally wet years from 1999 to 2002. Similar conditions occurred between 2003 and 2006, with only the record breaking rainfall of 2007 and 2008 replenishing the water levels.

Some major worldwide droughts since 1900 are:

- 1900 in India, killing between 250,000 and 3.25 million.
- 1921–2 in the Soviet Union, in which over 5 million perished from starvation due to drought.
- 1930s in south-west North America (see next section).
- 1928–30 in north-west China, resulting in over 3 million deaths by famine.
- 1936 and 1941 in Sichuan Province, China, resulting in 5 million and 2.5 million deaths respectively.
- 1997–2009 in Australia (the Millennium Drought), which led to a water supply crisis across much of the country; as a result many desalination plants were built for the first time. The drought devastated south-west Western Australia, south-east South Australia, Victoria and northern Tasmania, and was said by a Bureau of Meteorology spokesperson to be 'very severe and without historical precedent'.
- 2006 in Sichuan Province, China, which experienced its worst drought in modern times, with nearly 8 million people and over 7 million cattle facing water shortages.
- 2011 in the state of Texas, which lived under a drought emergency declaration for the entire calendar year.
- 2010 in Ethiopia, which killed 300,000 people and overall affected some 57 million people.

11.7   Examples of the consequences of drought

In the 1930s, drought covered virtually the entire Plains of the United States for almost a decade (Warrick, 1980). The drought's direct effect is most often remembered as agricultural. Many crops were damaged by deficient rainfall, high temperatures and high winds (Figure 11.3) causing dust storms, as well as insect infestations. The resulting

Figure 11.3 A dust storm approaching Rolla, Kansas, USA, 6 May 1935 (image: Franklin D. Roosevelt Library Digital Archives)

agricultural depression contributed to the Great Depression's bank closures, business losses, increased unemployment, and other physical and emotional hardships. Although records focus on other problems, the lack of precipitation would also have affected wildlife and plant life, and would have created water shortages for domestic needs.

Although the 1930s drought is often referred to as if it were one episode, there were at least four distinct drought events: 1930–1, 1934, 1936, and 1939–40 (Riebsame et al., 1991). These events occurred in such rapid succession that affected regions were not able to recover adequately before another drought began.

Paleoclimatic evidence in the south-west of North America suggests (Woodhouse et al., 2009) that drought in the mid 12th century far exceeded the severity, duration and extent of subsequent droughts. The driest decade of this drought was anomalously warm, though not as warm as the late 20th and early 21st centuries. The convergence of prolonged warming and arid conditions suggests the mid 12th century may serve as a conservative analogue for severe droughts that might occur in the future. The severity, extent and persistence of the 12th century drought that occurred under natural climate variability have important implications for water resource management. The causes of past and future drought will not be identical but warm droughts, inferred from paleoclimatic records, demonstrate the plausibility of extensive, severe droughts, provide a long term perspective on the ongoing drought conditions in the south-west United States, and suggest the need for regional sustainability planning for the future.

By far the largest part of Australia is desert or semi-arid lands. A 2005 study by Australian and American researchers investigated the desertification of the interior, and suggested that one explanation was related to human settlers who arrived about 50,000 years ago (Williams, 2014). Regular burning by these settlers could

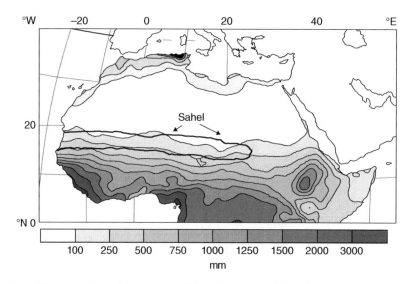

Figure 11.4   Mean annual precipitation over West Africa (in mm). The Sahel is indicated, showing the main area impacted by the drought of 2012 (Nicholson, 2013)

have prevented monsoons from reaching interior Australia. Recently it became evident that severe ecological damage for the whole Murray–Darling basin could occur if it did not receive sufficient water.

Recurring droughts have caused desertification in East Africa and have resulted in major food shortages. During the 2011 drought up to 150,000 people were reported to have died. In 2012, a severe drought occurred in the western Sahel (Figure 11.4).

11.8   Strategies for drought protection, mitigation or relief

Continuous observation of rainfall levels and comparisons with current usage levels can help prevent man-made drought and mitigate naturally occurring droughts. A number of other actions may also help as follows:

- Using dams: many dams and their associated reservoirs can supply additional water in times of drought.
- Cloud seeding, that is intentional weather modification to induce rainfall. This may be effective in only limited regions (for example Williams and Elliott, 1985).
- Desalination of sea water, producing water for irrigation or consumption.
- Planned crop rotation to minimize erosion, allowing farmers to plant fewer water-dependent crops in drier years.
- Regulation of the use of sprinklers, hoses or buckets on outdoor plants, the filling of pools, and other water-intensive home maintenance tasks.
- Collection and storage of rainwater from roofs or other suitable catchments, known as rainwater harvesting.
- Recycling water from wastewater (sewage) after treatment and purification.
- Building canals or redirecting rivers for irrigation in drought-prone areas (for example Shiklomanov, 1985).

Appendix 11.1   Defining aridity

The terms *arid* and *semi-arid* refer to areas where rainfall will not support agriculture requiring regular water. Hence the following index has been used

$$\alpha = P / (10 + T) \tag{11.2}$$

where P is mean annual precipitation (mm) and T is mean annual air temperature (°C). When $\alpha < 15$ the area is considered arid.

A more precise definition of aridity was given in Rodier (1985), known as the radiation index of aridity RIA:

$$RIA = R / I_p \tag{11.3}$$

where R is the radiation balance of the land surface and I_p is the heat necessary for evaporation of the annual precipitation p. When $2 < RIA < 3$ a region is regarded as semi-arid, and when $RIA > 3$ it is considered to be desert.

The Penman formula (section 3.5) has been used to estimate evaporation, although this approach was developed for humid countries and therefore requires some adjustment for use in arid countries. Nevertheless, world maps have been developed using this approach without adjustment (see for example UNESCO, 1979). Here

- $P / PET < 0.03$ indicates a hyper-arid zone
- $0.03 < P / PET < 0.20$ indicates an arid zone
- $0.20 < P / PET < 0.50$ indicates a semi-arid zone.

where P is mean annual precipitation depth and PET is potential evapotranspiration, i.e. the evapotranspiration from an area covered by vegetation with enough water to permit the evaporation of the maximum quantity of water demanded by the climatological conditions. For example south of the Sahara (Sahel), the arid zone corresponds to $50\,mm < P < 350$–$400\,mm$.

Summary of key points in this chapter

1. Four types of drought have been defined, namely meteorological, agricultural, hydrological and socioeconomic.
2. More droughts occur when regular weather patterns are interrupted, causing disruption to the water cycle.
3. El Niño/Southern Oscillation (ENSO) is a weather event which may be associated with drought in Indonesia, Australia and north-eastern South America.
4. El Niño events can occur every 2 to 7 years.
5. Many different indices have been developed to measure drought.
6. The PDSI index was developed to measure the cumulative departure in atmospheric moisture supply and demand at the surface.

7. One drawback to the Palmer index is that the values quantifying the intensity of drought, and indicating the beginning and end of a drought, are arbitrarily selected.

8. The Palmer index is widely used operationally and Palmer maps are published weekly in the United States.

9. Dendrochronology has been used to generate estimated Palmer index values for North America for the past 2000 years.

10. After a dry period caused by a long lasting high pressure, the ability of the Earth's surface to release moisture by evapotranspiration is limited.

11. When a low pressure system passes with its associated convergence zone, the air rises and cools; however it may not have a water content high enough to result in a condensation level efficiently low. As evapotranspiration moisture is missing, the disturbance will not be able to release the water it is transporting because the relative humidity level necessary for condensation has not been reached.

12. A self-supporting positive feedback mechanism is developed in which the less water that is present the less probable rain becomes, and water content diminishes further.

13. A simple general circulation model can be used to investigate the interaction of atmospheric and hydrologic processes during drought. In particular the feedback mechanism has been confirmed.

14. After the end of a drought growth period, the hydrologic system recovers water storage, mainly owing to decreased evapotranspiration, while rainfall resumes at normal levels.

15. More than 50% of the spatial and temporal variance in multi-decadal drought frequency over the United States is attributable to the Pacific decadal oscillation and the Atlantic multi-decadal oscillation.

16. There is growing promise for probabilistic climate forecasts one or two decades ahead based on quasi-periodic variations in sea surface temperatures, salinities and dynamic ocean topographies.

17. The high albedo of a desert contributes to a net radiative heat loss relative to its surroundings.

18. A frictionally controlled circulation over the desert imports heat aloft and maintains thermal equilibrium through sinking motion and adiabatic compression.

19. The reduction of vegetation, with the consequent increase in albedo, in the Sahel region of the southern margin of the Sahara could cause sinking motion and additional drying, and lead to arid conditions.

20. Comparison of numerical simulations has demonstrated the importance of capturing the dynamic nature of albedo and vegetation fields for persistent multi-year drought.

21. In the 1930s drought covered almost the entire Plains of the United States, resulting in an agricultural depression and contributing in a major way to economic depression.

22. Drought in the mid 12th century in south-west North America far exceeded the severity, duration and extent of subsequent droughts.
23. By far the largest part of Australia is desert or semi-arid lands.
24. Regular burning by early settlers in Australia may have prevented monsoons reaching the interior.
25. A number of strategies are possible for drought protection, mitigation and relief.

Problems

1. Define drought.
2. Give four types of drought.
3. How does the El Niño impact drought and where does it do so?
4. Define the PDSI drought index.
5. How does the Palmer moisture anomaly index (Z index) relate to the PDSI?
6. Name two commonly used drought indices other than the PDSI.
7. Describe the physics of how a drought occurs.
8. What role does evapotranspiration play in drought generation?
9. What are the important factors in at least half the droughts in the United States?
10. How does the El Niño impact drought?
11. What influences broad continental-scale weather patterns related to drought?
12. How does the albedo affect the margins of semi-arid regions such as the Sahel area of the southern Sahara?
13. Besides the albedo, what else may influence drought occurrence in the Sahel region?
14. How often do droughts occur on average in the United Kingdom?
15. How long can droughts last in the United States and Australia?
16. What were the consequences of the 1930s drought in the Plains area of the United States?
17. Historically, when did the most severe drought occur in the south-west of North America?
18. What might have caused the desertification of the interior of Australia?
19. Name six actions for mitigating the impact of droughts.

References

Alley, W.M. (1984) The Palmer drought severity index: limitations and assumptions. *J. Clim. Appl. Meteorol.*, 23, 1100–1109.
Blenkinsop, S. and Fowler, H.J. (2007) Changes in drought frequency, severity and duration for the British Isles projected by the PRUDENCE regional climate model. *J. Hydrol.*, 342, 50–71.

Bravar, L. and Kavvas, M.L. (1991a) On the physics of droughts. I: A conceptual framework. *J. Hydrol.*, 129, 281–297.

Bravar, L. and Kavvas, M.L. (1991b) On the physics of droughts. II: Analysis and simulation of the interaction of atmospheric and hydrologic processes during droughts. *J. Hydrol.*, 129, 299–330.

Charney, J.G. (1975) Dynamics of deserts and drought. *Q. J. R. Meteorol. Soc.*, 101, 193–202.

Dai, A., Trenberth, K.E. and Qian, T. (2004) A global dataset of Palmer drought severity index for 1870–2002: relationship with soil moisture and effects of surface warming. *J. Hydrometeorol.*, 5, 1117–1130.

Doesken, N.J., McKee, T.B. and Kleist, J.D. (1991) Development of a surface water supply index for the western United States: final report. Climatology Report 91-3, Colorado State University.

Fuchs, B. (2012) Palmer drought severity index (PSDI and scPDSI). Caribbean Drought Workshop, 22–24 May, National Drought Mitigation Center, University of Nebraska–Lincoln.

Gibbs, W.J. and Maher, J.V. (1967) Rainfall deciles as drought indicators. Bureau of Meteorology Bulletin 48, Commonwealth of Australia.

Hannaford, J., Lloyd-Hughes, B., Keef, C., Parry, S. and Prudhomme, C. (2011) Examining the large-scale spatial coherence of European drought using regional indicators of precipitation and streamflow deficit. *Hydrol. Proc.*, 25, 1146–1162.

Holton, J.R. (1979) *An Introduction to Dynamic Meteorology*, 2nd edn. Academic Press, New York.

Karl, T.R. (1986) Sensitivity of the Palmer drought severity index and Palmer's Z-index to their calibration coefficients including potential evapotranspiration. *J. Clim. Appl. Meteorol.*, 25, 77–86.

Lau, K.M., Shen, S.S.P., Kim, K.-M. and Wang, H. (2006) A multi-model study of the twentieth-century simulations of Sahel drought from the 1970s to 1990s. *J. Geophys. Res.*, 111, D07111.

McCabe, G.J., Palecki, M.A. and Betancourt, J.L. (2004) Pacific and Atlantic Ocean influences on multidecadal drought frequency in the United States. *Proc. Nat. Acad. Sci. USA*, 101, 4136–4141.

McKee, T.B., Doesken, N.J. and Kleist, J. (1993) The relationship of drought frequency and duration to timescales. Preprints 8th Conference on Applied Climatology, Anaheim, CA, pp. 179–184. American Meteorological Society, Boston.

Meng, X.H., Evans, J.P. and McCabe, M.F. (2011) South-east Australia's drought: numerical modelling and land–atmosphere feedback. Hydroclimatology: Variability and Change, Proceedings of Symposium at J-HO2 held at IUGG, Melbourne, Australia, July, IAHS Publ. 3XX.

Mu, Q., Zhao, M., Kimball, J.S., McDowell, N.G. and Running, S.W. (2013) A remotely sensed global terrestrial drought severity index. *Bull. Am. Meteorol. Soc.*, 94, 83–98.

Nicholson, S.E. (2013) The West African Sahel: a review of recent studies on the rainfall regime and its interannual variability. *International Scholarly Research Notices (ISRN) Meteorology*, 2013, Article ID 453521. dx.doi.org/10.1155/2013/453521.

Palmer, W.C. (1965) Meteorological drought. US Weather Bureau Research Paper 45.

Palmer, W.C. (1968) Keeping track of crop moisture conditions, nationwide: the new crop moisture index. *Weatherwise*, 21, 156–161.

Riebsame, W.E., Changnon, S.A. Jr. and Carl, T.R. (1991) *Drought and Natural Resources Management in the United States: Impacts and Implications of the 1987–89 Drought*. Westview, Boulder, CO.

Rodier, J.A. (1985) Aspects of arid zone hydrology. Chapter 8 in *Facets of Hydrology*, vol. II, ed. J. C. Rodda, pp. 205–247. Wiley, Chichester.

Shafer, B.A. and Dezman, L.E. (1982) Development of a surface water supply index (SWSI) to assess the severity of drought conditions in snowpack runoff areas. Proceedings 50th Annual Western Snow Conference, Reno, NV, pp. 164–175.

Sheffield, J., Wood, E.F., Chaney, N., Guan, K., Sadri, S., Yuan, X., Olang, L., Amani, A., Ali, A. and Demuth, S. (2014) A drought monitoring and forecasting system for sub-Sahara African water resources and food security. *Bull. Am. Meteorol. Soc.*, 95, 861–882.

Shiklomanov, I.A. (1985) Large scale water transfers. Chapter 12 in *Facets of Hydrology*, vol. II, ed. J.C. Rodda, pp. 345–387. Wiley, Chichester.

Steinemann, A. (2014) Drought information for improving preparedness in the Western States. *Bull. Am. Meteorol. Soc.*, 95, 843–847.

Svoboda, M., LeComte, D., Hayes, M., Hem, R., Gleason, K., Angel, J., Rippey, B., Tinker, R., Palecki, M., Stooksbury, D., Miskus, D. and Stephens, S. (2002) The drought monitor. *Bull. Am. Meteorol. Soc.*, 83, 1181–1190.

Tsakiris, G. and Vangelis, H. (2005) Establishing a drought index incorporating evapotranspiration. *Eur. Water*, 9–10, 3–11.

Tsakiris, G., Pangalou, D. and Vangelis, H. (2007) Regional drought assessment based on the reconnaissance drought index (RDI). *Water Resour. Manag.*, 21, 821–833.

Warrick, R.A. (1980) Drought in the Great Plains: a case study of research on climate and society in the USA. In *Climatic Constraints and Human Activities*, ed. J. Ausubel and A.K. Biswas, pp. 93–123. IIASA Proceedings Series, vol. 10. Pergamon, New York.

Werick, W.J., Willeke, G.E., Guttman, N.B., Hosking, J.R.M. and Wallis, J.R. (1994) National drought atlas developed. *Eos, Trans. Am. Geophys. Union*, 75, 89.

Wilhite, D.A. and Glantz, M.H. (1985) Understanding the drought phenomenon: the role of definitions. *Water Int.*, 10, 111–120.

Williams, M.C. and Elliott, R.D. (1985) Weather modification. Chapter 4 in *Facets of Hydrology*, vol. II, ed. J.C. Rodda, pp. 99–129. Wiley, Chichester.

Woodhouse, C.A., Meko, D.M., MacDonald, G.M., Stahle, D.W. and Cook, E.R. (2009) 1,200-year perspective of 21st century drought in southwestern North America. *Proc. Nat. Acad. Sci. USA*, 107, 21283–21288.

UNESCO (1979) Map of the world distribution of arid regions. Man and the Biosphere Programme (MAB) Technical Note 7.

Williams, M. (2014) Desertification: causes, consequences and solutions. Chapter 24 in *Climate Change in Deserts*, pp. 473–499. Cambridge University Press, Cambridge.

Zhao, C., Deng, X., Yuan, Y., Yan, H. and Liang, H. (2013) Prediction of drought risk based on the WRF model in Yunnan Province of China. *Adv. Meteorol.*, 2013, Article ID 295856.

12

Wind and the Global Circulation

12.1 Equations of motion

The atmospheric circulation is the large scale movement of air, and the means by which thermal energy is distributed on the surface of the Earth. The movement of air is governed by the equations of motion, where the mass of a parcel of air multiplied by its acceleration is equal to the vector sum of the forces acting on the parcel (see for example Hess, 1959). The forces acting on the parcel are the pressure gradient force, the gravitational force and the frictional force.

Consider a parcel of air as shown in Figure 12.1. The force in the x direction acting on the face ABCD is $p\delta y\delta z$, where p is the air pressure over the face. The force in the x direction acting on the face A′B′C′D′ is $-p'\,\delta y\delta z$, where p' is the air pressure over the face. Thus the net pressure force in the x direction is $(p-p')\delta y\delta z = -(\partial p/\partial x)(\delta x\delta y\delta z)$. Hence

$$\text{total pressure force} = -\nabla p\,\delta x\delta y\delta z \tag{12.1}$$

where ∇ is the Laplace operator. Taking g_a as the gravitational force per unit mass, then

$$\text{gravitational force on parcel} = \rho\delta x\delta y\delta z g_a \tag{12.2}$$

$$\text{frictional force on parcel} = \rho\delta x\delta y\delta z F \tag{12.3}$$

where ρ is the density of the air and F is the frictional force. Therefore the equation of motion in the inertial frame A is

$$\left(\frac{dv_a}{dt}\right)_A = -\frac{1}{\rho}\nabla p + g_a + F \tag{12.4}$$

where v_a is the velocity of the unit mass. Allowing for the spin of the Earth we have two more forces, the Coriolis force $-2\Omega v$ and the centrifugal force $\Omega^2 R$, where Ω is the angular rotation of the Earth and R is the radius of the Earth, measured positive

---

Hydrometeorology, First Edition. Christopher G. Collier.
© 2016 John Wiley & Sons, Ltd. Published 2016 by John Wiley & Sons, Ltd.
Companion website: www.wiley.com/go/collierhydrometerology

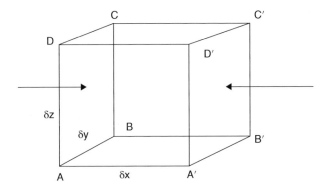

Figure 12.1   A parcel of air

outward from the axis of rotation. This may be simplified by linking g_a and $\Omega^2 R$ such that $g_a + \Omega^2 R = g$, the force of gravity. Therefore

$$\frac{dv}{dt} = -\frac{1}{\rho}\nabla p + g + F - 2\Omega v \tag{12.5}$$

On the synoptic scale, motion is essentially horizontal, i.e. vertical motion is small. Hence equation 12.5 can be simplified to

$$\frac{\partial p}{\partial z} = -\rho g \tag{12.6}$$

This is the *hydrostatic equation*, valid only for the synoptic scale.

When there is no acceleration and no friction we have horizontal motion under balanced forces, and the horizontal geostrophic wind velocity is

$$V_g = \frac{1}{\rho f}\left|\frac{\partial p}{\partial n}\right| \tag{12.7}$$

where f is the Coriolis parameter and n is the horizontal gradient of the vertical separation of the pressure and isobaric surfaces. This is the *geostrophic wind equation*. Note that the geostrophic wind equation is meaningless at the equator and not a very good representation near the equator. If we require a uniform easterly flow of 12 knots extending from 20°N to 20°S then $\partial p/\partial n$ must be zero at the equator because $f = 0$. However, away from the equator $\partial p/\partial n$ can take on a non-zero value.

The effect of friction is to reduce the wind speed, causing cross-isobar flow from high to low pressure. The friction velocity of the sea can be derived from the mean wind and the bulk stability (see Vickers et al., 2015). Isobars are not straight and this must be accounted for in deriving the wind velocity. Assuming that there is no friction and that the tangential acceleration is zero, then the flow is said to be *gradient flow*. This is given by

$$C - V^2 = f\left(V_g - V\right) \tag{12.8}$$

where V is the gradient wind, r is the radius of curvature of the isobars and $C = 1/r$ is the curvature. Taking h as the height of the pressure surface above some datum level, then $\partial p/\partial x = \rho g \partial h/\partial x$ and we may write the geostrophic wind equation as

$$V_g = \frac{g}{f}\left|\frac{\partial h}{\partial n}\right|$$
(12.9)

This is the *thermal wind equation*, where h is the thickness of the layer between two pressure surfaces. The thermal wind is the difference between the geostrophic winds at two levels (see for example McIlveen, 1992). Large scale air motion around both low and high pressure systems is discussed in Appendix 12.1.

12.2   Atmospheric Ekman layer

The atmospheric Ekman layer is similar to the planetary boundary layer described in the oceans by Ekman (1905). The balance of forces at different levels in the atmospheric boundary layer is shown in Figure 12.2. Consider a quasi-steady wind aloft in geostrophic balance between a pressure gradient force and the Coriolis force. The acceleration is therefore weak within the Ekman layer. The frictional force becomes increasingly important at lower and lower levels, whereas the Coriolis force is expected to become weaker towards the surface. The equations representing the Ekman layer are:

$$u = u_g\left[1-\exp(-\gamma z)\cos(\gamma z)\right]-v_g\exp(-\gamma z)\sin(\gamma z)$$
$$v = v_g\left[1-\exp(-\gamma z)\cos(\gamma z)\right]-u_g\exp(-\gamma z)\sin(\gamma z)$$
(12.10)

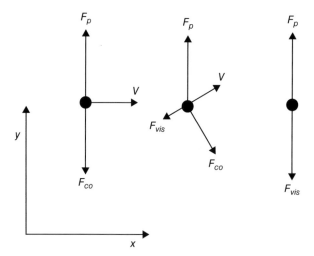

Figure 12.2   Schematic showing the balance of forces in an Ekman layer in the northern hemisphere at three levels from left to right: near the top, at mid-level and near the surface. F_p is pressure gradient force, F_{co} is Coriolis force, F_{vis} is frictional force and V is velocity (from Mak, 2011)

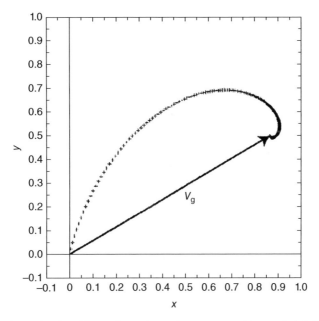

Figure 12.3 Hodograph of an Ekman layer driven by a geostrophic wind aloft $V_g = (u_g, v_g)$ (from Mak, 2011)

A curve $V = (u,v)$ at different vertical levels projected on an (x, y) plane is called a *hodograph*. Because of its spiral shape, it is called an *Ekman spiral*. An example of the structure of wind in an Ekman layer driven by a non-dimensional geostrophic flow aloft equal to $(u_g, v_g) = (\cos\phi, \sin\phi)$ with $\phi = 30°$ is shown in Figure 12.3. The key parameters are the Coriolis parameter $f = 10^{-4}\,\mathrm{s}^{-1}$ and the turbulent eddy coefficient $K = 5\,\mathrm{m}^2\mathrm{s}^{-1}$. The points refer to the values of the wind at heights of about 6 m apart. The wind is seen to diminish towards the surface, directed 45° to the left relative to the geostrophic wind aloft.

12.3 Fronts

There is a discontinuity of wind across a stationary or moving front with a fixed configuration. On either side of a sloping discontinuity of temperature and therefore density, the y component of the geostrophic wind is given by equation 12.7. Following Scorer (1997), Figure 12.4 shows a vertical section normal to the wind component, with the plane of the front sloping upwards at angle of α; the velocity across the front is v_G. The pressure gradient just below the discontinuity Δp exceeds that just above the front by an amount $(\partial p/\partial x)(g\Delta\rho\Delta z)$, and ρ is replaced by $\rho + \Delta\rho$. Neglecting the higher order terms in the discontinuity with temperature T,

$$\frac{1}{\rho + \Delta\rho} = \frac{1}{\rho}\left(1 + \frac{\Delta\rho}{\rho}\right)^{-1} = \frac{1}{\rho}\left(1 - \frac{\Delta\rho}{\rho}\right) = \frac{1}{\rho}\left(1 + \frac{\Delta T}{T}\right) \tag{12.11}$$

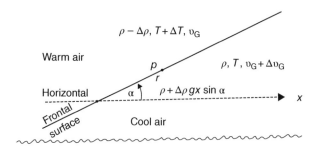

Figure 12.4 A discontinuity of density $\Delta\rho$ inclined at an angle α to the horizontal (from Scorer, 1997)

and therefore below the discontinuity where $\Delta z = \alpha x$, and when $\alpha = 10^{-2}$ say,

$$v_G + \Delta v_G = v_G\left(1 + \frac{\partial T}{T}\right) - \frac{g}{f}\frac{\Delta T}{T} \qquad (12.12)$$

where the term $\Delta T \Delta \rho$ is neglected. Using the Boussinesq approximation, i.e. where the effect of density variations on the inertia represented by $(\Delta T/T)v_G$ is small compared with the final term in equation 12.12 which contains g, gives

$$\nabla\Delta v_G = \frac{\Delta T}{T}\frac{g\alpha}{f} \qquad (12.13)$$

which is the change in the geostrophic wind in passing up through the frontal surface.

12.4   Jet streams

The warm air above a front is subject to the thermal wind equation 12.9, and therefore usually blows almost along the front with the cold air on the poleward side of the front. A typical cross-section of a frontal zone is shown in Figure 12.5 in which the vertical scale is exaggerated. It shows the order of magnitude of the wind maximum and its horizontal and vertical extent. The jet stream is the name given to the wind maximum. This phenomenon may be reproduced in rotating fluids in which temperature gradients are generated (see for example Hide, 1966). In Figure 12.5 the front is depicted as a zone of large gradient rather than a discontinuity. Above a jet stream the temperature gradient is reversed, usually in the stratosphere, and the wind decreases with height. The strength of a jet stream varies along a front, and has a maximum to the west of the centre of a cyclone which is active on a wave of a front. The warm air in the jet stream therefore overtakes the cyclone, and is accelerated into the maximum at the jet entrance, and decelerated as it emerges at the jet exit.

The energy required to produce the acceleration is derived from a direct circulation in which the cold air sinks and the warm air rises over it, decreasing the potential energy of the air. The motion in this circulation is ageostrophic, and is largely accounted for by the mechanism of the acceleration of the geostrophic wind (ageostrophic motion is discussed in Appendix 12.2). At a jet exit the deceleration is affected by an

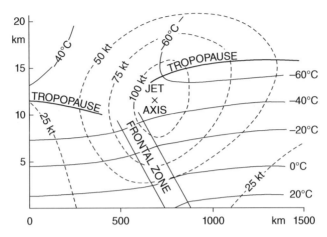

Figure 12.5 Typical dimensions and simplified structure of a jet stream and wind speed blowing across it (into the paper); wind speeds are in knots (from Scorer, 1997)

indirect circulation in which the warm air sinks and moves ageostrophically towards higher pressure. Here kinetic energy is converted into potential energy and moves ageostrophically towards higher pressure. The jet maximum is usually about 1 km higher at the entrance than at the exit because the maximum is further downwind at the lower levels.

12.5 Hurricanes

The feature of hurricanes which makes them different from larger cyclones is that they depend on a continuous supply of heat to maintain the circulation. This heat source is the rain whose latent heat of condensation is given to the air in which the rain condenses. This heat source exists mainly because the sea level pressure at the centre is lower than in the surrounding air, and therefore the air temperature is lower by between 3 and 5 °C at the centre because the air expands adiabatically as it moves inwards. Much more violent convection is supported here.

Because of the strong winds and rough sea, the air spiralling inwards at the sea surface is subject to considerable retardation due to friction. In the absence of friction the tangential velocity (relative to axes fixed in space) would increase as r^{-1}, where r is the distance from the centre. However, usually it is observed to increase inwards roughly as $r^{-\frac{1}{2}}$. Near the centre the rotation of axes fixed to the Earth is a small contribution which can be neglected. The air does not penetrate to the very centre of the storm but only to the outer edge of the eye, which is a region free from rain clouds and is composed of air sucked down from the stratosphere by the low pressure in the vortex.

Hurricanes may become mid-latitude cyclones if they move to a high enough latitude (say 37°) and are regenerated by the inflow of a cold front. The sea is not warm enough to provide the heat source needed to maintain a hurricane except at low latitudes in the autumn. The heat flows from the sea largely as latent heat in the form of evaporated water; the heat is released when the rain falls out.

Hurricanes are called typhoons in the Pacific and Indian Oceans. The general name is tropical cyclone. They occur in the Arabian Sea, the Bay of Bengal, the western Pacific, the Atlantic in low latitudes, northern Australia and the China Sea. They move from the east in the trade winds at first, then they curve along the east side of the continents.

12.6   Lee waves

Most usually standing waves are lee waves resulting from flow over a hill somewhere upstream. If a linear velocity is added to the coordinate system the waves could be seen as travelling over level ground (or sea), but in this case an impulse is required to start them which may be a downdraught of air from a higher level storm moving with a different velocity, the explosion of a volcano or the arrival of a meteorite.

Scorer (1997) describes the use of a simplified equation to represent standing waves as follows:

$$\frac{\partial^2 \zeta}{\partial z^2} + S\frac{\partial \zeta}{\partial z} + \left(l^2 - k^2\right) = 0 \tag{12.14}$$

in which ζ is the displacement of the flow; $S = 2\alpha - \beta - g/c^2$, where $\alpha = (1/u)(du/dz)$ is the vertical velocity gradient, $\beta = (1/\theta)(\partial\theta/\partial z)$, θ is potential temperature, and c is the velocity of light; $l^2 = g\beta/u_0^2$ is known as Scorer's parameter; and k is the wave number. The variations in l^2 with height are very important in determining the character of lee waves (Scorer, 1997).

12.7   Land and sea breezes

Latitudinal circulation is the consequence of the fact that incident solar radiation per unit area is highest at the heat equator, and decreases as the latitude increases, reaching its minimum at the poles. Longitudinal circulation, on the other hand, comes about because water has a higher specific heat capacity than land and thereby absorbs and releases more heat, but the temperature changes less than land. Even at mesoscales (a horizontal range of 5 to several hundred kilometres), this effect is noticeable. As shown in Figure 12.6, this causes the sea breeze (air cooled by the water) ashore in the day, and carries the land breeze (air cooled by contact with the ground) out to sea during the night (see for example Zhang and Zheng, 2004; Mak, 2011).

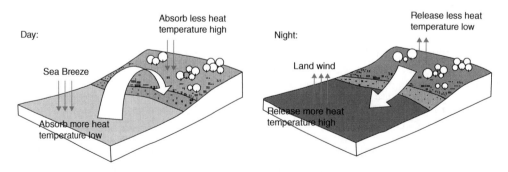

Figure 12.6   Diurnal wind change in coastal area (from Wikipedia Ambassador Program, author Yue Guan, 11 February 2011)

12.8   The wind structure of the atmospheric circulation

The atmosphere transports heat, but present-day atmospheric characteristics prevent heat from being carried directly from the equator to the poles (Trenberth et al., 2009). Currently, there are three distinct wind cells – Hadley cells, Ferrel cells and polar cells – and they divide the troposphere into regions of essentially closed wind circulations (see for example Kushnir, 2007). The main bulk of the vertical motion occurs in the Hadley cell. There is one discrete Hadley cell that may split, shift and merge in a complicated process over time. Low and high pressures on the Earth's surface are balanced by opposite relative pressures in the upper troposphere.

12.9   Hadley cell

The atmospheric circulation pattern of the Hadley cell (Hadley, 1735) provides an explanation for the trade winds. It is a closed circulation loop, which begins at the equator with warm, moist air lifted aloft in equatorial low pressure areas (the Inter-Tropical Convergence Zone, ITCZ) to the tropopause and carried poleward. At about 30°N/S latitude, it descends in a high pressure area. Some of the descending air travels equatorially along the surface, closing the loop of the Hadley cell and creating the trade winds.

Though the Hadley cell is described as lying on the equator, it follows the Sun's zenith point, and undergoes a semi-annual north–south movement. The Hadley system provides an example of a thermally direct circulation. The thermodynamic efficiency and power of the Hadley system enables it to be considered as a heat engine.

Heat from the equator generally sinks around 30° latitude where the Hadley cells end, and as a result the warmest air does not reach the poles. If atmospheric dynamics were different, however, it is possible that one large overturning circulation per hemisphere could exist and that wind from the low latitudes could transport heat to the high latitudes. Farrell (1990) suggested that during equable climates, the Hadley cells extended from the equator to the poles.

Held and Hou (1980) outlined the dynamics of this circulation through a simplified model of the Hadley cell. For the model, there are three main assumptions. First, the Hadley cell circulation is constant. Second, the air moving toward the poles in the upper atmosphere conserves its axial angular momentum, and the surface air moves towards the equator slowed down by friction. Third, the thermal wind balance is valid for the circulation (Vallis, 2006). For simplicity, the model is also symmetric around the equator.

Angular momentum is defined as the cross product of a particle's distance from the axis of rotation r and the particle's linear momentum p. In the Hadley cell, as an air particle moves toward high latitudes it comes closer to the Earth's spin axis, so r becomes smaller. If angular momentum is conserved in the Hadley cell, as Held and Hou (1980) assume, p must become larger to balance the decrease in r; p equals mass times velocity. Since the mass of the air particle cannot change, the velocity of the particle must increase. In the case of the Hadley cell, the velocity is the zonal (east–west) velocity; so as the particle moves towards the pole, the velocity must increase in the eastward direction. Eventually, the zonal velocity is so strong that the particle stops moving towards the pole and only travels to the east. At this latitude, air sinks, and then to close the loop, it returns to the equator along the surface. Therefore, because of the conservation of angular momentum, Hadley cells exist only from the equator to the mid-latitudes.

Heaviside and Czaja (2012) discuss the mechanisms responsible for poleward atmospheric heat transport at low latitudes using the ECMWF reanalysis (ERA-40) data set. Deep meridional overturning circulations are found in regions experiencing frequent convection (i.e. the Indo-Pacific 'warm pool'), and it is suggested that these are the primary reasons why atmospheric heat transport seems dominated by axisymmetric motions referred to as 'Hadley cell heat transport'. In contrast, the more complex distribution of meridional mass transport by the circulations such as the eastern Atlantic area, and the presence of a pronounced minimum in moist static energy at mid-levels, constrain these regions to contribute little to heat transport towards the pole throughout the year. Regions experiencing frequent convection do not, however, account for all the annual mean atmospheric heat transport towards the pole. At the Equator, an annual net southward transport of heat of about 0.4 peta-watts (PW: 1 petawatt $= 10^{15}$ watts) is found, but the deep overturning cells found in the most convectively active regions contribute only 25% of this.

During December–February, most of the cross-equatorial heat transport is attributable to the convective regions, whereas in June–August, only half is attributable to these regions. In this season, there is a significant amount of heat transport by the strong Somali jet associated with the Asian summer monsoon (section 12.14). Rather than attributing the net southward cross-equatorial heat transport to asymmetries in the climatology associated with the northerly position of the Inter-Tropical Convection Zone throughout the year over the eastern Pacific and Atlantic oceans, it is suggested that the Somali jet is a significant contributor to the annual net southward atmospheric heat transport at the Equator.

12.10   Polar cell

The polar cell is cool and dry relative to equatorial air, although it is still sufficiently warm and moist to instigate convection and drive a thermal loop. Air circulates within the troposphere up to about 8 km. Warm air rises at lower latitudes and moves towards the poles through the upper troposphere at both the north and south poles. When the air reaches the polar areas, it has cooled considerably, and descends as a cold, dry, high pressure area, moving away from the pole along the surface but turning westward as a result of the Coriolis effect to produce the polar easterlies.

The outflow from the cell creates quasi-horizontal harmonic waves in both hemispheres of the atmosphere, known as Rossby waves (Rossby, 1938). These ultra-long waves play an important role in determining the path of the jet stream, which travels within the transitional zone between the tropopause and the Ferrel cell. By acting as a heat sink, the polar cell also balances the Hadley cell in the Earth's energy system.

The polar cell, the orography and the katabatic winds in Antarctica can create very cold conditions at the surface. The coldest temperature recorded on Earth is −89.2 °C at Vostok Station in Antarctica, measured in 1983 (Greenemeier, 2008).

12.11   Ferrel cell

Some air rising at the polar fronts diverges at high altitude towards the poles to create the polar cell. The rest of the air moves in the opposite direction to the high level zones of convergence and subsidence at the subtropical ridges on each side of the Equator. These mid-latitude counter-circulations create the Ferrel cells (Ferrel, 1856) that encircle the globe in the northern and southern hemispheres. The Ferrel cell is a secondary circulation feature, dependent for its existence upon the Hadley cell and the polar cell. It is sometimes known as the 'zone of mixing'. At its southern extent (in the northern hemisphere) it overrides the Hadley cell, and at its northern extent it overrides the polar cell. Just as the trade winds can be found below the Hadley cell, the westerlies can be found beneath the Ferrel cell. The strong, high pressure areas divert the prevailing westerlies, such as a Siberian high. In contrast to the Hadley and polar systems, the Ferrel system provides an example of a thermally indirect circulation. The Ferrel system acts as a heat pump with a coefficient of performance of 12.1, and consumes kinetic energy.

While the Hadley, Ferrel and polar cells are major factors in global heat transport, they do not act alone. Differences in temperature also drive a set of longitudinal circulation cells, and the overall atmospheric motion is known as the zonal overturning circulation.

12.12   Walker circulation

Warm air rises over the equatorial, continental and western Pacific Ocean regions, and flows eastward or westward, depending on its location. When it reaches the tropopause, it subsides in the Atlantic and Indian Oceans and in the eastern Pacific. The Pacific Ocean cell plays a particularly important role in the Earth's weather. This ocean-based cell results from a difference in the surface temperatures of the western and eastern Pacific. Under ordinary circumstances, the western Pacific is warm and the eastern part is cool. The process begins when strong convective activity over equatorial East Asia and subsiding cool air off South America's west coast create a wind pattern which pushes Pacific water westward and piles it up in the western Pacific. (Water levels in the western Pacific are about 60 cm higher than in the eastern Pacific, a difference due entirely to the force of moving air: Hein et al., 2004.)

The Pacific cell is referred to as the Walker circulation (Walker, 1923) and arose after work undertaken to discover when the monsoon winds would fail. This led to the identification of a link between periodic pressure variations in the Indian Ocean and the Pacific, which is termed the 'Southern Oscillation'. Under 'normal' circumstances, the weather behaves as expected; but every few years the winters become unusually warm or unusually cold, or the frequency of hurricanes increases or decreases, and the pattern sets in for an indeterminate period. Meanwhile in the Atlantic, the westerlies are unable to reach their usual intensity. These winds remove the tops of developing hurricanes and reduce the number able to reach full strength.

12.13   El Niño/Southern Oscillation

The behaviour of the Walker cell is the key to understanding the El Niño (more accurately, the El Niño/Southern Oscillation or ENSO) phenomenon (Wang and Picaut, 2004). If convective activity slows in the western Pacific for some reason (this reason is not currently known), the climate begins to change. First, the upper level westerly winds fail. This cuts off the source of cool subsiding air, and therefore the surface easterlies cease. The consequence of this is twofold. In the eastern Pacific, warm water moves in from the west since there is no longer a surface wind to constrain it. This and the corresponding effects of the Southern Oscillation result in long term unseasonable temperatures and precipitation patterns in North and South America, Australia, and south-east Africa, and disruption of ocean currents.

El Niño and La Niña are opposite surface temperature anomalies in the Southern Pacific, which heavily influence the weather on a large scale. In the case of El Niño, warm water approaches the coasts of South America, which results in the blocking of the upwelling of nutrient-rich deep water. This has serious impacts on the fish populations. In the La Niña case, the convective cell over the western Pacific strengthens inordinately, resulting in colder than normal winters in North America, and a more robust cyclone season in South East Asia and eastern Australia. There is increased upwelling of deep cold ocean waters and more intense uprising of surface air near South America, resulting in increasing numbers of drought occurrences, although it is often argued that fishermen reap benefits from the more nutrient-filled eastern Pacific waters. Over the last 10 years numerical predictions of the ENSO have slowly improved (Barnston et al., 2012).

12.14   Monsoons

Monsoon winds are associated with excessive rainfall in many parts of the world including Asia, North America, South America, and Africa. The primary mechanism behind a monsoon is a shift in global wind patterns (Webster, 2006), although regional, seasonal and diurnal variations of convection are also important (Romatschke et al., 2010). During most of the year, winds blow from land to ocean, making the air dry. However, during certain months of the year, the winds begin to blow from the ocean to the land, making the air moist. This moist ocean air is what causes monsoonal rains over many countries.

Differential heating occurs when the Sun heats the land and oceans, as in section 12.7 but on a larger scale. Incoming solar radiation heats land masses faster than large bodies of water. In tropical and subtropical climates, solar heating is most intense in the summer months. As the land heats throughout the summer, a large low pressure system builds over the land. The heat from the Sun also warms the surrounding ocean waters, but the effect happens much more slowly due to the high heat capacity of water. Therefore the ocean, as well as the layer of air above the ocean, stays cooler for longer. The cooler air above the ocean is moist and more dense, creating a high pressure zone relative to the pressure above the land mass.

Winds flow from high pressure areas to low pressure areas due to the pressure gradient. Once the temperature conditions on the land and oceans change, the resultant pressure changes cause the winds to change from a land-to-ocean direction to an ocean-to-land direction. Monsoon season does not end as abruptly as it begins. While it takes time for the land to heat up, it also takes time for that land to cool in the autumn. This makes monsoon season a time of rainfall that diminishes rather than ends.

One important synoptic feature associated with the onset of the Indian monsoon is the existence of a strong cross-equatorial low level jet (LLJ), with its core around 850 hPa over the Indian Ocean and South Asia (Findlater, 1969). This LLJ generally supports the large scale moisture and momentum transport from ocean to atmosphere and the consequent rainfall over India. The Somali monsoon wind is deflected to the north as it crosses the Equator, and is further deflected to the east by the mountains of Africa as it passes over Kenya, Somalia and the Sahel.

Swapna and Kumar (2002) analysed twice-daily zonal (u) and meridional (v) winds at 10 m height and at 850 hPa and total columnar water vapour derived from the National Centers for Environmental Prediction (NCEP) reanalysis project for the years 1987 and 1988. Figure 12.7 shows the Indian summer monsoon flow for the pentad (5 day average) mean winds at 850 hPa for 21–25 July 1988. For this year India had more than normal rainfall. During the active monsoon periods, the core of the jet is directed to the Indian subcontinent, producing heavy rainfall over India. However, during the break period (less active) the core of the low level jet is directed south of the Indian peninsula, leading to a weakening of monsoon conditions over India.

Analysis of numerical model simulations with mountains reveals (Hahn and Manabe, 1975) that the large scale circulation associated with the South Asian monsoon may be well simulated. However, the onset of the monsoon is approximately 10–15 days later than normal, and the atmosphere over the western Pacific seems to be dynamically too active, whereas the atmosphere over the northern parts of the Bay of Bengal and northern India is relatively inactive. More recently Krishnamurti

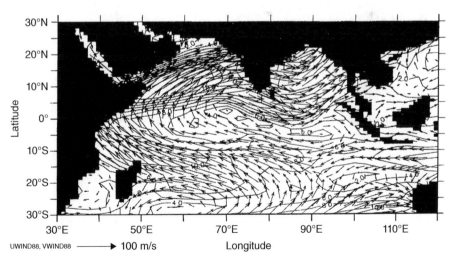

Figure 12.7 Pentad mean winds at 850 hPa over the Indian Ocean, 21–25 July 1988 (from Swapna and Kumar, 2002)

et al. (1983) have demonstrated, using a mesoscale fine mesh numerical model, the important role of advective accelerations in the near-equatorial balance of forces. Work continues using numerical models.

Appendix 12.1   Large scale air motion

When the isobars are curved around a low pressure, the pressure gradient acts towards the centre of the low and in the northern hemisphere the wind blows anticlockwise. Assume the wind is blowing in a circle of radius r. Then the Coriolis force is directed outwards to the direction of the pressure gradient dp/dr, and for steady motion

$$\frac{V^2}{r} = \frac{1}{\rho}\frac{dp}{dr} - 2V\omega\sin\phi \quad (\text{cyclone}) \tag{12.15}$$

where ρ is the density of the air moving with a velocity V; ϕ is the latitude; and ω is the angular velocity of the Earth.

   If the isobars are curved round the centre of a high pressure, dp/dr is negative and the pressure gradient gives an acceleration $-(dp/dr)/\rho$ outwards. Then in the northern hemisphere the equation for steady state motion is

$$\frac{V^2}{r} = 2V\omega\sin\phi + \frac{1}{\rho}\frac{dp}{dr} \quad (\text{anticyclone}) \tag{12.16}$$

The equation for the cyclone is a quadratic equation in V having two roots:

$$V = -r\omega\sin\phi \pm \sqrt{\left[r^2\omega^2\sin^2\phi + (r/\rho)(dp/dr)\right]} \tag{12.17}$$

The positive sign before the square root gives a positive value for V corresponding to anticlockwise motion.

Appendix 12.2   Ageostrophic motion

Take V to be the actual velocity; V' to be the geostrophic velocity appropriate for the pressure gradient; and $V - V_g = V'$ to be the ageostrophic wind (also known as the geostrophic departure). Then

$$\frac{dV}{dt} = -\frac{\nabla p}{\rho} + fv\hat{k} + F \tag{12.18}$$

Here "^" means a vector product (see for example Stephenson, 1965). Assuming unaccelerated, frictionless motion, i.e. geostrophic flow, then

$$0 = -\frac{\nabla p}{\rho} + fV_g\hat{k} + 0 \tag{12.19}$$

Subtracting these two equations gives

$$\frac{dV}{dt} = f(V - V_g)\hat{k} + F$$

$$\frac{dV}{dt} = fV'\hat{k} + F$$

(12.20)

Multiplying by $\hat{k}$ and rearranging:

$$fV' = \hat{k}\frac{dV'}{dt} + \hat{k}\frac{\partial V_g}{\partial t} + V\hat{k}\frac{\partial V_g}{\partial s} + \omega\hat{k}\frac{\partial V_g}{\partial z} - \hat{k}F$$

(12.21)

This is an expression for the ageostrophic wind. Consider the isallobaric wind (change of pressure pattern) V_I', the ageostrophic wind due to downwind changes V_h', the ageostrophic wind due to vertical advection V_v' and the ageostrophic wind due to friction V_F'. In the special case when $V_I' = V_h' = V_v' = V_F' = 0$, the equation reduces to

$$fV_a' = \hat{k}dV_a'/dt$$

(12.22)

where V_a' is the ageostrophic wind due to the acceleration of the wind. This is the inertial motion showing that this is part of the ageostrophic wind. Hence

$$V_I' = -\nabla_H(\partial p/\partial t)/\rho f^2$$

(12.23)

An isallobar is a line of constant rate of change, i.e. $\partial p/\partial t$, and the isallobaric wind blows down the isallobaric gradient into low pressure and out of high pressure. It aids ascending motion and increases cloud and therefore precipitation, but in high pressure systems it aids subsidence, so removing cloud.

Summary of key points in this chapter

1. The atmospheric circulation is the large scale movement of air and is the means thermal energy is distributed on the surface of the Earth.
2. There are a number of forces acting on a parcel of air: total pressure, gravitational, frictional, the Coriolis and the centrifugal forces.
3. The hydrostatic equation is $\partial p/\partial z = -\rho g$ and is applicable on the synoptic scale.
4. The geostrophic wind represents horizontal motion under balanced forces when there is no acceleration acting on the parcel of air and no friction.
5. When there is no friction and the tangential acceleration on a parcel of air is zero, the flow is said to be gradient flow.
6. The thermal wind equation is the difference between the geostrophic winds at two levels.

7. The Ekman layer represents flow in the planetary boundary layer where the pressure gradient force, the Coriolis force and the frictional force are balanced.

8. The structure of the wind profile in the Ekman layer is a spiral.

9. A discontinuity of wind (and temperature) across a stationary or moving boundary with a fixed configuration is known as a front.

10. The Boussinesq approximation is where the effect of density variations on the inertia is small compared to other variations.

11. The warm air above a front is subject to the thermal wind equation.

12. The jet stream is the name given to the wind maximum above a frontal zone.

13. The energy required to produce the acceleration in a jet stream is derived from a direct circulation in which cold air sinks and the warm air rises over it, reducing the potential energy of the air. The motion in this circulation is ageostrophic.

14. Hurricanes depend on a continuous supply of heat to maintain their circulation. The heat source is the rain whose latent heat of condensation is given to the air in which the rain condenses.

15. Air spiralling inwards at the sea surface is subject to considerable retardation due to friction.

16. Hurricanes may become mid-latitude cyclones if they move to a high enough latitude.

17. Hurricanes are called typhoons in the Pacific and Indian Oceans, but generally may be referred to as cyclones.

18. Lee waves are standing waves resulting from flow over a hill somewhere upstream, but may also be caused by an impulse from a downdraught of air from a higher level storm, a volcano or a meteorite.

19. Land and sea breezes result from temperature differences between the atmosphere over the land and sea.

20. Three distinct wind cells account for the structure of the atmospheric circulation, namely Hadley, Ferrel and polar cells.

21. The Hadley cell provides an explanation for the trade winds. It is a closed circulation loop beginning at the equator with warm moist air lifted aloft in equatorial low pressure areas. As air moves towards the poles, axial angular momentum is conserved. Heat is transported poleward.

22. The polar cell in both hemispheres results from warm air rising at low latitudes and moving polewards through the upper troposphere. The outflow creates quasi-horizontal harmonic waves known as Rossby waves which play an important role in determining the path of the jet stream.

23. The Ferrel cells are the mid-latitude counter-circulations that encircle the globe in both hemispheres. They depend for their existence upon the Hadley and polar cells.

24. The circulation over the Pacific Ocean is known as the Walker cell, and is associated with monsoon winds and the Southern Oscillation.

25. The El Niño/Southern Oscillation results from convection activity in the western Pacific causing the climate to 'domino', resulting in unseasonable temperatures and precipitation patterns in North and South America, Australia and south-east Africa and disruption of ocean currents.

26. El Niño and La Niña are opposite surface temperature anomalies in the southern Pacific.

27. Seasonal monsoon winds are associated with excessive rainfall in many parts of the world, resulting from differential heating of the atmospheric layer above the oceans. Pressure changes cause the winds to change from a land-to-ocean direction to an ocean-to-land direction.

28. The equatorial low level jet stream over Kenya, Somalia and the Sahel, having a core at a height of about 850 hPa, is associated with the onset of the Indian monsoon.

29. During the active monsoon periods the core of the jet is directed to the Indian subcontinent, producing heavy rainfall over India, but in less active periods the jet is directed south of the Indian peninsula.

Problems

1. Specify the net pressure force in the x direction on a parcel of air using a schematic drawing and a mathematical representation.

2. Name the forces acting on a parcel of air.

3. What are the hydrostatic and geostrophic wind equations?

4. Give a formula for gradient wind flow.

5. Describe the thermal wind and give the mathematical representation.

6. Show a schematic giving the balance of forces in an Ekman layer.

7. Give equations representing the Ekman layer.

8. Give the structure of wind in an Ekman layer assuming geostrophic flow $(u_g, v_g) = (\cos\phi, \sin\phi)$ with $\phi = 30°$, $f = 10^{-4} s^{-1}$ and $K = 5 m^2 s^{-1}$.

9. What is the Boussinesq approximation?

10. Give a formula for the geostrophic wind passing up through a frontal surface inclined at an angle α and assuming the effect of density variations on the inertia are small.

11. Draw a diagram showing a simplified structure of a jet stream above a frontal zone, giving typical dimensions.

12. What is the heat source maintaining a hurricane?

13. In the absence of friction, how would the tangential velocity in a hurricane increase?

14. What causes lee waves?

15. Give a simplified equation representing standing waves.

16. What is Scorer's parameter?

17. Define how land and sea breezes differ.

18. Describe the Hadley cell.

19. Give three assumptions in modelling the Hadley cell.

20. What is angular momentum, and describe how this impacts the Hadley cell?

21. What are Rossby waves?

22. Describe how the polar cell creates Rossby waves.

23. What is the Ferrel cell?

24. Describe how the Walker cell links to the Southern Oscillation.

25. What is the El Niño/Southern Oscillation, and how and where does it influence the weather?

26. What causes the monsoons?

27. Where do monsoons occur?

28. Describe the equatorial low level jet stream.

29. Describe how the equatorial low level jet stream influences the monsoon over the Indian peninsula.

30. What influences the occurrence of the monsoon break period over India?

References

Barnston, A.G., Tippett, M.K., L'Heureux, M.L., Li, S. and DeWitt, D.G. (2012) Skill of real-time seasonal ENSO model predictions during 2002–11: Is our capability increasing? *Bull. Am. Meteorol. Soc.*, 93, 631–651.

Ekman, V.W. (1905) On the influence of the Earth's rotation on ocean currents. *Arch. Math. Astron. Phys.*, 2, 1–52.

Farrell, B.F. (1990) Equable climate dynamics. *J. Atmos Sci.*, 47, 2986–2995.

Ferrel, W. (1856) An essay on the winds and currents of the oceans. *Nashville J. Med. Surg.*

Findlater, J. (1969) A major low-level current near the Indian Ocean during the northern summer. *Q. J. R. Meteorol. Soc.*, 95, 362–380.

Greenemeier, L. (2008) Welcome to the coldest town on Earth. *Sci. Am.*, 24 December.

Hadley, G. (1735) On the cause of the general trade winds. *Phil. Trans. Roy. Soc.*, 34, 58–62.

Hahn, D.G. and Manabe, S. (1975) The role of mountains in the South Asia monsoon circulation. *J. Atmos. Sci.*, 32, 1515–1541.

Heaviside, C. and Czaja, A. (2012) Deconstructing the Hadley cell heat transport. *Q. J. R. Meteorol. Soc.*, 139, 2181–2189.

Hein, Z., Appledoorn, G., Burgers, G. and Van Oldenborgh, G.J. (2004).Relationship between sea surface temperature and thermocline depth in the eastern equatorial Pacific. *J. Phys. Ocean*, 34, 643.

Held, I.M. and Hou, A.Y. (1980) Nonlinear axially symmetric circulations in a nearly inviscid atmosphere. *J. Atmos. Sci.*, 37, 515–533.

Hess, S.L. (1959) *Introduction to Theoretical Meteorology*. Holt, New York.

Hide, R. (1966) Review article on the dynamics of rotating fluids and related topics in geophysical fluid mechanics. *Bull. Am. Meteorol. Soc.*, 47, 873–885.

Krishnamurti, T.N., Wong, V., Pan, H.-L., Pasch, R., Molinari, J. and Ardanuy, P. (1983) A three-dimensional planetary boundary layer model for the Somali jet. *J. Atmos. Sci.*, 40, 894–908.

Kushnir, Y. (2007) The climate system: general circulation and climate zones. Lectures EESC 2100 2007, Department of Earth and Environmental Sciences, Columbia University. eesc. columbia.edu/courses/ees/climate/lectures/gen_circ/.

Mak, M. (2011) *Atmospheric Dynamics*. Cambridge University Press, Cambridge.

McIlveen, R. (1992) *Fundamentals of Weather and Climate*. Chapman & Hall, London.

Romatschke, U., Medina, S. and Houze, R.A. Jr (2010) Regional, seasonal and diurnal variations of convection in the South Asian Monsoon Region. *J. Clim.*, 23, 419–439.

Rossby, C.-G. (1938) On the mutual adjustment of pressure and velocity distributions in certain simple current systems II. *J. Marine Res.*, 1, 239–263.

Scorer, R.S. (1997) *Dynamics of Meteorology and Climate*. Wiley, Chichester.

Stephenson, G. (1965) Mathematical Methods for Science Students, Longmans, London.

Swapna, P. and Kumar, M.R.R. (2002) Role of low level flow on the summer monsoon rainfall over the Indian subcontinent during two contrasting monsoon years. *J. Ind. Geophys. Union*, 6, 123–137.

Trenberth, K.E., Fasullo, J.T. and Kiehl, J.B. (2009) Earth's global energy budget. *Bull. Am. Meteorol. Soc.*, 90, 311–322.

Vallis, G.K. (2006) The overturning circulation: Hadley and Ferrel cells. Chapter 11 in *Atmospheric and Oceanic Fluid Dynamics*. Cambridge University Press, New York.

Vickers, D., Mahrt, L. and Andreas, E.L. (2015) Formulation of the sea surface friction velocity in terms of the mean wind and bulk stability. *J. Appl. Meteorol. Climatol.*, 54, 691–703.

Walker, G.T. (1923) A preliminary study of world weather. *Mem. India Met. Dept.*, 24, 75–131.

Wang, C. and Picaut, J. (2004) Understanding ENSO physics: a review. In *Earth's Climate: The Ocean–Atmosphere Interaction*, ed. C. Wang, S.-P. Xie and J.A. Carton, pp. 21–48. American Geophysical Union.

Webster, P.J. (2006) The coupled monsoon system. In *The Asian Monsoon*, ed. B. Wang, pp. 3–66. Springer, New York.

Zhang, D.-L. and Zheng, W.-Z. (2004) Diurnal cycles of surface winds and temperatures as simulated by five boundary layer parameterizations. *J. Appl. Meteorol.*, 43, 157–169.

13

Climatic Variations and the Hydrological Cycle

13.1   An introduction to climate

Climate may be defined as the sequence of weather which is experienced at a given locality. This sequence is usually taken to extend for between 50 and 100 years, the average lifespan of a human being. Unpredictable climatic fluctuations may have major social and economic impacts on humankind, and the need for an increased scientific effort in this area has been recognized by the World Meteorological Organization and the United Nations in setting up the World Climate Programme in 1979 (WMO, 1975) and the Intergovernmental Panel on Climate Change (IPCC) in 1988.

Among the most serious Earth science and environmental policy issues confronting society are the potential changes in the Earth's water cycle due to climate change. The science community now generally agrees that the Earth's climate is undergoing changes in response to natural variability, including solar variability, and increasing concentrations of greenhouse gases and aerosols. Furthermore, agreement is widespread that these changes may profoundly affect atmospheric water vapour concentrations, clouds, precipitation patterns, and runoff and stream flow patterns. Global climate change will affect the water cycle, possibly creating perennial droughts in some areas and frequent floods in others. For example, as the lower atmosphere becomes warmer, evaporation rates will increase, resulting in an increase in the amount of moisture circulating throughout the troposphere (lower atmosphere). An observed consequence of higher water vapour concentrations is the increased frequency of intense precipitation events, mainly over land areas. Furthermore, because of warmer temperatures, more precipitation is falling as rain rather than snow. This is discussed further later in this chapter.

It is widely recognized (WMO, 1979) that water resources, together with food and energy, have the largest dependence on changes of climate. Part of the reason for this is that many water resources development projects are designed to function for around 100 years or even more in some cases. Therefore any changes in climate which may affect design parameters could endanger lives through increased likelihood of sudden structural failures.

Hydrometeorology, First Edition. Christopher G. Collier.
© 2016 John Wiley & Sons, Ltd. Published 2016 by John Wiley & Sons, Ltd.
Companion website: www.wiley.com/go/collierhydrometerology

Hare (1985) considers a number of idealized time series of some variables such as air temperature or humidity (Figure 13.1). In this figure, curves A to D illustrate *modes of variation* typical of these types of time series as follows:

- *Curve A*: a periodic variation about a *mean value* or *central tendency*. Such variations as these are rare in atmospheric time series, except for daily or annual solar parameters.
- *Curve B*: short term variations typical of atmospheric time series. The series changes almost impulsively in a short period to a new regime within which *quasi-periodic* variations occur.
- *Curve C*: a series in which a downward trend occurs up to a particular point, and thereafter the series is *stationary*.
- *Curve D*: a series with a constant central tendency, but with *short term variations* which increase in amplitude as time progresses.

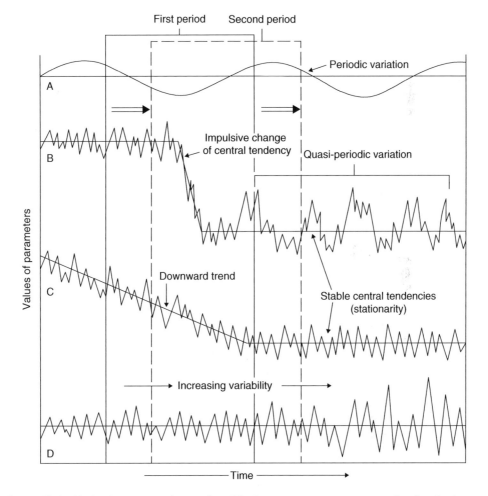

Figure 13.1 Idealized time series (curves A to D) of a representative parameter of a climatic element that is continuous in time (such as temperature or pressure). Vertical bars indicate arbitrary averaging or integrating periods (usually 30 years) which are recalculated each decade (see dashed bars) (after Hare, 1985)

In order to define a particular climate, a number of time series for several parameters are analysed over a long period. In general 30 year periods, updated every decade, are used, as shown in Figure 13.1 by the annotation for the first and second periods.

Climate noise is that part of the variance of climate attributable to short term weather changes, and *climatic variability* is the manner of variation in the climatic parameters within the typical averaging period (Hare, 1985). Hence the climate of an averaging period must be defined in terms of the central tendency and estimates of the variability. *Climatic change* occurs when the differences between successive averaging periods exceed the differences that can be accounted for by noise. Short term changes lasting a few decades are referred to as *climatic fluctuations*.

Climate may be regarded as one element in a global system which includes other geophysical and biochemical cycles. Proposal for interdisciplinary studies of the total system have been made (Roederer, 1985), which aim to clarify various main linkages classified as:

1. The *natural energy system*, the main element of which is the flux of solar energy, but which includes terrestrial and atmospheric radiation, heat fluxes in the soil and ocean, and energy transformations in the oceans and atmosphere.
2. The *hydrological cycle*, the subject of much of this book.
3. The *carbon cycle*, describing the exchange of carbon between atmosphere, oceans, the biosphere and the solid earth. This cycle is of major importance to climate, particularly aspects of atmospheric carbon dioxide.
4. The *other biochemical cycles*, such as for example the exchange of nitrogen and chlorofluoromethanes, and of both man-made and volcanic aerosols.

Perturbations of these linkages affect climate, but it is often difficult to separate out cause and effect. It is only with the help of numerical models that observed climatic change can be analysed.

13.2   Evidence of climate change

Instrumental records have existed in only a few places for as long as between 200 and 300 years, although in China temperature records have been extended for several thousand years (Chu, 1973; Yeh and Fu, 1985). Hence our knowledge of past climates depends largely on analyses of phenomena such as tree rings, fossils and ice cores. Table 13.1 is a summary of the data sources giving information on past climates.

These data have enabled the climate of the last million years, known as the Pleistocene period, to be documented for the northern hemisphere as shown in Figure 13.2. In this figure eight alternations between cold glacial epochs and relatively warm interglacial epochs occurring at intervals of about 100,000 years are evident. Large fluctuations in the northern hemisphere ice sheets and sea level variations of around 100 m occurred with these transitions. About 10,000 years before the present, extensive areas of the northern hemisphere were covered with continental ice sheets, the sea level had dropped by about 120 m, and the sea surface temperatures in the North Atlantic had dropped by up to 10 °C.

Since the Last Interglacial Maximum about 6000 years ago there has been a gradual cooling (as shown in Figure 13.3a), which increased in the late 18th century,

Table 13.1 Characteristics of paleoclimatic data sources (after Mason, 1976)

Proxy data source	Variable measured	Continuity of evidence	Potential geographical coverage	Period open to study (years BP)	Minimum sampling interval (years)	Usual dating accuracy (years)	Climatic inference
Layered ice cores	Oxygen isotope concentration, thickness (short cores)	Continuous	Antarctica, Greenland	10,000	1–10	±1–100	Temperature, accumulation
	Oxygen isotope concentration (long cores)	Continuous	Antarctica, Greenland	100,000+	Variable	Variable	Temperature
Tree rings	Ring-width anomaly, density, isotopic composition	Continuous	Mid-latitude and high latitude continents	1,000 (common) 8,000 (rare)	1	±1	Temperature, runoff, precipitation, soil moisture
Fossil pollen	Pollen-type concentration (varved core)	Continuous	Mid-latitude continents	12,000	1–10	±10	Temperature, precipitation, soil moisture
	Pollen-type concentration (normal core)	Continuous	50°S to 70°N	12,000 (common) 200,000 (rare)	200	±5%	Temperature, precipitation, soil moisture
Mountain glaciers	Terminal positions	Episodic	45°S to 70°N	40,000	—	±5%	Extent of mountain glaciers
Ice sheets	Terminal positions	Episodic	Mid-latitude to high latitudes	25,000 (common) 1,000,000 (rare)	—	Variable	Area of ice sheets
Ancient soils	Soil type	Episodic	Lower and mid-latitudes	1,000,000	200	±5%	Temperature, precipitation, drainage
Closed basin lakes	Lake level	Episodic	Mid-latitudes	50,000	1–100 (variable)	±5%	Evaporation, runoff, precipitation, temperature
Lake sediments	Varve thickness	Continuous	Mid-latitudes	5,000	1	±5%	Temperature, precipitation

(Continued)

Table 13.1 (Continued)

Proxy data source	Variable measured	Continuity of evidence	Potential geographical coverage	Period open to study (years BP)	Minimum sampling interval (years)	Usual dating accuracy (years)	Climatic inference
Ocean sediments (common deep sea cores, 2–5 cm per 1000 years)	Ash and sand accumulation rates	Continuous	Global ocean (outside red clay areas)	200,000	500 +		Wind direction
	Fossil plankton composition	Continuous	Global ocean (outside red clay areas)	200,000	500 +	±5%	Sea surface temperature, surface salinity, sea ice extent
	Isotopic composition of planktonic fossils; benthic fossils; mineralogic composition	Continuous	Global ocean (above $CaCO_3$ compensation level)	200,000	500 +	±5%	Surface temperature, global ice volume; bottom temperature and bottom water flux; bottom water chemistry
(rare cores > 10 cm per 1000 years)	As above	Continuous	Along continental margins	10,000+	20	±5%	As above
(cores < 2 cm per 1000 years)	As above	Continuous	Global ocean	1,000,000+	1000 +	±5%	As above
Marine shorelines	Coastal features, reef growth	Episodic	Stable coasts, oceanic islands	400,000	—	±5%	Sea level, ice volume

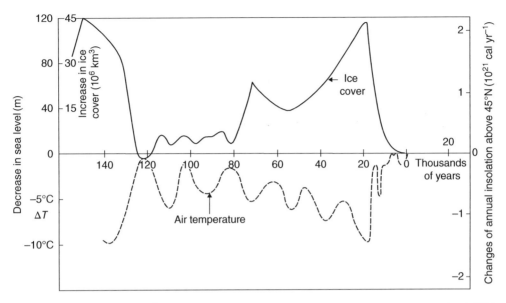

Figure 13.2   Changes in global ice cover and northern hemisphere air temperatures north of 45°N over the last 150,000 years (after Mason, 1976)

although from around 1940 cooling again occurred (as in Figures 13.3b and 13.3c). This central tendency was interrupted by the warm period of the Middle Ages from about 1150 to 1350 CE (Diaz et al., 2011), and the cooler period from about 1500 to 1850 known as the *Little Ice Age*.

One method of analysing time series, known as power spectrum analysis, can provide evidence of periodicity provided account is taken of any smoothing, filtering or removal of long term trends. Figure 13.4 shows the power spectrum of the Manley record of central England temperatures with the long term trend removed; the larger the area beneath the peaks, the more significant are the peaks. Significant peaks are evident at 2.1, 3.1, 5.2, 7.6, 14.5, 23 and 76 years. The first peak is associated with the quasi-biennial oscillation; the 3.1 year peak and perhaps the 5.2 and 7.6 year peaks could result from the Southern Oscillation (Philander, 1983); and the 23 year peak may be associated with the double sunspot cycle. The number of sunspots oscillates on roughly an 11 year cycle, but this periodicity is not particularly significant in Figure 13.4, although other analyses (for example King, 1973) have indicated the significance of this frequency. This illustrates the dangers of drawing conclusions from limited or smoothed data sets, which may or may not reveal a range of periodicities.

The periodicities discussed relate to short term climatic fluctuations. Further fluctuations may also be introduced by the activities of humankind, for example in increasing atmospheric carbon dioxide, mainly by burning fossil fuels (Figure 13.5), or in releasing pollutants into the atmosphere. These effects may manifest themselves as either atmospheric warmings, as is thought to be the case for carbon dioxide (for a review see Bach, 1976), or atmospheric coolings, as may be the case with aerosol input to the stratosphere produced by volcanic activity (Lamb, 1977; Kondratyev et al., 1985). However, the mechanisms are not yet fully understood, and require

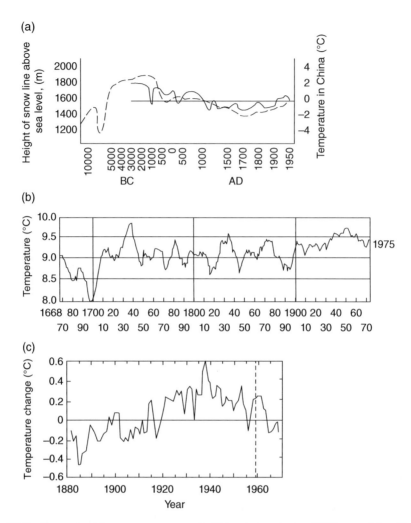

Figure 13.3 Illustrating (a) temperature curve in China during the last 5000 years (after Chu, 1973); (b) 10 year running means of central England temperatures from 1650 to 1975 (after Manley, 1974); and (c) recorded changes in the annual mean temperature of the northern hemisphere since 1880 (after Budyko, 1969)

detailed modelling before the total effects on climate can be defined under all circumstances. There is only limited evidence that the variability of weather is changing (Ratcliffe et al., 1978).

Local changes in the air–sea interaction produce further short term climatic fluctuations. Global sea surface temperature (STT) anomalies – one example being the *El Niño*, a current of warm water which occurs off the coast of South America in certain years, and may be associated with the Southern Oscillation – are thought to affect atmospheric systems (Namias and Cayan, 1981). Similar associations may explain rainfall deficiencies in the Sahel (sub-Saharan North Africa) (Folland et al., 1986; see Chapter11). These types of relationships are sometimes referred to as *teleconnections*.

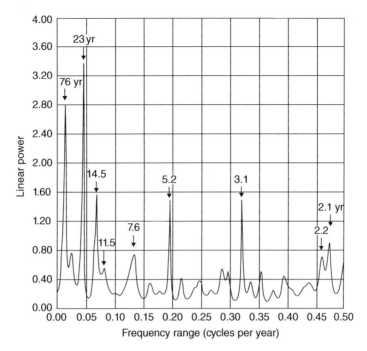

Figure 13.4 Power spectrum of the Manley record of central England temperatures with the long term trend removed (after Mason, 1976)

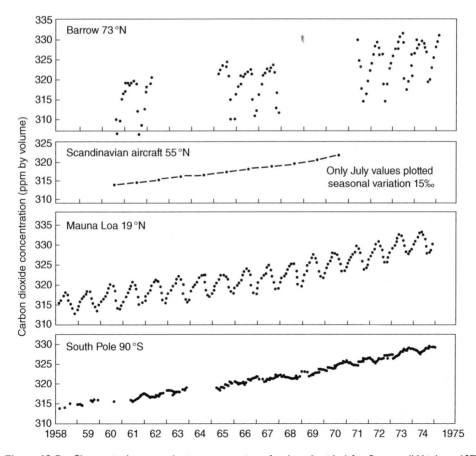

Figure 13.5 Changes in the atmospheric concentration of carbon dioxide (after Rotty and Weinberg, 1977)

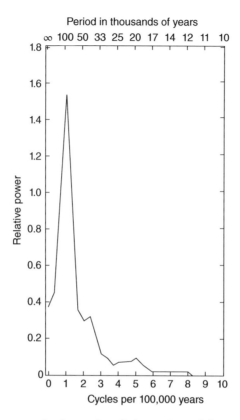

Figure 13.6 Power spectrum of a time series of observations of the oxygen isotope content in a deep sea core from the equatorial Pacific which indicates fluctuations in global ice volume over the last 600,000 years (US National Academy of Sciences, 1975)

Longer term climatic change is identified by power spectrum analysis of paleoclimatological records for the last million years (Figure 13.6). Significant peaks occur at periods of around 100,000, 40,000 and 20,000 years. Changes on these timescales are responsible for major changes in world climate.

13.2.1   Climatology of the last ice age

The climatology of the last ice age has been discussed by Burroughs (2005). The global average temperature at the height of the last ice age was about 5 °C lower than current values. Humans had to deal with the challenges caused by the conditions during the ice ages arising from the variations in periods of cooling and warming. As the ice sheets advanced, weather patterns and climatic zones were pushed further south in the northern hemisphere. The strength of the Inter-Tropical Convergence Zone (ITCZ) was reduced as temperatures in the tropics were reduced, causing rainfall in equatorial regions to decrease and the ITCZ to narrow. The last ice age was marked by a number of large ice sheet melting events with temperatures rising to around modern-day values by about 10,000 years ago. The timeline from the UK Natural Environment

Research Council (NERC) (www.esd.ornl.gov/projects/qen/nerc130k.html) is outlined in what follows.

The timespan of the last 130,000 years has seen the global climate system switch from warm interglacial to cold glacial conditions, and back again. This broad interglacial–glacial–interglacial climate oscillation has been recurring on a similar periodicity for about the last 900,000 years, though each individual cycle has had its own idiosyncrasies in terms of the timing and magnitude of changes. Winograd et al. (1997) describe limited outlines of the earliest cycles. Even for the most recent oscillation beginning around 130,000 years ago there is still too much ambiguity in terms of the errors in geological dating techniques, in the gaps in the record, and in the slowness of responses by indicator species, to know precisely when certain events occurred and whether the climate changes were truly synchronous between different regions. The following is a general summary reflecting the present consensus gained from ice cores, deep ocean cores, and terrestrial and lake sediments around the world.

Warmth.   Around 130,000–110,000 years ago (known as the Eemian Interglacial), the Earth's climates were generally much like those of today. The climate record derived from long ice cores taken through the Greenland ice cap suggested that the warm climate of this period might have been interrupted by many sudden and short lived cold phases. However, it is now thought that the lower layers of the ice sheet have become disrupted. There has been at least one major cold and dry event which is supported by the pollen records from Europe and China (see An and Porter, 1997).

Cooling.   The sediment records that cross the interglacial boundary indicate that cooling was a relatively sudden event and not a gradual transition to colder conditions taking many thousands of years. The Atlantic sediment record of Adkins et al. (1997) suggests that the move from interglacial to many glacial conditions occurred over a period of less than 400 years.

Conditions often changed suddenly, followed by several thousand years of stable climate or even a temporary reversal to warmth, but overall there was a decline. Northern forest zones retreated as the summers and winters grew colder. Large ice sheets began to grow in the northern latitudes when the snow that fell in winter failed to melt, until it reached thousands of metres in thickness.

As it grew colder, the Earth's climate also became drier. In areas that were not directly affected by the ice sheets, aridity began to cause forests to die and to give way to dry grassland. Eventually, much of the grassland retreated to give way to deserts and semi-deserts, as global conditions reached a cold, dry low point around 70,000 years ago (called the Lower Pleniglacial) when most of northern Europe and Canada were covered by thick ice sheets.

In-between.   By around 60,000–55,000 years ago, conditions around the world became warmer, although still colder than today. The ice melted back partially, and there followed a phase in which the climate oscillated between warmer and colder conditions, often in sudden jumps. Conditions in the tropics may have been moister than at present, and at other times they were drier. Generally, the mid-latitude zones seem to have been drier than present, with cold steppe and wooded steppe instead of forests.

Cooling followed.   After about 30,000 years ago, the Earth's climate system entered another freeze state when temperatures fell, deserts expanded and ice sheets spread across the northern latitudes much as they had done 70,000 years ago. This cold and arid phase is known as the Late Glacial Cold Stage (and is also sometimes called the Upper Pleniglacial).

The point at which the global ice extent was at its greatest, about 21,000 years ago (18,000 radiocarbon or 14C years ago) is known as the Last Glacial Maximum. The Last Glacial Maximum was much more arid than the present almost everywhere, with desert and semi-desert occupying large areas of the continents and with forests shrunk back. The greatest global *aridity* (rather than ice extent) may have been reached slightly after the Last Glacial Maximum.

Interstadials. Sudden warm and moist phases occurred intermittently during the timespan of the Last Glacial phase, and Greenland and Europe were as warm as at present. Twenty-four of these short lived warm events have been recognized from the Greenland ice core data (Dansgaard–Oeschger events: Dansgaard et al., 1993). From the speed of the climate changes recorded in the Greenland ice cap (Dansgaard et al., 1989), and by observation of the speed of change in sedimentation conditions on land, it is thought that the complete 'jump' in climate occurred over only a few decades. The interstadials usually lasted a few centuries to about 2000 years, before an equally rapid cooling period. Recent study of high resolution deep sea cores (Bond et al., 1997) suggests that for at least the last 30,000 years, interstadials tended to occur at the warmer points of a north Atlantic temperature cycle which had a periodicity of around 1500 years. The same basic 1500 year climate cycle has continued into the Holocene (see later).

The interstadials are associated with brief peaks in atmospheric methane concentration, suggesting that the biological activity of swamps and herbivores around the world increased as a result of moisture and warmth.

Heinrich events. Opposite in sign to the interstadials were sudden intense cold and dry phases which occasionally affected Europe and the north Atlantic region, and possibly many other parts of the world. These Heinrich events were first recognized as the traces of 'ice surges' into the north Atlantic, but they also appear in the Greenland ice cores and are detectable in the European pollen records and Antarctic ice cores. They are also evident as pine pollen peaks in Florida, and environmental changes in the Middle East, China, New Zealand and South America. Heinrich events are the most extreme of a spectrum of sudden, brief cold events which seem to have occurred very frequently over the last 115,000 years. The Greenland and North Atlantic record (Bond and Lotti, 1995) suggests that during the last 50,000 years, Heinrich events occurred around 41,000, 35,000, 23,000, 21,000 and 17,000–15,000 years ago. Each Heinrich event lasted between several hundred and several thousand years, with the 21,000 year event and the 17,000–15,000 year events representing the 'extreme' Last Glacial Maximum conditions.

Warming, then a cold snap. Around 14,000 years ago (about 13,000 radiocarbon years ago) there was a rapid global warming and moistening of climates, occurring within the space of only a few years or decades. In many respects, this phase seems to have resembled some of the earlier interstadials that had occurred so many times before during the glacial period. Conditions in many mid-latitude areas appear to have been about as warm as they are today. Forests began to spread back, and the ice sheets began to retreat. However, after a few thousand years of recovery, the conditions returned to a very short lived ice age known as the Younger Dryas. Although the Younger Dryas did not affect everywhere in the world, it destroyed the returning forests in the north and led to a brief resurgence of the ice sheets (Alley et al., 1993).

After about 1300 years of cold and aridity, the Younger Dryas seems to have ended in the space of only a few decades (various estimates from ice core climate indicators range from 20 to 70 years for this sudden transition) when conditions became as warm as they are today (Taylor et al., 1997).

The start of the present warm phase, the Holocene. Following the sudden ending of the Younger Dryas, about 11,500 years ago, forests extended. Ice sheets again began melting, though because of their size they took about 2000 more years to disappear completely. The Saharan and Arabian deserts almost completely disappeared under a vegetation cover, and in the northern latitudes forests grew slightly closer to the poles than they do at present. This phase, known as the Holocene Optimum, occurred between about 9000 and 5000 years ago, though the timing of the warmest and moistest conditions probably varied somewhat between different regions. The optimum may have been interrupted by a severe cold and dry phase that affected climates across north Africa, southern Asia, Europe, the Americas and Antarctica about 8200 years ago, perhaps lasting for a century or two before a return to warmer and wetter conditions (Stager and Mayewski, 1997). In Africa the climate does not seem to have returned to the moist warm state that existed before this sudden drought, but it was significantly moister than at present. A widespread cool event associated with relatively wet conditions seems to have occurred in many parts of the world around 2600 years ago (van Geel et al., 1996). It seems that at least in the North Atlantic region, and possibly globally, there was a warm–cold cycle with a periodicity of around 1500 years (Bond et al., 1997). In the north Atlantic region, and probably adjacent oceanic areas of Europe, the change from peak to trough of each period was about 2 °C. Several cold phases seem to have been sudden, lasting several centuries before a return to warmer conditions. These changes had periodicities of 11,100, 10,300, 9400, 8100, 5900, 4200, 2800 and 1400 years ago.

The unstable nature of the Earth's climate history suggests that it may be liable to change suddenly in the future. By putting large quantities of greenhouse gases into the atmosphere, humans are exerting pressure on the climate system which might produce a drastic change without much prior warning.

13.2.2 Intergovernmental Panel on Climate Change (IPCC)

The recognition that the world's climate was changing, possibly due to the activities of humankind, led to the United Nations Intergovernmental Panel on Climate (IPCC). This panel produced its first report in 1990 which served as the basis for the United Nations Framework Convention on Climate Change. This was followed by subsequent reports: the Fourth Assessment Report (AR4) was published in 2007, and the Fifth Assessment Report (AR5) in 2014.

These reports provide a review of the scientific, technical and socioeconomic aspects of climate change; each one updates our knowledge of climate change. Evidence of global warming has continued to grow since the Fourth Assessment Report in 2007, suggesting wetter winters and drier summers in the United Kingdom, although the summers might experience more heavy rainfall events. The state of the climate in 2010 was reviewed by the Met Office Hadley Centre, and published by the UK Departments of Environment Food and Rural Affairs (Defra) and Energy and Climate Change (Defra/DECC, 2010). The Working Group I contribution to the Fifth

Assessment Report (IPCC, 2013) summarizes the current situation as follows (expressed in terms of qualitative confidence and quantified likelihood):

> Warming of the climate system is unequivocal, and since the 1950s, many of the observed changes are unprecedented over decades to millennia. The atmosphere and ocean have warmed, the amounts of snow and ice have diminished, sea level has risen, and the concentrations of greenhouse gases have increased ... Each of the last three decades has been successively warmer at the Earth's surface than any preceding decade since 1850 ... In the Northern Hemisphere, 1983–2012 was *likely* the warmest 30-year period of the last 1400 years (*medium confidence*). ...
>
> *Confidence* in precipitation change averaged over global land areas since 1901 is *low* prior to 1951 and *medium* afterwards. Averaged over the mid-latitude land areas of the Northern Hemisphere, precipitation has increased since 1901 (*medium confidence* before and *high confidence* after 1951). For other latitudes area-averaged long-term positive or negative trends have *low confidence* [see Figure 13.7]. ...
>
> Changes in many extreme weather and climate events have been observed since about 1950 ... It is *very likely* that the number of cold days and nights has decreased and the number of warm days and nights has increased on the global scale. It is *likely* that the frequency of heat waves has increased in large parts of Europe, Asia and Australia. There are *likely* more land regions where the number of heavy precipitation events has increased than where it has decreased. The frequency or intensity of heavy precipitation events has *likely* increased in North America and Europe. In other continents, *confidence* in changes in heavy precipitation events is at most *medium*.

The IPCC report also noted that evidence continues to accumulate, strengthening the link between human activity and a wide range of indicators of a changing climate both globally and regionally. The 12 month running mean of global average temperatures continues to rise, as shown in Figure 13.8. Natural variability within the climate system was suggested as an explanation for the recent slowdown in this rise. Lubchenco and Karl (2012) noted that destructive weather is growing more prevalent in the United States (Figure 13.9). According to the insurance company Munich Re

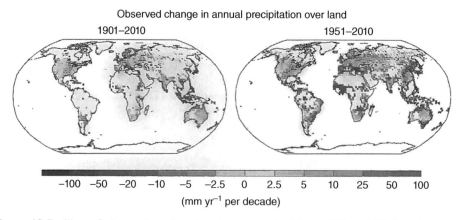

Observed change in annual precipitation over land
1901–2010   1951–2010

−100 −50 −20 −10 −5 −2.5 0 2.5 5 10 25 50 100
(mm yr⁻¹ per decade)

Figure 13.7 Maps of observed precipitation change over land from 1901 to 2010 (trends in annual accumulation) from one data set (from IPCC Fifth Assessment Report)

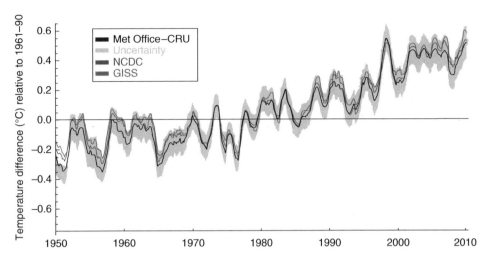

Figure 13.8 Illustrating 12 month running mean of global average temperatures from three data sets: HadCRUT3 (black and grey area) produced by the Met Office CRU; NCDC (dark grey) produced by the National Climate Data Center; and GISS (medium grey) produced by the Goddard Institute for Space Studies. The grey shaded area shows the approximate 95% confidence range for the HadCRUT3 data; the true global average is expected to lie outside this range around 5% of the time (from Defra/DECC, 2010)

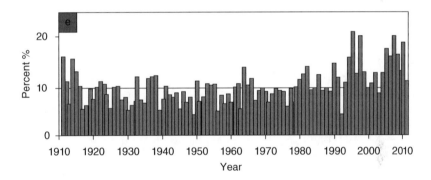

Figure 13.9 Extremes in 1 day heavy precipitation, defined as the monthly averages that rank in the top or bottom 10th percentile of all data on record (after Lubchenco and Karl, 2012)

(2012), the number of extreme meteorological and hydrological events has more than doubled over the past 20 years.

13.3 Causes of climatic change

The causes of climatic change lie in the linkages listed in section 13.1. However, the natural energy system is thought to control major changes of climate, with the other cycles producing fluctuations within the major changes or perhaps accelerating or decelerating these changes. Each system is examined separately here, although, as we shall see, their combined effects can only be assessed using a global numerical model.

13.3.1 The natural energy system

This system is concerned not just with solar energy but also with the whole spectrum of terrestrial and atmospheric radiation, and with energy transfers within both the oceans and the atmosphere. Climatic changes depend upon modification to the fluxes of energy and the amount of energy stored in the atmosphere, the Earth and the oceans.

Changes in the oceans, manifest as sea surface temperature anomalies, generate climatic fluctuations, as we have noted in the previous section. However, whilst these influences are extremely important, particularly for long range forecasting (periods of about a month ahead) (Gilchrist, 1986), it is less certain how much the oceans influence climatic change.

Changes in solar input and long term changes in insolation seem to offer one explanation of some of the observed periodicities. Wetherald and Manabe (1975) suggest that a 2% increase in the solar constant (total solar radiation) would produce a rise of 3°C in the mean global surface temperature. A decrease of 2°C, however, would produce an average temperature drop of 4.3°C. It was also concluded that a 6% change (from −4% to +2%) in the solar constant would produce a 27% increase in the area mean rates of precipitation. Such changes as these would not be uniform from equator to pole because of changes in snow cover and in albedo (reflection coefficient). Indeed, changes in the solar constant of this magnitude have not occurred in the last few hundred years, and, although there is some evidence of atmospheric periodicities which coincide with the sunspot cycle and which are reflected in small changes of solar constant, the relationship is still not uniformly accepted (for a review see Bonnet, 1985).

Changes of solar radiation are also brought about on very long timescales by changes in the orbit of the Earth about the Sun. The eccentricity of the orbit has a periodicity of 9600 years, changes in the obliquity of the axis of the Earth have a periodicity of 40,000 years, and changes due to the precession of longitude of the perihelion equivalent to precession of the solstices and equinoxes have a periodicity of 21,000 years (Mason, 1976). All these periodicities appear in the power spectrum analysis of the oxygen isotope content of fossil plankton (Figure 13.6). The changes of incident solar radiation over the last million years which are implied by this orbit geometry were calculated as a function of season and latitude by Milankovitch (1930; 1938).

Mason (1976) compared the variations in annual insolation received northwards of 45°N due to changes in the orbit of the Earth with changes in air temperature and ice cover over the last 150,000 years (Figure 13.10). There is a close correlation between the major advances and recessions of the ice and the 40,000 year cycle of variations in solar insolation known as the Milankovitch cycle: the maxima of the ice cover are nearly coincident with the minima in the radiation received at about 45°N. The oceans play a significant part in this process as they act as stores of heat which may be released as the ice melts, so aiding further melting. This type of feedback process is essential to explain the rather rapid transition from an ice age which cannot be explained entirely by the increased solar radiation. Changes in albedo caused by either increased or decreased snow cover may also generate complex feedback mechanisms.

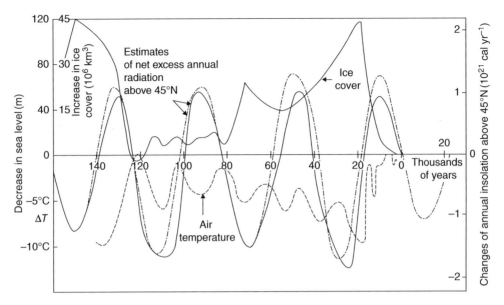

Figure 13.10   Changes in global ice cover, northern hemisphere air temperature (as in Figure 13.2), and total insolation north of 45°N over the last 150,000 years (after Mason, 1976)

13.3.2   The hydrological cycle

Although, as we shall discuss later, climatic change may have significant effects upon the hydrological cycle, the converse is also true. Changes in the vegetation, brought about by natural events such as mountain building, earthquakes, volcanoes or the climate itself, modify evapotranspiration. However, the activities of humankind have had by far the largest impact upon natural vegetation. It has been suggested (Flohn, 1973) that over the last 8000 years about 11% of the land area has been converted to arable land, and 31% of the forest land is not as it originally was. The loss of forests is particularly important as they play a major role in the interception and evaporation of rainwater.

As well as modifying evapotranspiration, changes of land use result in changes of albedo which, particularly in areas around the edges of deserts, may cause instabilities in local climate. Charney (1975) suggested that a reduction of vegetation, with a consequent increase in albedo, in the Sahel region on the southern edge of the Sahara would lead to a perpetuation of arid conditions. It was shown, by numerical experiment, that increasing the albedo from 14% to 35% had the effect of decreasing the rainfall in the Sahel by about 40% during the rainy season.

Reductions of vegetation may result from human activities or reductions of soil moisture. Soil moisture content may also directly affect evaporation, and consequently the proportion of the net radiation available as latent heat. Walker and Rowntree (1977) demonstrated that deficiencies in soil moisture could, without necessarily significantly reducing the existing vegetation, cause deserts to persist. Anomalies of soil moisture may also propagate and affect wider areas (Rowntree and Bolton, 1983). Spracklen et al. (2012) (see also Aragao, 2012), using numerical models of atmospheric transport with satellite observations of rainfall and vegetation cover,

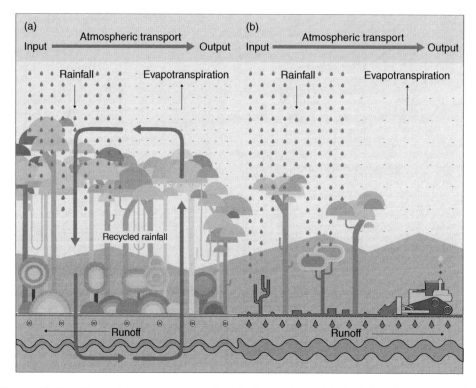

Figure 13.11　Effects of deforestation on rainfall in the tropics. (a) Much of the rainfall over tropical forests comes from water vapour that is carried by the atmosphere from elsewhere. However, a large component is 'recycled' rain, that is water pumped by trees from soil into the atmosphere through evapotranspiration. Water exits from forests either as runoff into streams and rivers, or as vapour that is carried away by the atmosphere. The atmospheric transport of water vapour into the forest is balanced by exit of water in the form of vapour and runoff. (b) The analysis of Spracklen et al. (2012) suggests that deforestation reduces evapotranspiration and so inhibits water recycling. This decreases the amount of moisture carried away by the atmosphere, reducing rainfall in regions to which the moisture is transported. Decreasing evapotranspiration may also increase localized runoff and raise river levels (from Aragao, 2012) (see plate section for colour representation of this figure)

found that deforestation can greatly reduce tropical rainfall through the mechanism illustrated in Figure 13.11.

Changes in soil moisture may be brought about by irrigation which results in increased evaporation. However, irrigation actually produces a rise of global temperature as the water causes a decrease in the reflection of incoming solar radiation. Budyko (1974) estimated that present-day irrigation generates an increase in the Earth's mean surface temperature of about 0.07°C.

Increased use of water by humankind for irrigation, domestic and industrial processes increases the rate at which water passes through the hydrological cycle over land. Whilst this may not in itself lead to climatic change, it is likely to cause a redistribution of water resources, particularly if large scale water transfers are involved (Shiklomanov, 1985), reducing runoff, depleting ground water resources, and increasing the amounts of water in the oceans and the polar ice fields. It is these changes which could affect climate.

13.3.3   The carbon cycle

Carbon dioxide (CO_2) is a very small constituent of the atmosphere (0.03%). However, it transmits short wavelength radiation from the Sun, yet absorbs a portion of the longer wavelength radiation emitted by the Earth, and this property causes it to have a profound effect on climate. The combination of the transmission of short wavelength radiation and the absorption of long wavelength radiation is called the *greenhouse effect*.

The biosphere extracts carbon dioxide from the atmosphere in spring and summer, and returns carbon dioxide to the atmosphere during the rest of the year. Likewise, carbon dioxide is transferred between the atmosphere and the oceans. In addition to these natural circulations, humankind has injected carbon dioxide into the atmosphere through the burning of fossil fuels, the cutting and burning of forests, and the conversion of land for agricultural or urban development. A steady increase in the atmospheric carbon dioxide has been measured (Figure 13.5), although the size of the yearly increase is only about half of what would be expected if all the carbon dioxide produced by burning fossil fuel remained in the atmosphere. It is thought that most of the rest of the carbon dioxide, about 35% (Crane and Liss, 1985), is absorbed by the oceans, the sea water reacting with the carbon dioxide to form a solution of carbonic acid (H_2CO_3).

Since it is thought that the surface layer of the oceans can take up only about 10% of the carbon dioxide injected into the atmosphere each year, it would appear that water deeper in the oceans must be absorbing significant amounts. The mechanism by which this is accomplished is not yet fully understood. Recently it has been found that tropical forests are removing much higher quantities of carbon dioxide from the atmosphere than realized. Stephens et al. (2007) have shown that tropical forests are absorbing about one billion tonnes more carbon than previously thought, and that northern mid-latitude forests are absorbing 0.9 billion tonnes, or 38%, less than assumed.

The effects of the steady increase in atmospheric carbon dioxide have been studied using numerical models (for example Schneider, 1975; Manabe and Wetherald, 1975; 1980; Mitchell, 1983; and by the IPCC). It has been found that a doubling of carbon dioxide would lead to an average global temperature increase of about 2°C in 70 years, although the temperature increase in the polar regions might be three to five times larger. Such changes, if confirmed, are significant, and could lead to important changes in global climate. However, increased warming could produce increased cloudiness which would reduce solar radiation, causing cooling, and so one effect could negate another. These complex interactions remain an area of active research.

13.3.4   Other biochemical cycles

A variety of particles are found in the atmosphere. These particles are either man-made or are produced by natural processes, such as from sea salt spray, wind-blown dust, forest fires, meteoric debris or volcanic emissions. Man-made pollutants include chlorofluoromethanes (gases used in spray cans) and other halocarbons, and sulphur compounds from the burning of fossil fuels (Bolin, 1979). Nuclear explosions, such as that of the Chernobyl accident in 1986 (ApSimon and Wilson, 1986), may generate unusual pollutants.

The man-made pollutants and oxides of nitrogen released by high flying aircraft (and perhaps from fertilizers) may interfere with the natural processes occurring in the atmospheric ozone layer (for example see Mitchell, 1984). Ozone is an unstable form of oxygen, and is constantly created and destroyed by incoming ultraviolet solar radiation (Crutzen and Andreae, 1985). These processes cause the atmosphere to be heated, and prevent much of the ultraviolet radiation from reaching the surface of the Earth. Changes in the ozone layer brought about by interactions with man-made pollutants could modify the heating in the stratosphere, or cause a health hazard to humans by increasing the amount of ultraviolet radiation reaching the Earth's surface.

Very small particles (aerosols) are also injected into the atmosphere either from industrial activities or from natural processes. The amount of aerosol affects visibility and turbidity (the ability to transmit solar radiation: see for example Mass and Schneider, 1977; Brinkman and McGregor, 1983), and could have a long term effect on global temperatures (see for example Lamb, 1970; Schneider and Mass, 1975). Figure 13.12 shows the mean residence times of aerosols in different layers of the atmosphere. Material close to the surface falls out in a matter of days, but material reaching the stratosphere may remain for several years. Volcanoes cause material to be placed directly in the stratosphere (Cronin, 1971), as do some nuclear explosions either deliberate or accidental. Such occurrences could have profound effects on the climate over many years, although it remains unclear to what extent climatic effects, such as the so-called 'nuclear winter', would really occur (Golding et al., 1986).

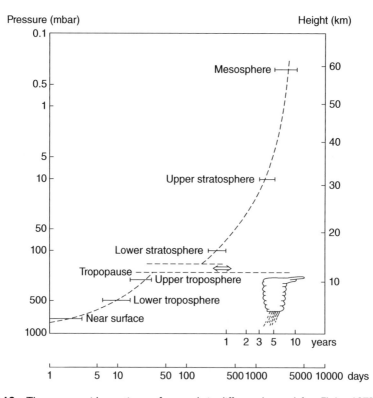

Figure 13.12   The mean residence times of aerosols in different layers (after Flohn, 1973)

13.4   Modelling climatic change

Climate models have continued to be developed and improved since the IPCC Fourth Assessment Report (AR4), and many models have been extended into Earth system models by including the representation of biogeochemical cycles important to climate change. The hydrological cycle has also been incorporated in these models (see for example Manabe et al., 1965). These models allow for policy-relevant calculations such as the carbon dioxide (CO_2) emissions compatible with a specified climate stabilization target. In addition, the range of climate variables and processes that have been evaluated has greatly expanded, and differences between models and observations are increasingly quantified using 'performance metrics'. Model evaluation covers simulation of the mean climate, of historical climate change, of variability on multiple timescales and of regional modes of variability. This evaluation is based on recent internationally coordinated model experiments, including simulations of historic climate and paleoclimate, specialized experiments designed to provide insight into key climate processes and feedbacks, and regional climate downscaling.

The processes which may cause climatic change must be studied using climate models, since the observations are often confusing, and it is often difficult to separate cause and effect. These models range from integral parameter, one-dimensional models to full four-dimensional representations of the global atmosphere. Much can be gained from simplistic models. For example the mean temperature T_r of outgoing radiation from the Earth may be estimated from

$$\varepsilon \sigma T_r^4 = I_0 (1 - A)/4 \qquad (13.1)$$

where I_0 is the solar constant, A is the mean planetary albedo (0.28 derived from satellite data), ε is the emissivity, and σ is the Stefan–Boltzmann constant ($5.67 \times 10^{-8}\,\mathrm{W\,m^{-2}K^{-4}}$). Equation 13.1 gives $T_r = 256\,\mathrm{K}$, a mean temperature which occurs in the atmosphere at a height of about 5 km.

In order to estimate the atmospheric circulation, at least the mean difference ΔT between the surface air temperature at the Equator and the poles must be added to equation 13.1. Hence, the total (over the depth of the atmosphere) meridional heat flux produced by air movements is assumed to balance the heat losses due to radiation:

$$\frac{C_p M U \Delta t}{\pi 0.5 R} = \epsilon \sigma T_r^4 \qquad (13.2)$$

where C_p is the specific heat at constant pressure, R is the radius of the Earth, M is the mass of a unit column of air, and U is the typical atmospheric wind velocity, around 10 m s⁻¹. From equation 13.2, Δt can be evaluated. This approach to analysing the circulation of a planetary atmosphere was developed by Golitsyn (1970: 1973). However, the real atmosphere of the Earth deviates considerably from the state of radiative equilibrium described by equation 13.1. Nevertheless, it is possible to represent aspects of climate by these simple models, and to carry out limited experiments to investigate the effects of changing the values of particular parameters (Monin, 1986).

So far we have discussed integral parameter climate models, but more exact formulations of the vertical structure of the atmosphere also provide a useful framework for studies of climatic change. In a column of air a balance may be assumed between heat fluxes due to turbulent and radiative vertical heat exchanges:

$$\frac{d\lambda}{dz}\frac{dT}{dz} + \alpha\rho_{ab}\left(F_D + F_U - 2f\sigma T^4 + \gamma S\right) = 0 \tag{13.3}$$

where λ is the vertical turbulent heat conductivity; α is the coefficient of absorption of long wave radiation; ρ_{ab} is the density of absorbing substances; F_D and F_U are the downward (atmosphere) and upward Earth's surface and atmospheric fluxes of long wave radiation; T is the black body temperature; f is a correction factor to allow for the fact that the substances are not perfect black bodies; S is the downward flux of short wave solar radiation; and γ is the ratio of the absorption coefficients for short wave and long wave radiation. Equation 13.3 may be solved with equations for the fluxes of radiative energy and appropriate boundary conditions. Whilst these one-dimensional models produce results which fit the thermal structure of the atmosphere quite well, the neglect of the divergence of the heat fluxes caused by large scale atmospheric motions restricts their applicability.

This problem can be partially overcome by constructing a model from the zonally averaged hydrostatic equations and the continuity equation as follows:

$$\frac{\partial \overline{p}}{\partial z} = -g\rho \frac{\partial \overline{v}_z}{\partial z} + \frac{1}{R\sin\theta}\frac{\partial\left(\overline{v}_\theta \sin\theta\right)}{\partial\theta} = 0 \tag{13.4}$$

where θ is the co-latitude and $\overline{v}$ is the zonally meaned velocity component. These models can reproduce quite well the observed zonal temperature and velocity fields together with the meridional circulation and the energy cycle (Corby et al., 1972; Stone, 1972; Kirichkov, 1978). Consequently they have been used to study aspects of climatic change (Stone, 1973).

A rationale for the use of large scale numerical models for the study of climate and climatic change has been given by Gilchrist (1978). Controlled experiments cannot be carried out on the atmosphere, and the interactions of systems on a global scale are so complex that the assumptions of simpler models are not comprehensive enough to produce reliable results. The results of large scale experiments with these models are impressive (see for example Kutzbach and Street-Perrott, 1985; Rind, 1986; Chervin, 1986), and tend to support the theories of climatic change discussed in section 13.3. As an illustration of the data produced by integrations of these models, Table 13.2 from Rind (1986) shows the changes (compared with the present conditions) produced in mean global climate under conditions of the ice ages (10,000 years before the present), of the Mesozoic period (65 million years before the present), and of doubling the level of carbon dioxide in the atmosphere.

In mid-2009 the Athena supercomputer, a Cray XT-4 with 4512 quad-core nodes, was due to be decommissioned. The US National Science Foundation agreed to meet the operating and maintenance costs for Athena for an additional year to provide 6 months of dedicated use for climate simulation and prediction. Kinter et al. (2013) describe the results, which confirmed that dedicated computational resources can 'substantially accelerate progress in climate simulation and prediction'.

Table 13.2   Global average (a)–(e) January values and (f) August values for four climate simulations: Ice Age I (18,000 years before present); Ice Age II (as Ice Age I with sea surface temperature reduced by 2 °C); Double CO_2 (carbon dioxide); Mesozoic (65 million years before present) (from Rind, 1986)

Climate variable	Ice Age II	Ice Age I	Current	2×CO$_2$	Mesozoic
(a) Temperature					
Surface air temperature (°C)	6.3	8.4	11.8	17.0	19.2
Sea surface temperature (°C)	17.9	19.9	19.2	23.1	23.3
Vertically integrated air temperature (°C)	−26.4	−24.0	−22.9	−17.8	−17.3
(b) Hydrologic cycle					
Precipitation (mm d^{-1})	2.8	3.0	3.1	3.5	3.6
Evaporation (mm d^{-1})	2.8	3.0	3.1	3.5	3.6
Evaporation (W m^{-2})	81.0	88.1	89.7	102.7	104.0
Sensible heat flux (W m^{-2})	26.0	24.8	24.5	20.4	21.1
Low level cloud (%)	40.6	39.8	38.2	34.7	30.1
High level cloud (%)	27.9	28.7	28.7	30.3	29.2
Total cloud cover (%)	54	54	53	52	48
Ground albedo (%)	15.8	15.4	12.2	10.9	7.6
Planetary albedo (%)	33.6	33.1	31.3	29.3	25.3
Solar radiation absorbed at ground (W m^{-2})	168	169	177	181	195
Relative humidity (%) at 959 hPa	76	76	78	78	80
Relative humidity (%) at 633 hPa	52	53	54	55	55
Relative humidity (%) at 321 hPa	40	40	41	44	45
Moist convective heating (10^{14}W)	371	400	410	478	498
Moist convective cloud depth (hPa)	378	384	385	409	436
(c) General circulation					
Vertically integrated stream function (10^{9}kg s^{-1})	−17	−15	−15	−13	−22
200 hPa meridional wind (m s^{-1})	0.4	0.3	0.3	0.3	0.7
200 hPa zonal wind (m s^{-1})	13.2	14.4	14.8	16.0	9.7
200 hPa eddy poleward transport of angular momentum (10^{17}J)	40.6	37.3	30.7	39.2	19.2
200 hPa divergence of E–P flux* (10^{16}J)	−178	−251	−258	−259	−261
(d) Eddy energy					
Tropospheric APE† (10^{19}J)	483	477	362	348	178
Tropospheric zonal APE† (10^{19}J)	346	346	269	266	146
Tropospheric eddy APE† (10^{19}J)	137	131	93	82	32
Topography (m)	348	348	233	233	89
633 hPa baroclinic eddy generation (W m^{-2})	0.62	0.59	0.49	0.42	0.25
Eddy kinetic energy (10^{6}J m^{-2})	127	124	107	96	71
Transient eddy kinetic energy (10^{4}J m^{-2})	91	91	78	69	52
Stationary eddy kinetic energy (10^{4}J m^{-2})	36	32	28	27	19
Tropospheric lapse rate (°C km^{-1})	5.7	5.6	5.7	5.6	5.7
Moist adiabatic lapse rate (°C km^{-1})	7.4	7.2	7.0	6.4	6.4
Dynamic lapse rate (°C km^{-1})	6.5	6.5	6.6	6.5	7.9
(e) Energy transports					
Poleward transport sensible heat by eddies (10^{14}W)	30.8	29.2	24.1	20.6	13.9
Poleward transport sensible heat by transient eddies (10^{14}W)	17.3	16.9	16.0	13.6	8.7
Poleward transport sensible heat by stationary eddies (10^{14}W)	13.5	12.4	8.1	7.0	5.2

(Continued)

Table 13.2   (Continued)

Climate variable	Ice Age II	Ice Age I	Current	$2\times CO_2$	Mesozoic
Poleward transport latent heat by eddies (10^{14}W)	21.9	23.5	22.7	27.4	16.6
Poleward transport static energy by eddies (10^{14}W)	52.3	52.3	46.7	47.9	30.4
Atmospheric poleward transport static energy (10^{14}W)	66.9	68.8	61.4	65.5	37.1
(f) August results					
Surface air temperature (°C)	9.0	11.1	14.9	19.9	21.3
Total cloud cover (%)	50.2	50.8	51.4	48.4	44.3
Relative humidity (%) at 633 hPa	51.3	51.3	52.8	54.6	55.9
Eddy kinetic energy (10^6J m^{-2})	100.5	100.8	91.3	78.6	62.9
Poleward transport static energy by eddies (10^{14}W)	39.5	42.1	36.6	36.1	26.0
Atmospheric poleward transport static energy (10^{14}W)	52.1	56.5	53.1	53.6	33.7

*E–P flux: the Eliassen–Palm flux is a vector quantity with non-zero components in the latitude–height plane, the direction and magnitude of which determine the relative importance of the eddy heat flux and momentum flux.
†APE: available potential energy.

   The AR5 report concludes that there is very high confidence that models repro-duce the general features of the global-scale annual mean surface temperature increase over the historical period, including the more rapid warming in the second half of the 20th century, and the cooling immediately following large volcanic eruptions. Most simulations of the historical period do not reproduce the observed reduction in global mean surface warming trend over the last 10 to 15 years. There is medium confidence that the trend difference between models and observations during 1998–2012 is to a substantial degree caused by internal variability, with possible contributions from forcing error and from some models overestimating the response to increasing greenhouse gas (GHG) forcing. Most, though not all, models overestimate the observed warming trend in the tropical troposphere over the last 30 years, and tend to underestimate the long term lower stratospheric cooling trend.
   The simulation of large scale patterns of precipitation has improved somewhat since the AR4 report, although models continue to perform less well for precipitation than for surface temperature. The spatial pattern correlation between modelled and observed annual mean precipitation has increased from 0.77 for models available at the time of the AR4 to 0.82 for current models. At regional scales, precipitation is not simulated as well, and the assessment remains difficult owing to observational uncer-tainties. The simulation of clouds in climate models remains challenging. There is very high confidence that uncertainties in cloud processes explain much of the spread in modelled climate sensitivity. However, the simulation of clouds in climate models has shown modest improvement relative to models available at the time of the AR4, and this has been aided by new evaluation techniques and new observations for clouds. Nevertheless, biases in cloud simulation lead to regional errors in cloud radiative effect of several tens of watts per square metre. Models are able to capture the general characteristics of storm tracks and extratropical cyclones, and there is some evidence of improvement since the AR4. Storm track biases in the North Atlantic have improved slightly, but models still produce a storm track that is too zonal and underestimate cyclone intensity.

Many models are able to reproduce the observed changes in upper ocean heat content from 1961 to 2005 with the multi-model mean time series falling within the range of the available observational estimates for most of the period. The ability of models to simulate ocean heat uptake, including variations imposed by large volcanic eruptions, adds confidence to their use in assessing the global energy budget and simulating the thermal component of sea level rise. The simulation of the tropical Pacific Ocean mean state has improved since the AR4, with a 30% reduction in the spurious westward extension of the cold tongue near the equator, a pervasive bias of coupled models. The simulation of the tropical Atlantic remains deficient with many models unable to reproduce the basic east–west temperature gradient.

13.5   Possible effects of climate change upon the hydrological cycle and water resources

Since the hydrological cycle is an integral part of the climate system, any changes in climate must be expected to affect water resources. This relationship is shown schematically in Figure 13.13. As one might expect, many factors interact with each other and with the water resources system, which may be modified in the ways listed by Schwarz (1977). However, as Nemec (1985) points out, it is not an easy matter to model stream flow over the lifespan of a water resource system, and consequently it is difficult to assess the impact of climatic change with any certainty. Nevertheless it is instructive to use river models to estimate the change in stream flow which might result from changes in precipitation and evapotranspiration. Nemec (1985) shows the results of this type of analysis for the Pease and Leaf Rivers in the United States (Figure 13.14). For the dry river basin (Pease River) an increase of 25% in precipitation and a decrease of 1 °C in temperature (corresponding to a 4% change in evapotranspiration) increases the runoff by 250%. This compares with a runoff increase of 70% for the humid basin (Leaf River).

Changes in rainfall and evapotranspiration producing changes in runoff could cause problems for existing water supply systems, particularly if over the lifetime of the systems other changes occur. One example of a system, the Colorado River, United States, which is susceptible to climatic change as a consequence of physical encroachment into the lower part of the floodplain and reservoirs being full, has been discussed by Rhodes et al. (1984). Likewise Diaz et al. (1985) examined the impact of rainfall fluctuations on water consumption in the south-western United States (see also Stockton and Boggess, 1982). Changnon (1983) reported an increase in winter precipitation of 12% since 1930 in northern and eastern Illinois with a corresponding increase in winter floods.

Although observational studies of short term climatic fluctuations can provide useful information on likely changes necessary to design parameters and operating procedures (McCullough, 1983), it is likely that major impacts on water resources can only be studied by use of numerical models. These models will take the form either of large scale atmospheric models, as discussed in the previous section, or of more specialized hydrological models, such as those discussed by Deschesnes et al. (1985a; 1985b) and Sharma (1985).

Hagemann et al. (2013) report that multiple global climate (three) and hydrological models (eight) assessed systematically the hydrological response to climate change

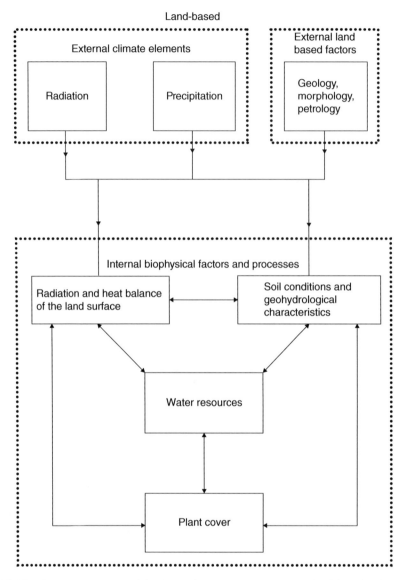

Figure 13.13   Conceptual scheme of the climate/water-resources relationship (after Novaky et al., 1985)

and projected the future state of global water resources. This multi-model ensemble allowed them to investigate how the hydrology models contribute to the uncertainty in projected hydrological changes compared to the climate models. Due to their systematic biases, general circulation model (GCM) outputs cannot be used directly in hydrological impact studies, so a statistical bias correction was applied. The results showed a large spread in projected changes in water resources within the climate–hydrology modelling chain for some regions. They clearly demonstrated that climate models are not the only source of uncertainty for hydrological change, and that the spread resulting from the choice of the hydrology model is larger than the spread originating from the climate models over many areas. However, there were also areas

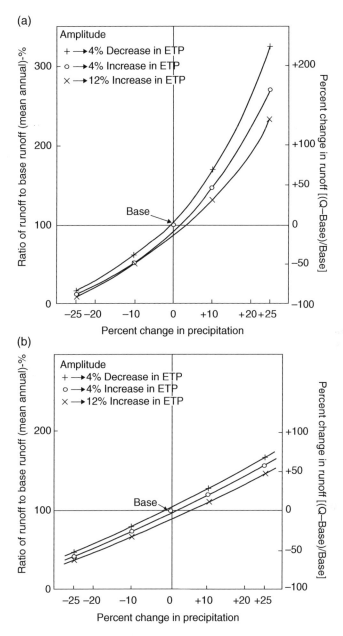

Figure 13.14 Changes in stream flow as a function of changes in precipitation and potential evapotranspiration (ETP). (a) Pease river at Vernon, Texas; drainage area 9034 km²; mean precipitation base 540 mm; mean runoff base 11 mm. (a) Leaf river near Collins, Mississippi; drainage area 1949 km²; mean precipitation base 1314 mm; mean runoff base 409 mm. (after Nemec, 1985)

showing a robust change signal, such as at high latitudes and in some mid-latitude regions, where the models agreed on the sign of projected hydrological changes, indicative of higher confidence in this ensemble mean signal. In many catchments an increase of available water resources is expected but there were some severe decreases in central and southern Europe, the Middle East, the Mississippi River basin, southern Africa, southern China and south-eastern Australia.

Appendix 13.1   Estimating return times for events in a long term climate record

Following Chapman et al. (2013), consider that in time period Δt there are N events which are in ascending order $x_1, ..., x_k, ..., x_n$ and have a mean time period between observations of $\delta t = \Delta t / N$. Within the period Δt there are $N - k + 1$ events of size $x \geq x_k$. Hence the return time R for an event at least as large as $x = x'_k$ (Gilchrist, 2000) is

$$R(x) = \frac{\Delta t}{N - k + 1} \approx \frac{\Delta t}{N - k} \tag{13.5}$$

for $(N - k)$ large when the cumulative density function of the record $C(x = x_k) = k/N$, giving

$$R(x) = \frac{\delta t}{1 - C(x)} \tag{13.6}$$

Defining the change ΔR in return time due to change in the climate state alone, then it may be shown that

$$\Delta R = \frac{R \Delta C}{1 - C} \tag{13.7}$$

This is not applicable at the largest events $k \to N$.

Summary of key points in this chapter

1. Climate may be defined as the sequence of weather which is experienced at a given locality.
2. Unpredictable climatic fluctuations may have major social and economic impacts on humankind.
3. Among the most serious Earth science and environmental policy issues confronting society are the potential changes in the Earth's water cycle due to climate change.
4. These changes may profoundly affect atmospheric water vapour concentrations, clouds, precipitation patterns, and runoff and stream flow patterns.
5. Water resources development projects are designed to function for around 100 years or even more in some cases.
6. Modes of variation typical of climatic time series are a mean value or central tendency, a quasi-periodic variation, a stationary trend and short term variations.

7. Climate noise is that part of the variance of climate attributable to short term weather changes.

8. Climatic variability is the manner of variation of the climatic parameters within the typical averaging period.

9. Short term changes lasting a few decades are referred to as climatic fluctuations.

10. Climate may be regarded as one element in a global system which includes geophysical and biochemical cycles.

11. Perturbations in the natural energy system (the main element of which is solar energy), the hydrological cycle and other biochemical cycles affect climate.

12. Instrumental records have existed in only a few places for as long as between 200 and 300 years, although in China records extend for several thousand years.

13. The climate of the last million years, known as the Pleistocene period, has been documented in the northern hemisphere.

14. Eight alternations between cold glacial epochs and relatively warm inter-glacial epochs have occurred at intervals of about 100,000 years.

15. About 10,000 years before the present, extensive areas of the northern hemisphere were covered with continental ice sheets, the sea level dropped by about 120 m and the sea surface temperature in the North Atlantic dropped by up to 10°C.

16. Since the Last Interglacial Maximum about 6000 years ago there has been a gradual cooling. The cooler period from about 1500 to 1850 is known as the Little Ice Age.

17. Power spectrum analysis provides evidence of periodicity, with a peak at 2.1 years being associated with the quasi-biennial oscillation, and the 23 year peak being associated with the double sunspot cycle.

18. The periodicities relate to short term climatic fluctuations, with others being related to the activities of humankind such as the burning of fossil fuels or the release of pollutants into the atmosphere.

19. Local changes in the air–sea interaction produce further short term climatic fluctuations, e.g. the El Niño.

20. The associations between atmospheric circulation anomalies and, say, rainfall deficiencies are sometimes referred to as teleconnections.

21. The global average temperature at the height of the last ice age was about 5°C lower than current values. The last ice age was marked by a number of large ice sheet melting events.

22. The Last Glacial Maximum coincided with the greatest global aridity.

23. The sudden warm and moist phases are known as interstadials. They often took Greenland and Europe from a full glacial climate to conditions about as warm as at present.

24. Heinrich events are of opposite sign to interstadials. They are characterized by sudden intense cold and dry phases.

25. Around 14,000 years ago there was a rapid global warming and moistening of climate. This was followed after a few thousand years by a sudden ice age known as the Younger Bryas.

26. The present warm phase is known as the Holocene.

27. The unstable nature of the Earth's climate history suggests that it may be liable to change suddenly in the future.

28. The Intergovernmental Panel on Climate Change (IPCC) produced its first report in 1990. The Fourth Assessment Report (AR4, 2007) suggested wetter winters and drier summers in the United Kingdom, although the summers might experience more heavy rainfall events. The Fifth Assessment Report (AR5) was finally published in 2014.

29. According to the insurance company Munich Re, the number of extreme meteorological and hydrological events has more than doubled over the past 20 years.

30. The natural energy system is thought to control major changes of climate, with other cycles producing fluctuations within the major changes.

31. Changes in the oceans manifest as sea surface temperature anomalies generating climatic fluctuations.

32. Changes in solar input and long term changes in solar insolation seem to offer an explanation of some observed periodicities.

33. There is a close correlation between major advances and recessions of the ice and the 40,000 year cycle of variations in solar insolation known as the Milankovitch cycle.

34. Changes of land use including deforestation result in changes in albedo and evapotranspiration.

35. The combination of the transmission of short wavelength radiation and the absorption of long wavelength radiation is called the *greenhouse effect*.

36. The biosphere extracts carbon dioxide from the atmosphere in spring and summer and returns carbon dioxide to the atmosphere during the rest of the year.

37. A variety of particles, either man-made or produced by natural processes, are found in the atmosphere. This could have a long term impact on global temperatures.

38. Climate models allow for policy-relevant calculations such as the impact of carbon dioxide emissions.

39. Exact formulation of the vertical structure of the atmosphere provides a useful framework for studies of climate change.

40. The IPCC AR5 report concludes that there is very high confidence that models reproduce the general features of the global-scale annual mean surface temperature increase over the historical period.

41. It is not an easy matter to model stream flow over the lifespan of a water resource system. Nevertheless the use of multi-model ensembles does allow investigation of how hydrology models contribute to the uncertainty in projected hydrological changes compared to climate models.

Problems

1. Define climate.
2. What do changes in climate impact?
3. List the models of variation in time series.
4. List the main linkages in the global climate system.
5. What does the natural energy system comprise?
6. List the data sources providing information on past climates.
7. Draw a graph illustrating changes in global ice cover and northern hemisphere air temperatures.
8. List the peaks shown in the power spectrum of the Manley record of central England temperatures.
9. When did the Little Ice Age occur?
10. What is the double sunspot cycle?
11. How does the burning of fossil fuels manifest itself in the atmosphere?
12. What are teleconnections?
13. What was the average temperature at the height of the last ice age?
14. Over what period did the global climate system switch from warm interglacial to cold glacial conditions and back again?
15. What characterized the Last Glacial Maximum?
16. What are interstadials?
17. What are Heinrich events?
18. What was the impact of the Younger Dryas and when did it end?
19. How long ago was the Holocene Optimum?
20. What is the Intergovernmental Panel on Climate Change (IPCC), and what was the date of its first report?
21. Draw a graph of the 12 month running mean of global average temperatures, starting in 1950.
22. Give three causes of climate change in the natural energy system.
23. What was suggested by Milankovitch?
24. How do changes in vegetation impact climate change?
25. How might major changes in soil moisture be brought about?
26. What is the greenhouse effect?
27. What impact does the biosphere have on atmospheric composition?
28. How do atmospheric aerosols impact climate?
29. Give an equation for the mean temperature of outgoing radiation from the Earth.
30. Give a simple equation for estimating the atmospheric circulation.
31. Give a rational for the use of large scale numerical models for the study of climate and climate change.
32. What is suggested for the spread in modelling climate sensitivity reported in the literature?
33. Draw a conceptual scheme for the climate/water-resource relationship.

References

Adkins, J.F., Boyle, E.A., Keigwin, L. and Cortijo, E. (1997) Variability of the North Atlantic thermohaline circulation during the last interglacial period. *Nature* 390, 154–156.

Alley, R.B., Meese, D.A., Shuman, C.A., Gow, A.J., Taylor, K.C., Groots, P.M., White, J.W.C., Ram, M., Waddington, E.D., Mayewski, P.A. and Zielinski, G.A. (1993) Abrupt increase in snow accumulation at the end of the Younger Dryas event. *Nature*, 362, 527–529.

An, Z.S. and Porter, S.C. (1997) Millennial-scale climatic oscillations during the last interglaciation in central China. *Geology*, 25, 603–606.

ApSimon, H. and Wilson, J. (1986) Tracking the cloud from Chernobyl. *New Scientist*, 17(July), 42–45.

Aragao, L.O.C. (2012) Environmental science: the rainforest's water pump. *Nature*, 489(13 Sept.), 217–218.

Bach, W. (1976) Global air pollution and climatic change. *Rev. Geophys. Space Phys.*, 14, 429–474.

Bolin, B. (1979) Global ecology and man. Proceedings World Climate Conference, WMO Publication 537, pp. 88–111. World Meteorological Organization, Geneva.

Bond, G.C. and Lotti, R. (1995) Iceberg discharges into the North Atlantic on millennial time scales during the last glaciation. *Science* 267, 1005–1010.

Bond, G., Showers, W., Cheseby, M., Lotti, R., Almasi, P., deMenocal, P., Priore, P., Cullen, H., Hajdas, I. and Bonani, G. (1997) A pervasive millennial-scale cycle in North Atlantic Holocene and Glacial climates. *Science* 278, 1257–1265.

Bonnet, R.M. (1985) Solar terrestrial relations. In *Global Change*, ed. T.F. Malone and J.G. Roederer. Proceedings Symposium sponsored by International Council of Scientific Unions (ICSU) 20th General Assembly, pp. 397–419. Cambridge University Press, Cambridge.

Brinkman, A.W. and McGregor, J. (1983) Solar radiation in dense Saharan aerosol in Northern Nigeria. *Q. J. R. Meteorol. Soc.*, 109, 831–847.

Budyko, M.I. (1969) The effect of solar radiation variations on the climate of the earth. *Tellus*, 21, 611–619.

Budyko, M.I. (1974) *Climate and Life*. Academic Press, New York.

Burroughs, W.J. (2005) *Climate Change in Prehistory: The End of the Reign of Chaos*. Cambridge University Press, Cambridge.

Changnon, S.A. (1983) Trends in floods and related climate conditions in Illinois. *Climate Change*, 5, 341–363.

Chapman, S.C., Stainforth, D.A. and Watkins, N.W. (2013) On estimating local long-term climate trends. *Phil. Trans. Roy. Soc. A*, 371(1991). doi 10.1098/rsta.2012.0287.

Charney, J.G. (1975) Dynamics of deserts and drought in the Sahel. *Q. J. R. Meteorol. Soc.*, 101, 193–202.

Chervin, R.M. (1986) Inter-annual variability and season climate predictability. *J. Atmos. Sci.*, 43, 233–251.

Chu, K. (1973) A preliminary study of the climatic fluctuations during the last 5000 years in China. *Scientia Sinica*, 6, 226–256.

Corby, G.A., Gilchrist, A. and Newson, R.L. (1972) A general circulation model of the atmosphere suitable for long period integrations. *Q. J. R. Meteorol. Soc.*, 98, 809–832.

Crane, A. and Liss, P. (1985) Carbon dioxide, climate and the sea. *New Scientist*, 21 November, 50–54.

Cronin, J.F. (1971) Recent volcanism and the stratosphere. *Science*, 172, 847–849.

Crutzen, P.J. and Andreae, M.D. (1985) Atmospheric chemistry. In *Global Change*, ed. T.F. Malone and J.G. Roederer. Proceedings Symposium sponsored by International Council of Scientific Unions (ICSU) 20th General Assembly, pp. 75–113. Cambridge University Press, Cambridge.

Dansgaard, W., White, J.W.C. and Johnsen, S.J. (1989) The abrupt termination of the Younger Dryas climate event. *Nature*, 339, 532–534.

Dansgaard, W., Johnsen, S.J., Clausen, H.B., Dahl-Jensen, D., Gundestrup, N.S., Hammer, C.U., Hvidberg, C.S., Steffensen, J.P., Sveinbjörnsdottir, A.E., Jouzel, J. and Bond G. (1993) Evidence for general instability of past climate from a 250-kyr ice-core record. *Nature*, 364, 218–220.

Defra/DECC (2010) The state of the climate. Met Office Hadley Centre evidence.

Deschesnes, J., Villeneuve, J.-P., Ledoux, E. and Girard, G. (1985a) Modelling the hydrologic cycle: the MC model. I: Principles and description. *Nord. Hydrol.*, 16, 257–272.

Deschesnes, J., Villeneuve, J.-P., Ledoux, E. and Girard, G. (1985b) Modelling the hydrologic cycle: the MC model. II: Modelling applications. *Nord. Hydrol.*, 16, 273–290.

Diaz, H.F., Holle, R.L. and Thorn, J.W. Jr (1985) Precipitation trends and water consumption related to population in the south western United States, 1930–83. *J. Clim. Appl. Meteorol.*, 24, 145–153.

Diaz, H.F., Trigo, R., Hughes, M.K., Mann, M.E., Xoplaki, E. and Berriopedro, D. (2011) Spatial and temporal characteristics of climate in medieval times revisited. *Bull. Am. Meteorol. Soc.*, 92, 1487–1500.

Flohn, H. (1973) Globale energiebilanz und Klimaschwankungen. In *Bonner Meteorologische Abhandlungen*, pp. 75–117. Westdeutscher.

Folland, C.K., Palmer, T.N. and Parker, D.E. (1986) Sahel rainfall and worldwide sea temperatures, 1901–85. *Nature*, 320(17 April), 602–607.

Gilchrist, A. (1978) Numerical simulation of climate and climatic change. *Nature*, 276, 342–345.

Gilchrist, A. (1986) Long-range forecasting. *Q. J. R. Meteorol. Soc.*, 112, 567–592.

Gilchrist, W. (2000) *Statistical Modelling with Quantile Functions*. CRC, New York.

Golding, B.W., Goldsmith, P., Machin, N.A. and Slingo, A. (1986) Importance of local mesoscale factors in any assessment of nuclear winter. *Nature*, 319, 301–303.

Golitsyn, G.S. (1970) Similarity theory for large-scale motions of planetary atmospheres. *Dokl. Ak. Nauk. SSSR*, 190(2), 323–326.

Golitsyn, G.S. (1973) *An Introduction to the Dynamics of Planetary Atmospheres* [in Russian]. Gidrometeoizdat, Leningrad.

Hagemann, S., Chen, C., Clark, D.B., Folwell, S., Gosling, S.N., Haddeland, I., Hanasaki, N., Heinke, J., Ludwig, F., Voss, F., and Wiltshire, A.J. (2013) Climate change impact on available water resources obtained using multiple global climate and hydrology models. *Earth Syst. Dynam.*, 4, 129–144.

Hare, F.K. (1985) Climatic variability and change. In *Climate Impact Assessment*, ed. R.W. Kates, J.H. Ausubel and M. Berberian, pp. 37–68. Wiley, Chichester.

IPCC (2013) Summary for Policymakers. In: *Climate Change 2013: The Physical Science Basis. Contribution of Working Group I to the Fifth Assessment Report of the Intergovernmental Panel on Climate Change*, ed. T.F. Stocker, D. Qin, G.-K. Plattner, M. Tignor, S.K. Allen, J. Boschung, A. Nauels, Y. Xia, V. Bex and P.M. Midgley. Cambridge University Press, Cambridge.

King, J.W. (1973) Solar radiation changes and the weather. *Nature*, 245, 443–446.

Kinter, J.L. III, Cash, B., Achuthavarier, D., Adams, J., Altshuler, E., Dirmeyer, P., Doty, B., Huang, B., Jin, E.K., Marx, L., Manganello, J., Stan, C., Wakefield, T., Palmer, T., Hamrud, M., Jung, T., Miller, M., Towes, P., Wedi, N., Satoh, M., Tomita, H., Kodama, C., Nasuno, T., Oouchi, K., Yamada, Y., Taniguchi, H., Andrews, P., Baer, T., Ezell, M., Halloy, C., John, D., Loftis, B., Mohr, R. and Wong, K. (2013) Revolutionizing climate modelling with Project Athena: a multi-institutional, international collaboration. *Bull. Am. Meteorol. Soc.*, 94(2), 231–245.

Kirichkov, S.E. (1978) Numerical experiments with a zonal atmospheric circulation model. *Izv. Akad. Nauk SSSR, Fiz. Atmos. Okeana*, 14(7), 691–702.

Kondratyev, K.Ya., Moskalenko, N.I., Parzhin, S.N. and Skvortsova, S.Ya. (1985) Volcanic activity and climates of the Earth and Mars. *Geofis. Int.*, 24, 217–243.

Kutzbach, J.E. and Street-Perrott, F.A. (1985) Milankovitch forcing of fluctuations in the level of tropical lakes from 18 to 0 kyr BP. *Nature*, 317(12 Sept.), 130–134.

Lamb, H.H. (1970) Volcanic dust in the atmosphere with a chronology and assessment of its meteorological significance. *Phil. Trans. Roy. Soc. Lond. A*, 266, 425–533.

Lamb, H.H. (1977) *Climate Present, Past and Future. Vol. II: Climatic History and Future.* Methuen, London.

Lubchenco, J. and Karl, T.R. (2012) Predicting and managing extreme weather events. *Phys. Today*, 65, 31–37.

Manabe, S. and Wetherald, R.T. (1975) The effect of doubling the CO_2 concentration on the climate of a general circulation model. *J. Atmos. Sci.*, 32, 3–15.

Manabe, S. and Wetherald, R.T. (1980) On the distribution of climatic change resulting from an increase in CO_2 content of the atmosphere. *J. Atmos. Sci.*, 37, 99–118.

Manabe, S., Smagorinsky, J. and Strickler, R.F. (1965) Simulated climatology of a general circulation model with a hydrologic cycle. *Mon. Weather Rev.*, 93, 769–798.

Manley, G. (1974) Central England temperatures: monthly means 1659 to 1973. *Q. J. R. Meteorol. Soc.*, 100, 389–405.

Mason, B.J. (1976) Towards the understanding and prediction of climatic variations. *Q. J. R. Meteorol. Soc.*, 102, 473–498.

Mass, C. and Schneider, S.H. (1977) Statistical evidence on the influence of sunspots and volcanic dust on long term temperature records. *J. Atmos. Sci.*, 33, 1995–2004.

McCullough, C.A. (1983) Short-term climatic predictions for water management. *Bull. Am. Meteorol. Soc.*, 64, 1273–1275.

Milankovitch, M. (1930) The astronomical theory of climate. In *Handbuch der Klimatologie I*, ed. A. Teil. Koppen and Keiger, Berlin.

Milankovitch, M. (1938) Neue Ergebrusse der astronomiechen Theorie der Klimaschwankungen. *Bull. Acad. Sci. Math Nat.*, 4, 41.

Mitchell, J.F.B. (1983) The seasonal response of a general circulation model to changes in CO_2 and sea temperatures. *Q. J. R. Meteorol. Soc.*, 109, 113–152.

Mitchell, J.F.B. (1984) The effect of pollutants on global climate. *Meteorol. Mag.*, 113, 1–16.

Monin, A.S. (1986) *An Introduction to the Theory of Climate.* Reidel, Dordrecht.

Munich Re (2012) *2011 Natural Catastrophe Year in Review.* www.iii.org/sites/default/files/docs/pdf/MunichRe-010412.pdf.

Namias, J. and Cayan, D.R. (1981) Large-scale air–sea interactions and short-period climatic fluctuations. *Science*, 214, 869–876.

Nemec, J. (1985) Water resources systems and climate change. In *Facets of Hydrology*, vol. II, ed. J.C. Rodda, pp. 131–152. Wiley, Chichester.

Novaky, B., Pachner, C., Szesztay, K. and Miller, D. (1985) Water resources. In *Climate Impact Assessment*, ed. R.W. Kates, J.H. Ausubel and M. Berberian, pp. 187–214. Wiley, Chichester.

Philander, S.G.H. (1983) El Niño Southern Oscillation phenomena. *Nature*, 302, 295–301.

Ratcliffe, R.A.S., Weller, J. and Collison, P. (1978) Variability in the frequency of unusual weather over approximately the last century. *Q. J. R. Meteorol. Soc.*, 104, 243–256.

Rhodes, S.L., Ely, D. and Dracup, J.A. (1984) Climate and the Colorado River: the limits of management. *Bull. Am. Meteorol. Soc.*, 65, 682–691.

Rind, D. (1986) The dynamics of warm and cold climates. *J. Atmos. Sci.*, 43, 3–24.

Roederer, J.G. (1985) The proposed International Geosphere–Biosphere Program: some special requirements for disciplinary coverage and program design. In *Global Change*, ed. T.F. Malone and J.G. Roederer. Proceedings Symposium sponsored by International Council of Scientific Unions (ICSU) 20th General Assembly, pp. 1–19. Cambridge University Press, Cambridge.

Rotty, R.M. and Weinberg, A.M. (1977) How long is coal's future? *Climate Change*, 1, 45–57.

Rowntree, P.R. and Bolton, J.A. (1983) Simulation of the atmospheric response to soil moisture anomalies over Europe. *Q. J. R. Meteorol. Soc.*, 109, 501–526.

Schneider, S.H. (1975) On the carbon dioxide climate confusion. *J. Atmos. Sci.*, 62, 2060–2066.

Schneider, S.H. and Mass, C. (1975) Volcanic dust, sunspots and temperature trends. *Science*, 190, 741–746.

Schwarz, H.E. (1977) Climatic change and water supply: how sensitive is the North East? In *Climate, Climatic Change and Water Supply*, pp. 111–120. National Academy of Sciences, Washington, DC.

Sharma, T.C. (1985) Large scale water transfers. In *Facets of Hydrology*, vol. II, ed. J.C. Rodda, pp. 345–387. Wiley, Chichester.

Shiklomanov, I.A. (1985) Large scale water transfers. Chapter 12 in *Facets of Hydrology*, vol. II, ed. J.C. Rodda, pp. 345–387. Wiley, Chichester.

Spracklen, D.V., Arnold, S.R. and Taylor, C.M. (2012) Observations of increased tropical rainfall preceded by air passage over forests. *Nature*, 489, 282–285.

Stager, J.C. and Mayewski, P.A. (1997) Abrupt early mid-Holocene climatic transition registered at the equator and the poles. *Science*, 276, 1834–1836.

Stephens, B.B., Gurrey, K.R., Trans, P.P., Sweeney, C., Peters, W., Bruhwiler, L., Cials, P., Ramonet, M., Bousquet, P., Nakazawa, T., Aoki, S., Machida, T., Inoue, G., Vinnichenko, N., Lloyd, J., Jarden, A., Helmann, M., Shibistova, O., Langenfelds, R.L., Steele, L.P., Francey, R.J. and Denning, A.S. (2007) Weak northern and strong tropical land carbon uptake from vertical profiles of atmospheric CO_2. *Science*, 316(22 June), 1732–1735.

Stockton, C.W. and Boggess, W.R. (1982) Climatic variability and hydrologic processes: an assessment for the southwestern United States. Proceedings International Symposium on Hydrometeorology, June, pp. 317–320. American Water Resources Association.

Stone, P.H. (1972) A simplified radiative dynamic model for the static stability of rotating atmospheres. *J. Atmos. Sci.*, 29, 405–418.

Stone, P.H. (1973) The effect of large-scale eddies on climatic change. *J. Atmos. Sci.*, 30, 521–529.

Taylor, K.C., Mayewski, P.A., Alley, R.B., Brook, E.J., Gow, A.J., Grootes, P.M., Meese, D.A., Saltzman, E.S., Severinghaus, J.P., Twickler, M.S., White, J.W.C., Whitlow, S. and Zielinski, G.A. (1997) The Holocene–Younger Dryas transition recorded at Summit, Greenland. *Science*, 278, 825–827.

US National Academy of Sciences (1975) *Understanding Climatic Change*. Washington, DC.

Van Geel, B., Burman, I. and Waterbolk, H.T. (1996) Archaeological and palaeoecological indications of an abrupt climate change in The Netherlands, and evidence for climatological teleconnections around 2650 BP. *J. Quaternary Sci.*, 11, 451–460.

Walker, J. and Rowntree, P.R. (1977) The effect of soil moisture on circulation and rainfall in a tropical model. *Q. J. R. Meteorol. Soc.*, 103, 29–46.

Wetherald, R.T. and Manabe, S. (1975) The effects of changing the solar constant on the climate of a general circulation model. *J. Atmos. Sci.*, 32, 2044–2066.

Winograd, I.J., Landwehr, J.M., Ludwig, K.R., Coplan, T.B. and Riggs, A.C. (1997) Duration and structure of the past four interglaciations. *Quaternary Res.*, 48, 141–154.

WMO (1975) *The Physical Basis of Climate and Climate Modelling*. GARP Publication Series 16. World Meteorological Organization, Geneva.

WMO (1979) *Proceedings of the World Climate Conference*. WMO Publication 537. World Meteorological Organization, Geneva.

Yeh, T. and Fu, C. (1985) Climatic change: a global and multidisciplinary theme. In *Global Change*, ed. T.F. Malone and J.G. Roederer. Proceedings Symposium sponsored by International Council of Scientific Unions (ICSU) 20th General Assembly, pp. 127–145. Cambridge University Press, Cambridge.

14

Hydrometeorology in the Urban Environment

14.1 Introduction

In the year 2000, 47% of the world's population lived in urban areas, and this is projected to rise to 60% by 2030 (UN, 2001). At least 75% of the population in the developed world lives in urban areas, and this is expected to rise to 83% by 2030. In the less developed regions of the world 18% of the population lived in urban areas in 1950, and this is expected to rise to 56% by 2030.

In 2000 only 3.7% of the world's population resided in cities of 10 million or more inhabitants, whereas 24.8% lived in urban settlements with fewer than 500,000 inhabitants. The number of cities with over 5 million inhabitants increased from 40 in 2001 to 58 in 2005. However the number of smaller urban areas has not decreased as the number of megacities has increased, and therefore the impact of urban areas generally on the weather and hydrology can be significant (Collier, 2006).

14.2 Urban boundary layer and the water cycle

Urban morphology impacts energy fluxes and airflow, leading to phenomena such as the urban heat island and convective rainfall initiation (see for example Collier, 2006). Horizontal and vertical flows over an urban area lead to a transition zone between the surface and the atmosphere known as the *urban boundary layer* (UBL). In most urban areas the surface is quite heterogeneous, there being significant differences arising from different land use (parks, rivers etc.) and building types (high rise, business quarters etc.). This has significant implications for the interpretation of measured energy budgets and weather phenomena. Temperature may be quite different in the centre of cities compared to the surrounding rural areas; this effect is known as the *urban heat island* (UHI) (see for example Oke, 1987). Figure 14.1 shows an example of the variation of the cloud base at the top of the boundary layer across the rural–urban boundary.

Most cities are anthropogenic sources of heat. Often there are large areas of asphalt and concrete with albedos and heat capacities that result in the conversion and storage of incoming radiation as sensible heat more effectively than land in surrounding

Hydrometeorology, First Edition. Christopher G. Collier.
© 2016 John Wiley & Sons, Ltd. Published 2016 by John Wiley & Sons, Ltd.
Companion website: www.wiley.com/go/collierhydrometerology

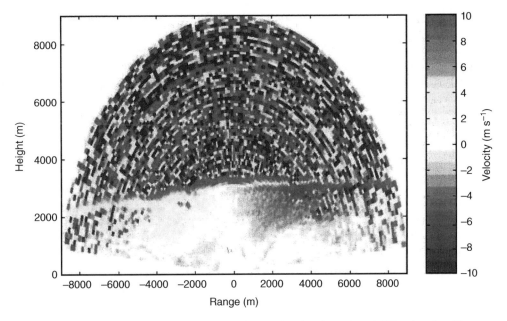

Figure 14.1   A vertical cross-section across the rural–urban boundary in West London, illustrating variation of the cloud base at the top of the boundary layer measured by Doppler lidar. The left-hand side of the image is over the rural area to the west, and the right-hand side of the image is over the urban area to the east. The colours represent the strength of Doppler radial velocities in m s⁻¹ towards (positive) and away from (negative) the lidar. Above the boundary layer the colours represent noise in the measurements (from Collier et al., 2005) (*see plate section for colour representation of this figure*)

rural areas. Thus the lower parts of the UBL are often warmer than the areas surrounding them, as noted above (Chandler, 1976; Oke, 1982). Harmon et al. (2004) and others have shown that the influence of building geometry on the radiation terms of the surface energy balance is a principal cause of surface temperature differences between rural and urban areas. UHIs have been identified in many cities.

Whilst the strength of the UHI has been linearly correlated with urban population (Oke, 1987), work in Lodz, Poland (Klysik and Fortuniak, 1999) has suggested that population magnitude may be a surrogate for the impact of building development density and the occurrence of different types of artificial surfaces. As pointed out by Arnfield (2003), over the last 20 years or so it has become evident that there are many types of UHIs displaying different characteristics and controlled by different interacting energy exchange processes. One of the dominant mechanisms that can explain the pronounced urban effects on rainfall during the warm season is the UHI-induced mesoscale convergence (which can significantly alter the boundary layer) and its interaction with the other local circulations (see for example Shepherd and Burian, 2003; Cotton and Pielke, 2007).

Figure 14.2 shows how impervious cover affects the water cycle. With natural ground cover, 25% of rain infiltrates into the aquifer and only 10% ends up as runoff. As imperviousness increases, less water infiltrates and more and more runs off. In highly urbanized areas, over one half of all rain becomes surface runoff.

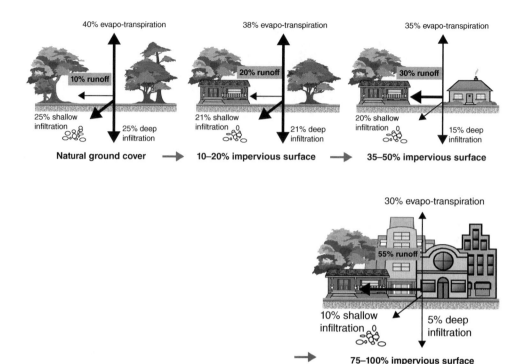

Figure 14.2 How impervious cover affects the water cycle (from California Water Land Use Partnership, WALUP: www.coastal.ca.gov/nps/watercyclefacts.pdf) (*see plate section for colour representation of this figure*)

14.3   Urban development and rainfall

Burian and Shepherd (2005) have pointed out that urbanization alters the appearance of the natural landscape and perturbs Earth system processes. The hydrological cycle, in particular, is changed during construction as vegetation is removed, the soil layer is modified, and built structures and drainage infrastructure are introduced. In general, development activities within a watershed will reduce infiltration and ground water recharge, increase surface runoff volumes and rates, reduce soil moisture, and modify the spatial distribution and magnitude of surface storage and fluxes of water and energy.

It has been known for many years that urban areas influence the occurrence of rainfall. Project METROMEX (Metropolitan Meteorological Experiment), summarized by Changnon (1981), demonstrated a 30–40% increase in rainfall at La Porte, Indiana, USA, located downwind of the Chicago–Gary area due to the growth of urban development during 1925–70 (Huff and Changnon, 1972). Similar impacts have been noted in Israel (Goldreich, 1987) and elsewhere. Around St Louis, Missouri, an afternoon maximum of rainfall 25 km east of the city has been found (Huff and Changnon, 1972), and a second maximum some 30 km north-east of the city during 2100–2400 Central Standard Time in the total summer rainfall shown in Figure 14.3 was described by Changnon and Huff (1986).

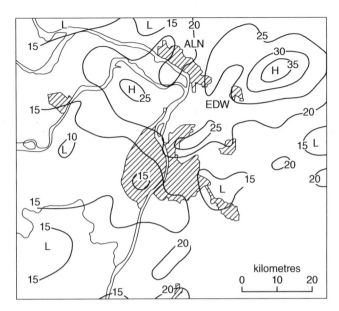

Figure 14.3 Total summer rainfall (in centimetres) around St Louis, Missouri, USA during 2100–2400 CST in 1971–5. The shaded areas are major urban areas (labelled also ALN, EDW). Areas of high and low (relative to average) rainfall are indicated by H and L respectively (from Changnon and Huff, 1986)

Recent work suggests that the greater frequency of rain initiations over urban and industrial areas is related to three factors originally proposed by Changnon et al. (1976):

- thermodynamic effects leading to more clouds and greater in-cloud instability
- mechanical and hydrodynamic effects that produce confluence zones where clouds initiate
- enhancement of the coalescence process due to giant nuclei.

Carraca and Collier (2007) studied the relative importance of the distribution and type of surface morphology and urban heating. A simple model of the surface sensible heat flux was used to explore the impact of urban heterogeneity. Sensitivity experiments are carried out to test the validity of the model, and experiments with a schematic urban morphology were used to investigate the impact of different types of building arrays. It was found that high rise buildings over relatively small areas may have just as much impact as somewhat lower buildings covering a much larger area. The urban area produces considerable spatial variation in surface sensible heat flux. Data from a C-band radar located to the north of Greater Manchester, United Kingdom, provided evidence that convective cells may be initiated by the sensible heat flux input generated by the high rise buildings in the city centre when the atmospheric boundary layer is unstable.

Burian and Shepherd (2005) found that the average annual and warm season diurnal rainfall distribution in the Houston urban area from 1984 to 1999 has an 18–28% higher fraction of the average daily rainfall occurring from noon to 4 p.m. compared with surrounding areas. This enhancement may be linked to the UHI (Shepherd and

Burian, 2003). A possible link between the trends of urbanization as detected by the remote sensing data and the patterns of precipitation over the Indian summer monsoon region was investigated by Kishtawal et al. (2009). Their analysis suggested that there was an adequate statistical basis to conclude that the observed increasing trend in the frequency of heavy rainfall events over the Indian monsoon region is more likely to be over the regions where the pace of land use and land cover change through urbanization is faster. Moreover, rainfall measurements from satellites also indicate that urban areas are more (less) likely to experience heavier (lighter) precipitation rates compared to non-urban areas.

To evaluate the impacts of the UHI effect on precipitation over a complex geographic environment in northern Taiwan, the next-generation mesoscale model, the Weather Research and Forecasting (WRF) model, coupled with the Noah land surface model and urban canopy model (UCM), was used by Lin et al. (2011). Based on a better land use classification derived from moderate resolution imaging spectroradiometer (MODIS) satellite data (the MODIS case), it significantly improved simulation results for the accumulation rainfall pattern as compared with the original US Geological Survey (USGS) 25-category land use classification (the USGS case). The precipitation system was found to develop later but stronger in the urban (MODIS) case than in the non-urban (USGS) case. In comparison with the observation by radar, simulation results predicted reasonably well. Not only was the rainfall system enhanced downwind of the city over the mountainous area, but it also occurred at the upwind plain area in the MODIS case. The simulation results suggested that the correct land use classification is crucial for an urban heat island modelling study. The UHI effect plays an important role in perturbing thermal and dynamic processes. It affects the location of thunderstorms and precipitation over the complex geographic environment in northern Taiwan.

14.4   Sewer flooding

In urban areas there are several types of flooding, namely pluvial flooding, flooding due to asset performance, fluvial flooding, flooding from coincident occurrence of natural phenomena and infrastructure failures, and ground water flooding.

When sewage escapes from a pipe through a manhole or drain, or by backing up through toilets, baths and sinks, this is known as sewer flooding. Sewer flooding can be caused by: a blockage in a sewer pipe; a failure of equipment; too much water entering the sewers from storm runoff (from roads and fields) and from rivers and water courses which have overflowed; or the sewer being too small to deal with the amount of sewage entering it. The cause of the problem may be some distance away from where the flooding is happening. If the sewage enters a building, it is called 'internal flooding'. If it floods gardens, or surrounding areas such as roads and public spaces, it is called 'external flooding'. Figure 14.4 illustrates combined and separate sewer systems. Many developments in old cities have combined sewers in which rainwater and sewage are moved around together.

Flows in combined sewers may be at or above capacity in some locations and below capacity in others. Therefore in-system storage is used to enable flood peaks to pass through the system without sewers surcharging or causing sewerage to be discharged into natural water courses. This storage may involve in-pipes, in-line storage

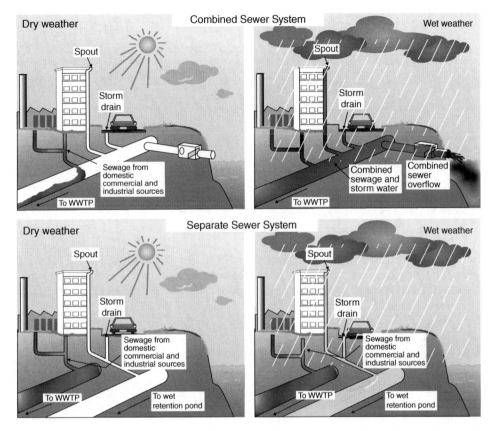

Figure 14.4 Types of sewer systems: WWTP is wastewater treatment plant (from Design of Stormwater Tanks, Grundfos)

tanks or off-line retention tanks. The operation strategy, known as real-time control, requires the ability both to control and to direct flows in the sewer system achieved by gates, dams and diversion regulators. This strategy requires detailed models of the system together with algorithms that direct control actions. As pointed out by Adams and Papa (2000), real-time control does not reduce the volume of combined sewage in a system, but rather changes the temporal distribution of this volume at the sewer outfall. This offers the potential for significant reduction in combined sewer overflows (see for example Labadie et al., 1975).

Simulation models are required to accurately describe the hydraulic phenomena of surcharged and flooded sewer systems. The distinct surface flow and its interaction with sewer flow is referred to as 'dual drainage modelling', as illustrated in Figure 14.5. This comprises

- areas linked completely via closed drains, e.g. roofs
- areas linked via surface inlets and closed drains, e.g. parking areas
- areas on private sites draining to the street
- areas not connected to the sewer system
- areas not subject to flooding
- areas where surface flow occurs.

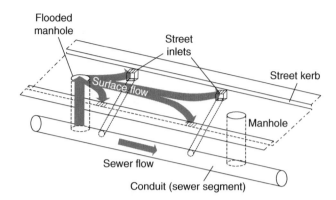

Figure 14.5   Interaction of surface and sewer flow (dual drainage concept) (from Schmitt et al., 2004)

Schmitt et al. (2004) describe hydraulic surface flow modelling based upon the Navier–Stokes equations. However the current state of knowledge regarding uncertainties in urban drainage models remains poor, as indicated by Deletic et al. (2012). In the next section we note progress in modelling surface runoff in urban areas.

14.5   Surface runoff from urban areas

Surface water flooding may be triggered or made worse in urban areas where the ground consists of mostly hard surfaces such as concrete or tarmac so that the rainwater flows straight off rather than soaks away into the ground. Figure 14.6 shows the impact of urban development on Thompson Creek in the Santa Clara Valley of California, USA. As the development grew, the main difference between pre- and post-development conditions was the significantly greater volume of runoff generated after the development (California Water and Land Use Partnership, WALUP).

Increases in extreme weather events across the United Kingdom and elsewhere, possibly brought about by climate change, are exacerbating the risks for properties thought to be in danger from pluvial flooding in low lying urban areas. In 2009, the Environment Agency estimated that around 3.8 million properties in England are susceptible to surface water flooding, compared to just 2.4 million properties that are at risk of fluvial or coastal flooding. Similar problems arise worldwide in which there are varied constituents in the surface runoff (see for example Choe et al., 2002).

Accurately predicting pluvial flood risks is difficult as there are no easily defined floodplains like there are for rivers and seas. Buildings, street furniture and how well the sewers and associated infrastructure all react to sudden intense rainfall need to be considered (see for example Schellart et al., 2011). In the summer of 2007 following a period of persistent wet weather, Hull was one of the many towns and cities across England to be affected after bursts of heavy rainfall triggered multiple flooding events and saw insured flood losses total £3 billion across the country – the largest amount ever paid out by UK insurers. This danger is only likely to get worse due to a combination of climate change and an expected increase in the United Kingdom's urban population, planning regulations that encourage building in flood risk areas and, as yet, limited investment by water and sewerage companies to update Victorian-era systems.

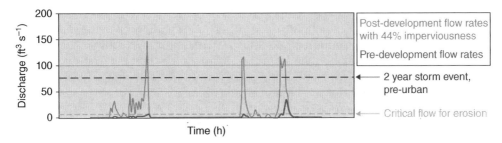

Figure 14.6   Comparison of pre- and post-development flow conditions, Thompson Creek, Santa Clara Valley, California, USA, modelled for a 714 acre development (from California Water Land Use Partnership, WALUP: www.coastal.ca.gov/nps/watercyclefacts.pdf)

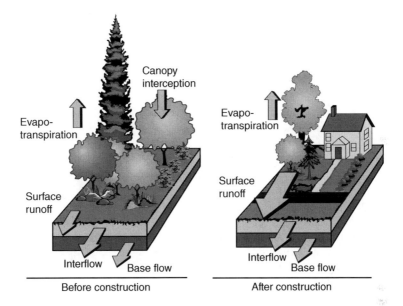

Figure 14.7   Impact of building on the local hydrologic cycle, illustrating the increase in surface runoff (after Maryland, USA, Department of the Environment)

The urban flooding problem is also exacerbated by fewer green spaces due to further urban development; the concreting over of driveways and the deregulation of planning for many home extensions means that water has nowhere to go. New residential developments with multiple occupation are putting pressure on already overburdened and ageing sewer and drainage infrastructure. Figure 14.7 shows the impact of building on the local hydrologic cycle.

Cole et al. (2013) present preliminary results from a study conducted as part of the UK Natural Hazards Partnership (NHP), which is addressing some of the limitations of existing surface water flooding (SWF) alert systems. The new NHP prototype SWF approach builds on the dynamic gridded surface runoff estimates from CEH's grid-to-grid (G2G) model (see Chapter 9) already employed for operational fluvial flood forecasting across Britain. This marks a potentially significant step forward beyond existing methods based on rainfall depth since runoff production in the model is shaped by spatial data sets on landscape properties – land cover (e.g. urban/suburban), terrain, soil and geology – along with dynamically and spatially changing antecedent soil moisture, calculated through continuous water accounting within G2G.

The CEH team has collaborated with the Health and Safety Laboratory (HSL) to produce dynamic maps of the possible impact of surface water flooding. These combine the dynamic hazard footprint with time-varying national impact data sets on population, infrastructure, property and transport such as the EA flood risk maps and the National Population Database from HSL.

14.6   Floodplain development

A floodplain is an area of land adjacent to a stream or river that stretches from the banks of its channel to the base of the enclosing valley walls and experiences flooding during periods of high discharge. Floodplains are made by a meander eroding sideways as they travel downstream. When a river breaks its banks and floods, it leaves behind layers of alluvium (silt). These gradually build up to create the floor of the floodplain.

Unfortunately local planning authorities continue to allow housing developments in areas at severe risk of flooding. In 2013, local councils in the United Kingdom allowed at least 87 planning developments involving 560 homes to proceed in England and Wales in areas at such high risk of flooding that the Environment Agency formally opposed them.

Historically, many towns have been built on floodplains, where they are highly susceptible to flooding, for a number of reasons as follows:

- access to fresh water
- the fertility of floodplain land for farming
- cheap transportation, via rivers and railroads, which often followed rivers
- ease of development of flat land.

More recently many developed countries have adopted floodplain construction regulations. However, in some tropical floodplain areas such as the Inner Niger Delta of Mali, annual flooding events are a natural part of the local ecology and rural economy, allowing for the raising of crops through recessional agriculture. Nevertheless in Bangladesh, which occupies the Ganges Delta, the advantages provided by the richness of the alluvial soil of floodplains are severely offset by frequent floods brought on by cyclones and annual monsoon rains, which cause severe economic disruption and loss of human life in this densely populated region.

Computer models provide one-dimensional (1D) and two-dimensional (2D) simulation techniques for the analysis and visualization of flood inundation areas (see Chapter 9). In the United States the Federal Emergency Management Agency (FEMA) has developed guides (FEMA, 1995) for use by community officials, property owners, developers, surveyors and engineers who may need to determine the base (100 year) flood elevations (BFEs) in special flood hazard areas designated as approximate zone A on FEMA's flood insurance rate maps published as part of the National Flood Insurance Program. One of the primary goals of these documents is to provide a means of determining BFEs at a minimal cost. Similarly in the United Kingdom the Environment Agency publishes flood risk maps for all areas on their web site.

14.7   Acid rain

14.7.1   Basics

Acid rain is caused by emissions of sulphur dioxide and nitrogen oxide, which react with the water molecules in the atmosphere to produce acids (Likens et al., 1972). Acid rain refers to the deposition of wet (rain, snow, sleet, fog, cloud water, and dew) and dry (acidifying particles and gases) acidic components. Acid rain was identified in Manchester, England in 1852, but public concern at its damage of vegetation only became widespread in the late 1960s (Gorham, 1984). The pollution emitted from various activities of humankind is subject to several physical and chemical processes before being deposited back to the surface of the Earth as shown in Figure 14.8.

Distilled water, once carbon dioxide is removed, has a neutral acidity level (pH) of 7. Liquids with a pH less than 7 are acidic, and those with a pH greater than 7 are alkaline. Unpolluted rain has an acidic pH (usually no lower than 5.7) because carbon dioxide and water in the air react together to form carbonic acid, a weak acid, according to the following reaction:

$$H_2O(l) + CO_2(g) \rightleftharpoons H_2CO_3(aq)$$

Carbonic acid then can ionize in water, forming low concentrations of hydronium and carbonate ions:

$$H_2O(l) + H_2CO_3(aq) \rightleftharpoons HCO_3^-(aq) + H_3O^+(aq)$$

However, unpolluted rain can also contain other chemicals which affect its pH. A common example is nitric acid produced by electric discharge in the atmosphere

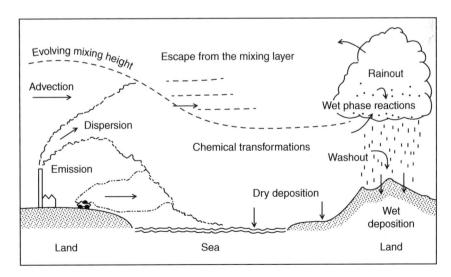

Figure 14.8   Processes involved in deposition of atmospheric pollutants including acid rain (Smith, 1984)

such as lightning (Likens et al., 1987) or forest fires. Nitric oxide (NO) is initially oxidized to nitrogen dioxide in the troposphere, mainly by reaction with ozone. Acid deposition as an environmental issue includes acids additional to H_2CO_3 (Likens and Bormann, 1974).

In Europe, wet deposition accounts for a large fraction of the total deposition over western Norway, but less than 15% of the total over central Europe (Beilke, 1983). The deposition measured over Europe is largely the result of sources within Europe (Levy and Maxim, 1987; Whelpdale et al., 1988); very little reaches Europe from North America. In Great Britain during 1990 Metcalfe et al. (1995) concluded that 34% of sulphur deposition came from European land-based sources and 46.7% from UK power stations and oil refineries, with shipping and natural sources such as algal blooms contributing only 4.6%. In China wet deposition of sulphur is 7–130 times higher than that from remote areas in the southern hemisphere (Galloway et al., 1987).

Wet deposition of radioactive nuclides was also identified in many European countries after the accident at Chernobyl, USSR. It has been shown that the levels of deposited caesium-137 can be closely related to rainfall intercepting the plume (see for example Clark and Smith, 1988).

14.7.2   *Modelling wet deposition*

Modelling wet deposition using a washout coefficient is implemented in the ADMS5 model (Apsley and CERC, 2012). This means that all the physical and chemical processes involved in the removal of pollutant from a plume are represented by simple proportionality between removal rates and the local airborne concentration of material. It is assumed that (i) removal processes act independently, and (ii) removal processes are irreversible.

The first assumption implies that plume strength can be written as a product of source strength and depletion factors corresponding to each removal process. Where species with significantly different deposition parameters are present in the release, a separate calculation must be performed for each. Variation of deposition parameters with the formation of new isotopes following radioactive decay cannot, however, be modelled and the values of dry deposition velocity and washout coefficient should reflect the contents of the initial isotope inventory. The first assumption precludes gases dissolving in raindrops only to come out of solution close to the ground. This *outgassing* is not described here.

Washout is assumed to occur in proportion to the local concentration throughout the depth of the plume, and the wet deposition is modelled in terms of a washout coefficient Λ. Washout coefficients are assumed to be uniform throughout the domain, except where limited by high concentrations of SO_2 or CO_2. When wet deposition is modelled using a washout coefficient, the mass of pollutant incorporated into rainfall is ΛC per unit volume per unit time, where C is the local airborne concentration. The washout coefficient is dependent on a large number of parameters, including rainfall rate, raindrop and aerosol size distributions, and concentrations in air and raindrops. It may be input directly by a user, or estimated by the computer model from other data such as rainfall rate.

Assuming irreversible uptake, the total wet deposition rate per unit horizontal area per unit time, F_{wet}, is found by integrating through a vertical column of air:

$$F_{wet} = \int \Lambda C \, dz \tag{14.1}$$

The result is that the plume strength diminishes with distance:

$$\left. \frac{dQ}{dx} \right|_{wet} = -\int F_{wet} \, dy \tag{14.2}$$

The following assumptions are made:

- All plume material lies in or below rain cloud, and no distinction is made between in-cloud scavenging (rainout) or below-cloud scavenging (washout).
- Uptake of pollutants is irreversible, and precipitation scavenging does not lead to a redistribution of material in the plume.
- Solution in raindrops does not lead to saturation.
- The rainfall rate is uniform over the domain.

The amount of pollutant remaining in the plume decays exponentially:

$$q_w = e^{-\Lambda t_r(x)} \tag{14.3}$$

where t_r is the travel time in rain.

The washout coefficient Λ may be specified as a constant value, in which case the wet deposition calculated will be independent of the precipitation rate. Wet deposition will then be predicted in the absence of precipitation. The default value of washout coefficient Λ may be taken as $1.0 \times 10^{-4} \, \text{s}^{-1}$. Alternatively the washout coefficient can be modelled as a function of pollutant species and rainfall rate only as follows:

$$\Lambda = a J^b \tag{14.4}$$

where J is the rainfall rate in mm h^{-1} and a and b are parameters depending on the type of pollutant, having default values of $a = 1.0 \times 10^{-4}$ and $b = 0.64$ (Jylha, 1991). The uptake of acidic pollutants is, in practice, limited at high concentrations by the limiting effect of the rain drop pH.

14.8   Urban air and water pollution

In the previous section wet deposition was discussed. As mentioned, this process is responsible for the impact not only of air pollution in urban and rural areas as acid rain, but also of a range of pollutants such as particulates, biological molecules or other harmful materials. The United Nations Environment Programme has estimated that more than 1 billion people are exposed to outdoor air pollution annually. Urban air pollution is linked to up to 1 million premature deaths each year. Also it is estimated to cost approximately 2% of GDP in developed countries and 5% in developing countries. According to a 2014 World Health Organization report, air pollution in 2012 caused the deaths of around 7 million people worldwide (WHO, 2014). Major air pollutants produced by human activity include: sulphur dioxide, nitrogen oxides, carbon monoxide, volatile organic compounds, particulates, persistent free radicals, toxic metals, chlorofluorocarbons, ammonia and radioactive pollutants.

Untreated sewage poses a major risk to human health since it contains waterborne pathogens that can cause serious human illness. Untreated sewage also destroys aquatic ecosystems, threatening human livelihoods, when the associated biological oxygen demand and nutrient loading deplete oxygen in the water to levels too low to sustain life. WHO has estimated that 2.6 billion people lacked access to improved sanitation facilities in 2008, with the lowest coverage in sub-Saharan Africa (37%), southern Asia (38%), and eastern Asia (45%) (WHO/UNICEF, 2008). Sewage can be intentionally discharged to waterways through pipes or open defecation, or unintentionally during rainfall events. When humans use these waterways for drinking, bathing or washing, they are exposed to the associated pathogens, many of which can live for extended periods in aquatic environments. Humans then become ill by ingesting contaminated water, by getting it on or in the skin, eyes or ears, or by preparing foods with contaminated water. Sometimes humans can even become ill from inhaling contaminated water droplets.

Appendix 14.1   Number of runoff events from an urban drainage system

As a result of urbanization, stormwater runoff flow rates and volumes are significantly increased due to increased impervious land cover and the decreased availability of depression storage (see Schueler, 1987). A runoff event will occur when the volume of a rainfall event exceeds the depression storage volume (see Adams and Papa, 2000). Therefore the probability of a runoff event occurring per rainfall event is equivalent to the probability that a rainfall event volume exceeds the depression storage, and is given by

$$prob[v > S_d] = \int_{v=S_d}^{\infty} f v(v)dv = \int_{v=S_d}^{\infty} \zeta e^{-\zeta v} = e^{-\zeta S_d} \qquad (14.5)$$

where v is average precipitation volume, S_d is depression storage, and $\zeta = 1/v$. The average annual number of runoff events n_r is then given by the product of the average annual number of rainfall events θ and the probability that a rainfall event causes a runoff event, i.e.

$$n_r = \theta\, prob[v > S_d] = \theta e^{-\zeta S_d} \qquad (14.6)$$

The volume of rainfall that does not become runoff is referred to as the loss volume l. Losses are of two types, namely (i) the volume of rainfall intercepted by depression storage and ultimately returned to the hydrologic cycle by evapotranspiration and subsequent infiltration, and (ii) the rainfall lost to infiltration during the storm event as described by the runoff coefficient. The average annual loss volume L is given by

$$L = \theta\, expected\, value\, E[l] = (\theta/\zeta)(1 - e^{-\zeta S_d}) \qquad (14.7)$$

The volume of rainfall that is lost to abstraction after the depression storage is filled is $v-\phi(v-S_d)$, where ϕ is a dimensional runoff coefficient and $(v-S_d)$ is the excess of rainfall over the depression storage.

Summary of key points in this chapter

1. In the year 2000, 47% of the world's population lived in urban areas.
2. In 2000 only 3.7% of the world's population resided in cities of 10 million or more inhabitants.
3. The number of cities with over 5 million inhabitants has increased from 40 in 2001 to 58 in 2005.
4. Urban morphology impacts energy fluxes and airflow, leading to phenomena such as the urban heat island (UHI).
5. Temperature may be quite different in the centre of cities compared to the surrounding rural areas, this being influenced by building geometry.
6. Most cities are anthropogenic sources of heat.
7. The strength of the UHI has been linearly correlated with urban population.
8. There are many types of UHIs controlled by different interacting energy exchange processes.
9. Impervious cover affects the water cycle. With natural ground cover, 25% of rain infiltrates into the aquifer and only 10% runs off, but as imperviousness increases less water infiltrates and more and more runs off.
10. In general, development activities (building) within a watershed will reduce infiltration and ground water recharge, increase surface runoff volumes and rates, reduce soil moisture, and modify the spatial distribution and magnitude of surface storage and fluxes of water and energy.
11. A 30–40% increase in rainfall has been measured downwind of the Chicago–Gary area in the United States due to urban development.
12. Rain initiations due to development result from thermodynamic, mechanical and hydrodynamic effects, and from enhancement of the coalescence process due to the presence of giant nuclei.
13. It has been concluded that the frequency of heavy rainfall events over the Indian monsoon region is more likely to be over regions where the pace of land use and land cover change through urbanization is faster.
14. In urban areas there are several types of flooding, namely pluvial, fluvial, flooding from infrastructure failures, and ground water flooding.
15. Sewer flooding can be caused by a blockage in a sewer pipe, a failure of equipment, rivers and water courses overflowing, or the sewer being too small to deal with the amount of sewage entering it.
16. Combined (natural and dirty water) and separate (sewage kept separate from naturally occurring stormwater) systems are implemented.

17. Simulation models are required to accurately describe the hydraulic phenomena of surcharged and flooded sewer systems, referred to as dual drainage modelling.

18. Surface water flooding may be triggered or made worse in urban areas where the ground consists of mostly hard surfaces, e.g. concrete, tarmac.

19. In 2009 the UK EA estimated that around 3.8 million properties in England are susceptible to surface water flooding, compared to just 2.4 million properties at risk of fluvial or coastal flooding.

20. A floodplain is an area of land adjacent to a stream or river that stretches from the banks of its channel to the base of the enclosing valley walls, and experiences flooding during periods of high discharge.

21. Historically many towns have been built on floodplains as there was access to fresh water, the floodplain was fertile for farming, transport was cheap, and development was possible on flat land. Unfortunately developments here are susceptible to flooding.

22. Acid rain is caused by emissions of sulphur dioxide and nitrogen oxide, which react with water molecules to produce acids. It causes damage to vegetation and buildings.

23. In Europe, wet deposition accounts for a large fraction of the total deposition over western Norway, but less than 15% of the total over central Europe.

24. Radioactive nuclides may be deposited as they were after the Chernobyl accident.

25. Wet deposition may be modelled numerically.

26. Urban air pollution is linked to up to 1 million premature deaths each year, and the World Health Organization reported that around 7 million deaths were caused worldwide by air pollution.

27. Untreated sewage poses a major risk to human health in many countries.

Problems

1. What proportion of the world's population lived in urban areas in the year 2000?

2. How many people lived in cities with over 5 million inhabitants in the year 2005?

3. What causes the urban heat island, and what are its characteristics?

4. Describe the impact of asphalt and concrete on the heat capacity in cities.

5. Describe one dominant mechanism explaining the urban effects on rainfall.

6. Describe how impervious cover affects the water cycle.

7. How does urbanization perturb the hydrological cycle?

8. What was the finding of the METROMEX Project?

9. What three factors account for the greater frequency of rain initiations over urban and industrial areas?
10. What might be responsible for an increased frequency of heavy rain over the Indian monsoon region?
11. Name several types of flooding in urban areas.
12. What causes a sewer to surcharge?
13. Describe a combined sewer system and a separate sewer system.
14. What are retention tanks used for?
15. What does dual drainage modelling comprise?
16. Draw a hydrograph showing flows before and after urban development.
17. Why are new residential developments putting pressure on the sewer system?
18. Define a floodplain.
19. Why have many towns been built on floodplains?
20. What is acid rain?
21. When and where was acid rain first identified?
22. Give a common example of a chemical which has an impact on lightning.
23. What sources provide the constituents of acid rain?
24. Give an equation for total wet deposition.
25. Give an equation for the decay of pollutants in a plume.
26. Give an equation relating washout coefficient to rainfall rate.
27. How many deaths does the World Health Organization estimate are caused by air pollution?
28. How many people worldwide lacked access to proper sanitation in the year 2008?

References

Adams, B.J. and Papa, F. (2000) *Urban Stormwater Management Planning with Analytical Probabilistic Models*. Wiley, Chichester.

Apsley, D.D. and CERC (2012) Modelling wet deposition. CERC, Cambridge. www.cerc.co.uk/environmental-software/assets/data/doc_techspec/CERC_ADMS5_P17_12.pdf.

Arnfield, A.J. (2003) Two decades of urban climate research: a review of turbulence exchanges of energy and water, and the urban heat island. *Int. J. Climatol.*, 23, 1–26.

Beilke, S. (1983) Acid deposition: the present situation in Europe. In *Acid Deposition*, Report EUR8307, pp. 3–30. Commission of the European Communities.

Chandler, T. (1976) The climate of towns. In *The Climate of the British Isles*, ed. T.J. Chandler and S. Gregory, pp. 307–329. Longman, London.

Burian, S.J. and Shepherd, J.M. (2005) Effect of urbanization on the diurnal rainfall pattern in Houston. *Hydrol. Proc.*, 19, 1089–1103.

Carraca, M.G.D. and Collier, C.G. (2007) Modelling the impact of high-rise buildings in urban areas on precipitation initiation. *Meteorol. Appl.*, 14, 149–161.

Changnon, S. (1981) *METROMEX: A Review and Summary*. Monograph 40, vol. 18. American Meteorological Society, Boston.

Changnon, S. and Huff, F.A. (1986) The urban-related nocturnal rainfall anomaly at St Louis. *J. Clim. Appl. Meteorol.*, 25, 1985–1995.

Changnon, S., Semonin, R.G. and Huff, F.A. (1976) A hypothesis for urban rainfall anomalies. *J. Appl. Meteorol.*, 15, 544–560.

Choe, J.S., Bang, K.W.and Lee, J.H. (2002) Characterization of surface runoff in urban areas. *Water Sci. Technol.*, 45, 249–254.

Clark, M.J. and Smith, F.B. (1988) Wet and dry deposition of Chernobyl releases. *Nature*, 322, 245–249.

Cole, S.J., Moore, R.J., Aldridge, T., Lane, A. and Laeger, S. (2013) Real-time hazard impact modelling of surface water flooding: some UK developments. Proceedings of the International Conference on Flood Resilience: Experiences in Asia and Europe (ICFR 2013), University of Exeter, ed. D. Butler, A.S. Chen, S. Djordjević and M.J. Hammond.

Collier, C.G. (2006) The impact of urban areas on weather. *Q. J. R. Meteorol. Soc.*, 132, 1–25.

Collier, C.G., Davies, F., Bozier, K.E., Holt, A., Middleton, D.R., Pearson, G.N., Siemen, S., Willetts, D.V., Upton, G.J.G. and Young, R.I. (2005) Dual Doppler lidar measurements for improving dispersion models. *Bull. Am. Meteorol. Soc.*, 86, 825–838.

Cotton, W. and Pielke, R.A. Sr (2007) *Human Impacts on Weather and Climate*, 2nd edn, Chapter 5. Cambridge University Press, Cambridge.

Deletic, A., Dotto, C.B.S., McCarthy, D.T., Kleidorfer, M., Freni, G., Mannina, G., Uhi, M., Henrichs, M., Fletcher, T.D., Rauch, W., Bertrand-Krajewski, J.L. and Tait, S. (2012) Assessing uncertainties in urban drainage models. *Phys. Chem. Earth*, 42–44, 3–10.

FEMA (1995) *Managing Floodplain Development in Approximate Zone A Areas: A Guide for Obtaining and Developing Base (100-Year) Flood Elevations*. FEMA 265. Federal Emergency Management Agency.

Galloway, J.N., Dianwu, Z., Jiling, X. and Likenb, G.E. (1987) Acid rain: China, United States, and a remote area. *Science*, 236, 1559–1562.

Goldreich, Y. (1987) Advertent/inadvertent changes in the spatial distribution of rainfall in the central coastal plain of Israel. *Climate Change*, 11, 361–373.

Gorham, E. (1984) Acid rain: an overview. In *Meteorological Aspects of Acid Rain*, ed. C.M. Bhumralkar. Butterworth, London.

Harmon, I.N., Best, M.J. and Belcher, S.E. (2004) Radiative exchange in an urban street canyon. *Boundary-Layer Meteorol.*, 110, 301–316.

Huff, F.A. and Changnon, S.A. (1972) Climatological assessment of urban effects on precipitation at St Louis. *J. Appl. Meteorol.*, 11, 823–842.

Jylha, K. (1991) Empirical scavenging coefficients of radioactive substances released from Chernobyl. *Atmos. Environ.*, 25A, 263–270.

Kishtawal, C.M., Niyogi, D., Tewari, M., Pielke, R.A. Sr and Shepherd, J.M. (2009) Urbanization signature in the observed heavy rainfall climatology over India. *Int. J. Climatol.*, 30, 1908–1916.

Klysik, K. and Fortuniak, K. (1999) Temporal and spatial characteristics of the urban heat island of Lodz, Poland. *Atmos. Environ.*, 33, 3885–3895.

Labadie, J.W., Grigg, N.S. and Bradford, B.M. (1975) Automatic control of large-scale combined sewer systems. *J. Environ. Eng. Div. ASCE*, 101, 27–39.

Levy, H. II and Maxim, W.J. (1987) Fate of US and Canadian combustion nitrogen emissions. *Nature*, 328, 414–416.

Likens, G.E. and Bormann, F.H. (1974) Acid rain: a serious regional environmental problem. *Science*, 184, 1176–1179.

Likens, G.E., Bormann, F.H. and Noye, M. (1972) Acid rain. In *Environment Science and Policy for Sustainable Development*, vol. 14, pp. 33–40. Published online 8 July 2010, Taylor & Francis. doi 10.1080/00139157.1972.

Likens, G.E., Keene, W.C., Miller, J.M. and Galloway, J.N. (1987) Chemistry of precipitation from a remote terrestrial site in Australia. *J. Geophys. Res.*, 92, 13299–13314.

Lin, C.-Y., Chen, W.-C., Chang, P.-L. and Sheng, Y.-F. (2011) Impact of the urban heat island effect on precipitation over a complex geographic environment in Northern Taiwan. *J. Appl. Meteorol. Climatol.*, 50, 339–353.

Metcalfe, S.E., Whyatt, J.D. and Derwent, R.G. (1995) A comparison of model and observed network estimates of sulphur deposition across Great Britain for 1990 and its likely source attribution. *Q. J. R. Meteorol. Soc.*, 121, 1387–1411.

Oke, T.A. (1982) The energetic basis of the urban heat island. *Q. J. R. Meteorol. Soc.*, 108, 1–24.

Oke, T.A. (1987) *Boundary Layer Climates*, 2nd edn. Routledge, London.

Schellart, A., Ochoa, S., Simoes, N., Wang, L.-P., Rico-Ramirez, M., Lignon, S., Duncan, A., Chen, A.S., Keedwell, E., Djordjevic, S., Savic, D.A., Saul, A. and Maksimovic, C. (2011) Urban pluvial flood modelling with real time rainfall information: UK case studies. 12th International Conference on Urban Drainage, Porto Alegre, Brazil, 10–15 September.

Schmitt, T.G., Thomas, M. and Ettrich, N. (2004) Analysis and modelling of flooding in urban drainage systems. *J. Hydrol.*, 299, 300–311.

Schueler, T. (1987) *Controlling Urban Runoff: A Practical Manual for Planning and Designing Urban Best Management Practices*. MWCOG, Washington, DC.

Shepherd, J.M. and Burian, S.J. (2003) Detection of urban-induced rainfall anomalies in a major coastal city. *Earth Interactions*, 7, 1–17.

Smith, F.B. (1984) The fate of airborne pollution. Section 1 in *Acid Rain*, Report 14, pp. 1–13. Watt Committee on Energy.

UN (2001) *World Population Monitoring 2001*. Department of Economic and Social Affairs, Population Division, UN E.01.xii.17. United Nations, New York.

Whelpdale, D.M., Eliassen, A., Galloway, J.N., Dovland, H. and Miller, J.M. (1988) The transatlantic transport of sulphur. *Tellus*, 408, 1–15.

WHO/UNICEF (2008) Progress on Drinking Water and Sanitation: Special Focus on Sanitation. International Year of Sanitation 2008. WHO, Geneva, UNICEF, New York. www.who.int/water_sanitation_health/monitoring/jmp2008.pdf.

WHO (2014) 7 million premature deaths annually linked to air pollution. World Health Organization, 25 March. www.who.int/mediacentre/news/releases/2014/air-pollution/en/.

Glossary

Accretion: The process which occurs if both water droplets and ice crystals exist, resulting in the formation of ice and snow pellets (graupel) or hail.

Acid rain: The deposition of wet (rain, snow, sleet, fog, cloud water and dew) and dry (acidifying particles and gases) acidic components.

Adiabatic: When no condensation or evaporation occurs, atmospheric processes may be regarded as adiabatic

Aerosol: Tiny particles in the atmosphere which may enhance scattering and absorption of solar radiation.

Aggregation: Rain formation; if only ice crystals exist in a cloud then snow results.

Agricultural drought: Drought can arise independently of any change in precipitation levels when soil conditions and erosion triggered by poorly planned agricultural activity cause a shortfall in water available to the crops.

Albedo: Coefficient of reflectance of radiation.

AMO: Atlantic Multi-decadal Oscillation.

AMSU: Advanced microwave sounding unit.

ANN: Artificial neural networks.

Anthroposphere: The effect of human beings on the Earth system.

AOFD: Automatic overland flow delineation.

Areal reduction factors (ARFs): Used in flood studies analysis.

ARMA: Autoregressive moving average error prediction with uncertainty estimation.

ATI: Area–time integral.

Atmosphere: The air surrounding the Earth, which is a mixture of gases, mainly nitrogen (about 80%) and oxygen (about 20%) with other minor gases.

Atmospheric stability: If a parcel of moist air has a vertical velocity w, the atmosphere is regarded as stable, neutral or unstable when dw/dt is <0, 0 or >0 respectively.

ATSR: Along-track scanning radiometer.

Attenuation: Loss of energy as it passes through a medium.

Avalanche: Rapid flow of snow, soil or rock down a hill or mountainside.

Hydrometeorology, First Edition. Christopher G. Collier.
© 2016 John Wiley & Sons, Ltd. Published 2016 by John Wiley & Sons, Ltd.
Companion website: www.wiley.com/go/collierhydrometerology

Avogadro's hypothesis: At normal temperature and pressure (273.15 K and 1013.25 mbar) the molecular volume is the same for all gases, namely $22.415 \times 10^3 \, cm^3$.

Biochemical cycles: Examples are the exchange of nitrogen and chlorofluoromethanes, and of both man-made and volcanic aerosols.

Biosphere: The global sum of all ecosystems, sometimes called the zone of life on Earth.

BMA: Bayesian model averaging.

BOREAS: Boreal Ecosystem–Atmosphere Study.

Bowen ratio: Ratio of sensible to latent heat.

Boyle's law: An equation reflecting the observations of experiments with gases, which states that pressure × volume is a constant at constant temperature.

BRDF: Bidirectional reflectance distribution function.

Bright band: Enhanced radar reflectivity where snow melts to form rain.

BSS: Brier skill score.

Carbon cycle: The exchange of carbon between atmosphere, oceans, biosphere and the solid earth.

CEH: Centre for Ecology and Hydrology, Wallingford, UK.

Charles's law: Any gas at a given pressure expands uniformly with temperature.

Clausius–Clapeyron equation: This gives the relationship between the equilibrium pressure and the temperature during a phase change.

Climate noise: That part of the variance of climate attributable to short term weather changes.

Climatic change: Occurs when the differences between successive averaging periods exceed the differences that can be accounted for by noise.

Climatic variability: The manner of variation of the climatic parameters within the typical averaging period.

Coalescence: The process of rain formation if a cloud contains only water.

Coastal flooding: Flooding caused by an increase in sea levels owing to particular weather and tidal conditions.

Condensation: The process whereby water vapour in the atmosphere is changed into liquid water as clouds and dew.

Conditional probabilities: Characterization of the level of persistence in the data record.

Conveyor belts: The warm conveyor belt and the cold conveyor belt occur in frontal systems.

COPE: Convective Precipitation Experiment (UK-led).

Copolar correlation coefficient ρ_{hv}: The successive estimates of correlation of the time series of horizontal and vertical reflectivity.

COSMOS-UK: COsmic-ray Soil Moisture Observing System UK.

Coupled models: Numerical models which couple different compartments of the Earth system.

CSIP: Convective Storm Initiation Project (UK-led).

CSO: Combined sewer overflow.

Dalton's law: States that in a mixture of gases, the total pressure is equal to the sum of the pressures which would be exerted by each gas if it filled the volume under consideration at the same temperature.

Darcy's law: Estimates the volume of water per unit area and time moving vertically through the soil or snow.

Data assimilation: A procedure to construct the best possible initial and boundary conditions from which to integrate forward the model.

DBM: Database mechanistic methodology a stochastic approach to data assimilation.

dBZ: Decibel relative to reflectivity Z; a unit used in weather radar.

Differential reflectivity Z_{DR}: $Z_{DR} = 10\log(Z_H/Z_V)$, where Z_H is the horizontally polarized reflectivity and Z_V is the vertically polarized reflectivity.

Direct inundation: The sea height exceeds the elevation of the land.

Discharge equation: For flow in open channels and pipes, $Q = VA$, where Q is the discharge in m^3s^{-1}, V is the velocity in m^{-1} and A is the wetted area in m^2.

Disdrometer: An instrument used to measure the drop size distribution and velocity of falling hydrometeors.

DOP: Degree of polarization.

Double mass curve: Analysis involving accumulating and plotting total precipitation throughout a data period, which reveals the long term sustained trends in the data.

Drain and sewer (urban) flooding: Flooding that occurs during heavy rain when drains have become blocked or full.

Drought recovery period: The time necessary to return to the climatological average conditions of water storage.

Dry adiabatic lapse rate: When a parcel of air rises dry adiabatically, its temperature will fall at a rate of about $10°C\ km^{-1}$; this is the dry adiabatic lapse rate (DALR).

Dual drainage: Interaction of surface and sewer flow.

EA: Environment Agency (UK).

ECMWF: European Centre for Medium Range Weather Forecasting.

EFAS: European Flood Awareness System.

EFFS: European Flood Forecasting System.

Ekman layer: Variation of wind in the planetary boundary layer; the Ekman spiral is the shape of the wind spiral in the Ekman layer.

El Niño: Unusually warm sea surface temperatures in the equatorial Pacific Ocean.

Emissivity: A measure of the emission of radiation over a range of wavelengths and directions from real materials; symbol ε.

Ensemble forecasting: System which samples the uncertainty inherent in weather prediction to provide more information about possible future weather conditions.

ENSO: El Niño/Southern Oscillation associated with the Walker cell.

EPS: ECMWF ensemble prediction system.

Equation of state for a perfect gas: The equation relating pressure p, specific volume V (volume per gram) and temperature T of a perfect gas.

Equatorial low level jet: A cross-equator jet stream associated with the onset of the Indian monsoon.

Equifinality: The concept that there are many possible parameter sets for hydrological models. Multiple acceptable parameter sets will result in a range of predictions, so allowing for a measure of uncertainty.

ESTOFS: Extratropical Surge and Tide Operational Forecast System (US).

ETSS: National Weather Service Extratropical Storm Surge Model (US).

Evaporation: The change of state from liquid water to vapour.

Evapotranspiration: The combination of evaporation from soil and water surfaces and transpiration from plants (evaporation of water from plant tissue), which is closely linked to photosynthesis.

Extreme value analysis: Construction of a series of extreme values by selecting the highest (or lowest) observation from a set of data.

FEH: Flood Estimation Handbook (UK).

FEMA: Federal Emergency Management Agency (US).

Ferrel cells: Mid-latitude counter-circulations encircling the globe in both hemispheres; an essential part of the general circulation.

Firn: Compacted snow having a density that exceeds 500 kg m^{-3}, turning into glacier ice when most of the gaps between snow crystals are filled.

First law of thermodynamics: The energy of a system in a given state relative to a fixed normal state (usually 0°C, one atmosphere pressure, and at rest in a given position) is equal to the algebraic sum of the mechanical equivalent of all the effects produced outside the system, when it passes in any way from the given state to the normal state.

Fisher–Tippett distribution types I, II and III: Used in extreme value analysis.

Flood forecasting techniques: adaptive approach: Rainfall–runoff modelling forecasts and control decisions in the current real-time period are based upon future anticipated inflows as well as on current conditions, so that available storage and flow capacities may be utilized in the best way, and warnings issued as quickly as possible.

Flood forecasting techniques: reactive approach: Upstream and downstream river level correlation in which real-time forecasting and river control are accomplished by using *a priori* derived operating rules.

Fluvial (river) flooding: Flooding that occurs when a river cannot cope with the amount of water entering it.

Fractal dimension: Unsmoothed shapes which show scaling behaviour.

Front: A discontinuity of wind (and temperature) across a stationary or moving boundary between air masses.

FRONTIERS system: Forecasting Rain Optimized using New Techniques of Interactive Radar and Satellite data (UK).

FSR: Flood Studies Report (UK).

G2G: Grid-to-grid hydrological model (CEH).

GANDOLF system: A system for generating automated nowcasts of convective precipitation (Met Office).

GCM: General circulation model.

GEM: Global environmental multiscale model.

Generalized likelihood uncertainty estimation (GLUE): A statistical method for quantifying the uncertainty of model predictions.

Geostationary satellite: Orbits at an altitude of 36,000 km.

Geostatistical techniques: Techniques including kriging.

Geostrophic wind equation: Expressed as $V_g = (1/\rho f)|(\partial p/\partial n)|$, where V_g is geostrophic wind velocity, ρ is air density, f is Coriolis parameter and $\partial p/\partial n$ is the horizontal pressure gradient; describes horizontal motion under balanced forces.

GEWEX: Global Energy and Water Cycle Exchanges Project; an integrated programme of research, observations and science activities that focuses on the atmospheric, terrestrial, radiative, hydrological, coupled processes and interactions that determine the global and regional hydrological cycle, radiation and energy transitions, and their involvement in climate change.

Glacier: Derived from the annual accumulation, ablation and change in mass of ice held in storage.

GPCP: Global Precipitation Climatology Project.

Gradient flow: Flow assuming no friction and zero tangential acceleration.

Graupel: Ice and snow pellets.

Ground water flooding: Flooding that results when rainfall causes the water that is naturally stored underground to rise to the surface, when it can flood low lying areas.

Gumbel distribution: A Fisher–Tippett distribution type I.

Hadley cell: A thermally direct circulation responsible for the trade winds; a main part of the atmospheric general circulation, usually between ±30° latitude.

Hallett–Mossop process: The process of splintering of ice nuclei during riming, resulting in large concentrations of ice particles in most clouds.

Harmonic (Fourier) analysis: Identification of the periodicity in data.

Heinrich event: Sudden intense cold and dry phases in past glacial periods which occasionally affected Europe and the north Atlantic region, and possibly many other parts of the world.

HSL: Health and Safety Laboratory (UK).

Hurricane: Wind system depending upon a continuous supply of heat to maintain its circulation.

Hydrodynamic models: Originally developed for flood mapping purposes, but now in some places adapted for use as a real-time forecasting tool.

Hydrological cycle: The circulation of water as liquid or vapour from the Earth's surface into the atmosphere, returning to the Earth's surface via clouds.

Hydrological drought: Brought about when the water reserves available in sources such as aquifers, lakes and reservoirs fall below the statistical average.

Hydrosphere: The combined mass of water found on, under and over the surface of a planet.

Hydrostatic equation: See *hydrostatic pressure law*.

Hydrostatic pressure law: Vertical gradient of pressure as a function of acceleration due to gravity and density, expressed as $\partial p/\partial z = -g\rho_a$.

Infiltration: Results when water passes into the land surface, occurring when the ground is not saturated or contains cracks or fissures.

Infoworks: Commercial hydrodynamic sewer network model.

Interstadials: Sudden warm and moist phases which occurred at various times during the timespan of the last glacial phase.

IPCC: Intergovernmental Panel on Climate Change.

ISO model: Input, storage, output of river flow numerical model.

Isohyets: Lines of constant rainfall.

ITCZ: Inter-Tropical Convergence Zone.

Jet stream: The name given to the wind maximum above a frontal zone or within a monsoon system.

Kriging: A statistical method that uses the variogram of the rainfall field, i.e. the variance between pairs of points at different distances apart, to optimize the raingauge weightings to minimize the estimation error.

La Niña: Unusually cool sea surface temperature in the equatorial Pacific Ocean.

Land and sea breezes: Caused by temperature differences over the land or the sea, resulting in low level atmospheric circulations.

Latent heat of evaporation: The amount of heat required to transform one gram of water into vapour at a constant temperature (597.3 cal g^{-1}, 2.50 MJ kg^{-1} at $0°C$).

Latent heat of fusion: The amount of heat required to transform one gram of ice to water (79.7 cal g^{-1}, 0.333 MJ kg^{-1}).

Latent heat of sublimation: The amount of heat required to transform one gram of ice to water vapour (677.0 cal g^{-1}, 2.83 MJ kg^{-1}).

LDR: Linear depolarization ratio.

Lee waves: A form of standing wave resulting from flow over a hill somewhere upstream (but may also result from other processes), causing an impulse on the atmosphere.

LHN: Latent heat nudging a procedure in numerical weather forecast models in which numerical values are moved gradually towards actual latent heat values.

LISFLOOD-FP: A two-dimensional hydrodynamic model specifically designed to simulate floodplain inundation in a computationally efficient manner over complex topography (Joint Research Centre, ISPRA, Italy).

Lithosphere: The rigid outermost shell of the Earth, comprising the crust and a portion of the upper mantle.

Little Ice Age: Cooler period from about 1500 to 1850 CE.

LPS: Land surface parameterization scheme.

Mesoscale convective complex (MCC): Several thunderstorms grouped together.

Mesoscale precipitation areas: Horizontal scale $10^{3/2}$km, vertical velocity scale 10 cm s^{-1}.

Meteorological drought: Brought about when there is a prolonged period with less than average precipitation.

METROMEX: Metropolitan Meteorological Experiment (US).

Mie theory: Microwave scattering for large drops or hail when particle diameters are larger than one-tenth of the radar wavelength.

Milankovitch theory: Close correlation between the major advances and recessions of polar ice and the 40,000 year cycle of variations in insolation; maximum ice cover is nearly coincident with minimum radiation received at about 45°N.

MODIS: Moderate resolution imaging spectroradiometer satellite instrument.

MOGREPS: Met Office Global and Regional Ensemble Prediction System (UK).

Monsoon: A seasonal global shift in wind patterns in Asia, North America, South America and Africa, often leading to rainy or dry weather periods of several months duration.

MOPS: Moisture Observation Processing System (UK).

MORECS: Meteorological Office Rainfall and Evaporation Calculation System.

MOS: Model output statistics.

MOSES: Met Office Surface Exchange Scheme.

MSG: METEOSAT second-generation satellites.

Natural energy system: The main element of the natural energy system is the flux of solar energy, but it includes terrestrial and atmospheric radiation, heat fluxes in the soil and ocean, and energy transformations in the oceans and atmosphere.

NCEP: National Centers for Environmental Prediction (US).

NFFS: National Flood Forecasting System (UK).

NHP: Natural Hazards Partnership (UK).

Nowcasting: Weather forecasts for lead times of 0–6 hours.

NWSRFS: The US NOAA's National Weather Service River Forecast System.

Optical depth: Gives the extent to which a given layer of material attenuates the intensity of radiation passing through it; sometimes called the opacity; symbol τ.

Palmer moisture anomaly index: A measure of surface moisture anomaly for the current month without consideration of the antecedent conditions; referred to as the Z index.

PDF: Probability density function.

PDSI: Palmer drought severity index used to measure the cumulative departure (relative to local mean conditions) in atmospheric moisture supply and demand at the surface.

Penman equation: Provides a solution to solving the aerodynamic or the energy balance equations using knowledge of the change with temperature of the saturated vapour pressure of water.

Permafrost or cryotic soil: Soil at or below the freezing point of water (0 °C) for two or more years.

Pleistocene period: The last million years of geological time.

PMM: Probability matching method of adjusting radar data with raingauge data.

POL: Proudman Oceanographic Laboratory (now the National Oceanography Centre, Liverpool).

Polar cell: Air circulation within the troposphere in the Arctic and the Antarctic regions associated with Rossby waves.

Polar orbit satellites: These fly at altitudes of 600–800 km.

Polarimetric radars: Provide information on the shape and orientation of the radar targets using parameters such as differential reflectivity Z_{DR}, the copolar correlation ρ_{hv}, specific differential phase shift K_{DP} and the linear depolarization ratio (LDR).

POLCOMS: Proudman Oceanographic Laboratory Coastal Modelling System (UK).

Potential temperature: Expressed as $\theta = T(1000/p)^K$, where $K = R/C_p$ and $R = C_p - C_V$, and where T is absolute temperature, R is the universal gas constant, C_p is specific heat at constant pressure and C_V is specific heat at constant volume.

Precipitation: The collective term for the forms that water takes in the atmosphere, including rain, drizzle, sleet (partly melted snowflakes, or rain and snow falling together), snow and hail.

Precision: Expressed quantitatively as the standard deviation of results from repeated trials under identical conditions.

PREVAH: ECMWF Precipitation–Runoff–Evapotranspiration–Hydrological response units model.

Probable maximum flood (PMF): Usually derived from the *probable maximum precipitation*.

Probable maximum precipitation (PMP): Defined as the theoretical greatest depth of precipitation for a given duration that is physically possible over a given size of storm area at a particular geographic location at a certain time of the year.

Pseudo-adiabatic process: Takes place when condensation occurs and the resulting water falls out of a sample; then the mass of the sample changes and heat is lost with the fallout of the water.

QPF: Quantitative precipitation forecast.

Radar equation: The relationship between the power received at the radar antenna from the sum of the contributions of the individual scatterers (hydrometeors) and the parameters of the radar system.

Radar reflectivity factor: Proportional to the drop size distribution and the sixth power of the raindrop diameters; symbol Z.

Radar wavelengths: Bands X (3 cm), C (5 cm) and S (10 cm) are commonly used for measuring precipitation.

Rainfall–runoff model types: Black box (or metric), conceptual (or parametric) and physics based (or mechanical).

Raingauge types: Non-recording, weighing, float, tipping bucket.

RANS-VOF: Reynolds-averaged Navier–Stokes volume of fluid model.

RAPIDS: The RAdar Pluvial flooding Identification for Drainage System is an artificial neural networks (ANNs) technique.

RASP: Risk Assessment for Flood and Coastal Defence for Strategic Planning.

Rayleigh theory: Microwave scattering for small drops.

Residence time: The average time a water molecule in parts of the hydrological cycle will spend in a particular area.

ROC: Relative operating characteristic a score used for assessing numerical forecast models.

Rossby waves: Quasi-horizontal ultra-long harmonic waves in both hemispheres determining the path of the jet stream.

Running mean: Mean taken over an odd number of values centred on the value of interest.

SAR: Synthetic aperture radar.

Saturated adiabatic lapse rate (SALR): Air in the troposphere becomes saturated as it rises (or dries out as it descends) and condensation (or evaporation) of water supplies heat to (or removes heat from) the air. This is known as an adiabatic process, and the SALR is the decrease of temperature with height as the parcel of air rises at a rate of about $5°C$ km^{-1}.

Saturated humidity mixing ratio: The mass of water vapour present in moist air, measured per gram of dry air, when the moist air is saturated.

SEBS: Surface energy balance system.

Seeder–feeder mechanism: Ice particles from generating cells fall into lower level frontal cloud and act as natural seeding particles, causing a rapid growth of precipitation particles.

Sensible and water vapour fluxes: Transport of momentum, heat and water vapour between the surface and the atmosphere.

Serial autocorrelation analysis: Identification of the periodicity in data.

Skin depth: A weighted average of the physical temperature over the depth of the surface layer of a material.

Sloping or baroclinic convection: Provides the basic mechanism for the development of large scale atmospheric weather systems.

SLOSH: The US NOAA Sea, Lake, and Overland Surges from Hurricanes model.

Socioeconomic drought: Relates to the supply and demand of various commodities under drought conditions.

Soil moisture: Measured using gravimetric, electrical resistance, dielectric, tensiometric, neutron, satellite and gamma ray attenuation methods.

Soil water content: Expressed as a ratio of the mass of water contained in a sample of soil either to the dry mass of the soil sample, or to the original volume of the sample.

Soil water potential: Expresses the potential energy of the water contained in the soil.

Specific differential phase shift K_{DP}: The rate of change of the difference between the horizontal and vertical phase returns (ϕ_h and ϕ_v) with range.

Specific heat at constant pressure C_p and specific heat at constant volume C_v: In the atmosphere C_p and C_v are treated as constant, with values derived experimentally of 240 cal kg^{-1} K^{-1} (10.04×10^6 erg g^{-1} K^{-1}) and 171 cal kg^{-1} K^{-1} (7.17×10^6 erg g^{-1} K^{-1}) respectively.

SSMI: Special sensor microwave imager.

SST: Sea surface temperature.

St Venant equations: Used in models to predict storm surge flooding and flood inundation.

Stage-discharge curve: Expressed as $Q = C(h + a)^n$, where Q is the discharge, h is the stage or gauge height, C and n are constants, and a is the value of the stage at zero flow.

Statistical interpolation (SI): Method for the spatial interpolation of precipitation; other methods include kriging, successive correction, and optimum interpolation.

Stefan–Boltzmann law: Relates the total radiant emittance from a black body to the fourth power of the absolute temperature; the Stefan–Boltzmann constant is $5.67 \times 10^{-8} \, \text{W m}^{-2} \text{K}^{-4}$.

STEPS: Short term ensemble prediction system (UK).

Sublimation: Change of phase of water from vapour to solid.

Surface water (pluvial) flooding: This happens when there is heavy rainfall on ground that is already saturated, or on paved areas where drainage is poor.

SVATs: Soil–vegetation–atmosphere transfer schemes.

Synoptic scale precipitation systems: Mid-latitude depressions.

Thermal inertia: The resistance of a material to temperature change; symbol P.

Thermal wind equation: Expressed as $V_g = (g/f)(|\partial h/\partial n|)$, where h is the thickness of the layer between two pressure surfaces, g is the acceleration due to gravity, and f is the Coriolis parameter. The thermal wind is the difference between the geostrophic winds at two levels, or the vertical shear of geostrophic wind.

Thiessen polygons: With this arrangement, individual gauge values are first weighted by the fractions of the catchment area represented by each gauge and then used to estimate areal precipitation.

Thunderstorms: Types are single cells, multi-cells and supercells.

TIROS: Television observation satellites.

Transpiration: The evaporation of water from plants through the small openings found on the underside of their leaves (known as stomata).

TRMM: Tropical Rainfall Measurement Mission.

Tropical storms: Known as hurricanes in the southern part of the North Atlantic, cyclones in the northern part of the Indian Ocean, typhoons in the south-western part of the North Pacific, and willy-willies in the north-east of Australia.

Tsunami: A series of enormous waves created by an underwater disturbance such as an earthquake, a landslide, a volcanic eruption, or possibly a meteorite.

UBL: Urban boundary layer.

UCM: Urban canopy model.

UHI: Urban heat island.

UM: Unified mesoscale numerical model (UK).

USGS: US Geological Survey.

Walker circulation: The Pacific Ocean atmospheric circulation pushing Pacific water westward.

Water balance at the land surface: Comprising terrestrial water balance, atmospheric water balance and combined atmospheric/land-surface water balance.

Water equivalent of snow: Measured from the depth of snow and the density of snow.

Water vapour: Vapour form of water, which is always present in the atmosphere in varying amounts, although it makes up less than 1% by volume.

WCRP: World Climate Research Programme, concerned with studying the dynamics and thermodynamics of the atmosphere.

Wegener–Bergeron–Findeisen process: A process of rain formation, sometimes shortened to the Bergeron process.

Wien's displacement law: States that the product of absolute temperature and wavelength is constant for a black body (perfect emitter at all wavelengths).

WMO: World Meteorological Organization.

WRF: Weather research and forecasting numerical model (NCAR).

Younger Dryas: Around 14,000 years ago the Earth was suddenly plunged back into a new and very short lived ice age, ending after about 1300 years.

ZPHI technique: A rain profiling algorithm based upon a set of three power law relationships between attenuation A and reflectivity Z, phase shift K_{DP} and A, and rainfall R and A, each of which is 'normalized' by the intercept parameter N_0^* of an exponential or gamma drop size distribution.

Z–R relationship: Power law between reflectivity Z and rainfall R, varying with precipitation type.

Index